T0328294

High-Risk Pollutants in Wastewater

High-Risk Pollutants in Wastewater

Edited by

Hongqiang Ren

Professor and the Dean
School of the Environment
Nanjing University
Nanjing, China

Xuxiang Zhang

Professor
School of the Environment
Nanjing University
Nanjing, China

ELSEVIER

Elsevier
Radarweg 29, PO Box 211, 1000 AE Amsterdam, Netherlands
The Boulevard, Langford Lane, Kidlington, Oxford OX5 1GB, United Kingdom
50 Hampshire Street, 5th Floor, Cambridge, MA 02139, United States

High-Risk Pollutants in Wastewater

Notices

Practitioners and researchers must always rely on their own experience and knowledge in evaluating and using any information, methods, compounds or experiments described herein. Because of rapid advances in the medical sciences, in particular, independent verification of diagnoses and drug dosages should be made. To the fullest extent of the law, no responsibility is assumed by Elsevier, authors, editors or contributors for any injury and/or damage to persons or property as a matter of products liability, negligence or otherwise, or from any use or operation of any methods, products, instructions, or ideas contained in the material herein.

ISBN: 978-0-12-816448-8

Publisher: Andre Gerhard Wolff
Acquisition Editor: Louisa Munro
Editorial Project Manager: Ruby Smith
Production Project Manager: Kiruthika Govindaraju
Cover Designer: Alan Studholme

Contents

Contributors

Jinju Geng, PhD
State Key Laboratory of Pollution Control and Resource Reuse, School of the Environment, Nanjing University, Nanjing, China

Xiwei He, PhD
State Key Laboratory of Pollution Control and Resource Reuse, School of the Environment, Nanjing University, Nanjing, China

Haidong Hu
State Key Laboratory of Pollution Control and Resource Reuse, School of the Environment, Nanjing University, Nanjing, China

Hui Huang, PhD
State Key Laboratory of Pollution Control and Resource Reuse, School of the Environment, Nanjing University, Nanjing, China

Kailong Huang, PhD
State Key Laboratory of Pollution Control and Resource Reuse, School of the Environment, Nanjing University, Nanjing, China

Shuyu Jia
State Key Laboratory of Pollution Control and Resource Reuse, School of the Environment, Nanjing University, Nanjing, China

Xiaofeng Jiang
State Key Laboratory of Pollution Control and Resource Reuse, School of the Environment, Nanjing University, Nanjing, China

Kan Li
State Key Laboratory of Pollution Control and Resource Reuse, School of the Environment, Nanjing University, Nanjing, China

Mei Li, PhD
State Key Laboratory of Pollution Control and Resource Reuse, School of the Environment, Nanjing University, Nanjing, China

Ruxia Qiao
State Key Laboratory of Pollution Control and Resource Reuse, School of the Environment, Nanjing University, Nanjing, China

Hongqiang Ren, PhD
State Key Laboratory of Pollution Control and Resource Reuse, School of the Environment, Nanjing University, Nanjing, China

Cheng Sheng
State Key Laboratory of Pollution Control and Resource Reuse, School of the Environment, Nanjing University, Nanjing, China

Jinfeng Wang
State Key Laboratory of Pollution Control and Resource Reuse, School of the Environment, Nanjing University, Nanjing, China

Bing Wu, PhD
State Key Laboratory of Pollution Control and Resource Reuse, School of the Environment, Nanjing University, Nanjing, China

Gang Wu
State Key Laboratory of Pollution Control and Resource Reuse, School of the Environment, Nanjing University, Nanjing, China

Ke Xu
State Key Laboratory of Pollution Control and Resource Reuse, School of the Environment, Nanjing University, Nanjing, China

Lin Ye, PhD
State Key Laboratory of Pollution Control and Resource Reuse, School of the Environment, Nanjing University, Nanjing, China

Jinbao Yin, PhD
State Key Laboratory of Pollution Control and Resource Reuse, School of the Environment, Nanjing University, Nanjing, China

Xuxiang Zhang, PhD
State Key Laboratory of Pollution Control and Resource Reuse, School of the Environment, Nanjing University, Nanjing, China

Yan Zhang, PhD
State Key Laboratory of Pollution Control and Resource Reuse, School of the Environment, Nanjing University, Nanjing, China

Huajin Zhao
State Key Laboratory of Pollution Control and Resource Reuse, School of the Environment, Nanjing University, Nanjing, China

CHAPTER

Introduction

1

Xuxiang Zhang, PhD, Hongqiang Ren, PhD
State Key Laboratory of Pollution Control and Resource Reuse, School of the Environment, Nanjing University, Nanjing, China

Chapter outline

1.1 Environmental high-risk pollutants

In recent years, in addition to large quantities of pollutants such as COD and ammonia nitrogen, new problems regarding the toxic substances in the environment have been highlighted. In China, nearly 50,000 kinds of chemicals are produced and used for different purposes, and this number is still increasing, as many new drugs, antibiotics, and pesticides are being continuously synthesized for use. Besides the chemicals, some biological substances (e.g., bacterial pathogen, viruses, and antibiotic resistance genes) coming from living organisms can also lead to adverse effects on both human and the environment. Notably, wastewater is an important environmental reservoir for both the chemical and biological substances, which can introduce them into receiving water body along with wastewater discharge. Many of the chemicals are structurally complex and are stable and difficult to degrade. Owing to the cumulative and persistent characteristics, the substances cannot be completely removed in the existing conventional wastewater treatment systems; thus, considerable amount of the substances or their transformed products enter water and soil environments through the discharge from industrial or urban point sources, as well as urban or agricultural nonpoint sources. The substances entering the environment are called "pollutants," and a variety of them have been frequently detected in natural waters such as rivers, lakes, oceans, and groundwater in recent years. It has been estimated that over 4000 kinds of trace organic pollutants are present in water environment in China, which have become an important factor endangering human and environmental health.

Although most of the pollutants have low concentrations or abundance in the environment, many of them can accumulate in organisms and spread and enrich through the food chain, leading to chronic poisoning of animals or human. The other

High-Risk Pollutants in Wastewater. https://doi.org/10.1016/B978-0-12-816448-8.00001-0

part of pollutants is not stable, but they are continuously discharged into natural water due to continuous production and extensive use for human living and animal husbandry. After long-term even full-lifecycle exposure, the phenomenon of "pseudopersistence" is formed in animal or human bodies to induce toxicities. Therefore, the chronic toxicity, microbial resistance, and synergistic toxicity caused by the substances have received growing concerns.

High-risk pollutants (HRPs) are defined as the highly diverse chemical and biological matters with high toxicity and complicated toxicological mechanisms that can pose serious risks to ecosystems and human health even at low concentrations. HRP pollution is regarded by the United Nations Environment Program as an urgent issue that needs to be dealt with through global cooperation, and since the 1990s many developed countries have introduced regulations to control HRPs. In 2014, China also issued the "High-Risk Pollutants Reduction Action Plan". European legislation follows the precautionary principle and has introduced stringent threshold limits (100 ng/L) for many chemical HRPs in water environments. Industrial and municipal wastewater contains a variety of HRPs including highly diverse chemical and biological substances with high toxicities, posing serious risks to ecosystems and human health even at low concentrations.

1.2 Control of HRPs in wastewater

Water shortage is one of today's grand challenges as many of the rivers or lakes in the world become polluted, and wastewater reuse or deep cleaning is considered effective in solving the problem. Presence of a variety of HRPs in industrial and municipal wastewater constrains the advanced treatment and reuse of the wastewater. Evaluation and control of HRPs in wastewater is an important issue for water pollution control, and only measuring quantitative integrated indices, such as COD_{Cr} and BOD_5, is not sufficient for assessing and controlling the risk induced by wastewater discharge, so biotoxicity indices and standards have been recommended as alternatives in many countries.

Recently, advanced treatment and reuse of wastewater have been increasingly implemented in many countries to solve the problems of serious water shortage and pollution. The advanced physicochemical technologies include membrane filtration, activated carbon absorption, and advanced oxidation, but they are neither environmentally nor economically sustainable, which seriously constrains their practical application. As a sustainable and cost-efficient alternative, biodegradation of HRPs either by endogenous or bioaugmented microbiota is considered as promising technologies. However, various bottlenecks hamper the implementation of biodegradation as a robust HRP removal technology that combines removal efficacy, energy efficiency, and risk reduction, and the key problem to be solved is how to eliminate the microbial constraints that limit biodegradation of HRPs present as mixtures in wastewater. Therefore, application of appropriate hazard/risk assessment and treatment technologies, which are effective, low cost, and simple to operate, is a key component in any strategy aimed at reducing ecological and health risks arising from wastewater discharge and reuse.

1.3 Objective and contents of this book

Currently, many books on the market have been focused on the treatment technology and engineering of wastewater to meet the discharge or reuse standards required by the governments of different countries, but provide little information regarding HRPs and their risk assessment and control technologies. Therefore, the major objectives of this book entitled "High-Risk Pollutants in Wastewater" are as follows:

(1) This book presents the basic knowledge regarding the diversity, concentrations, and health and environmental impacts of HRPs in wastewater. The book summarizes the information of the types (e.g., heavy metals, toxic organics, and pathogens) and toxicities of HRPs in wastewater, and also describes ecological and health hazards or risks arising from the living things' direct/indirect contacts with the HRPs during their full lifecycles (generation, disposal, discharge, and reuse) in wastewater or water environments.

(2) This book presents the concepts of appropriate technology for HRP hazard/risk assessment and wastewater treatment/reuse and the issues of strategy and policy for increasing the risk control coverage. The book focuses on the resolution of water quality monitoring and wastewater treatment and disposal problems in both developed and developing countries, and the concepts presented are wished to be valid and applicable in risk warning and control for wastewater discharge into the environment and reuse for any purposes.

This book mainly includes 11 chapters: (1) Introduction; (2) Chemical HRPs in wastewater; (3) Biological HRPs in wastewater; (4) Technologies for detection of HRPs in wastewater; (5) Human health hazards of wastewater; (6) Ecological safety hazards of wastewater; (7) Assessment technologies for hazards/risks of wastewater; (8) Physicochemical technologies for HRPs and risk control; (9) Biological technologies for HRPs and risk control; (10) Wastewater disinfection technologies; and (11) Risk management policy for HRPs in wastewater.

This book delivers the idea of recognition, assessment, and control of HRPs, including highly diverse chemical and biological matters with high toxicity at low concentrations. The book was technologically written, and does not involve traditional pollutants that also affect environmental and ecological health, such as nitrogen and phosphorus in wastewater. This book is expected to be used as a textbook or reference book for the graduate students majoring in environmental science, environmental engineering, and civil engineering, and will also provide a useful reference for wastewater treatment plant personnel, industrial wastewater treatment professionals, government agency regulators, environmental consultants, and environmental attorneys.

CHAPTER

Chemical HRPs in wastewater

2

Gang Wu, Jinfeng Wang, Jinju Geng, PhD
State Key Laboratory of Pollution Control and Resource Reuse, School of the Environment, Nanjing University, Nanjing, China

Chapter outline

In recent years, more and more chemicals are synthesized and used to meet people's rising living standards. It is reported that the chemicals registered under the Chemical Abstract Service (CAS) have been over 140 million (Li and Suh, 2019). Once they are used and discharged, these chemicals will exist in various environment medium. Human beings can be exposed to these chemicals. This raises a serious issue for people to pay more attention on the adverse effect of these chemicals on

High-Risk Pollutants in Wastewater. https://doi.org/10.1016/B978-0-12-816448-8.00002-2

ecosystem. However, it is impractical to focus on all the chemicals registered in CAS (Whaley et al., 2016). Some prioritized chemicals, especially for the chemicals with high risk, should be given enough attention firstly.

Chemical high-risk pollutants (HRPs) refer to the chemicals with high ecological risk to ecosystem (Rasheed et al. 2019; Zhou et al., 2019). These HRPs (anthropogenic or natural) include but are not limited to heavy metals, persistent organic pollutants (POPs), pharmaceuticals and personal care products (PPCPs), and endocrine disrupting chemicals (EDCs) (Li and Suh, 2019). Generally, wastewater is one of the major sink for these HRPs once they become consumer products. The occurrence of these HRPs in wastewater will pose some potential adverse effect on the downside ecosystem inevitably (Tran et al. 2017). Therefore, chemical HRPs in wastewater should be given enough attention. Understanding the environmental behavior and transformation characteristics of chemical HRPs is important for their removal and risk reduction in environment. In this chapter, the category, source, concentrations, migration, and transformation of chemical HRPs in wastewater are summarized in detail. This is beneficial to fill the knowledge gap of chemical HRPs in wastewater.

2.1 Heavy metals

Heavy metals refer to any metallic element that has a relatively high atomic mass (> 5 g/cm^3) and are toxic or poisonous even at a low concentration (Nagajyoti et al., 2010). Heavy metals, which are significantly toxic to environmental ecology, mainly include cadmium (Cd), chromium (Cr), nickel (Ni), mercury (Hg), lead (Pb), manganese (Mn), copper (Cu), and zinc (Zn). The source, occurrence and contamination level, and fate of heavy metals in wastewater, which will provide a clear understanding on ecological risk of metals in wastewater to readers, are summarized as follows.

2.1.1 Sources of heavy metals in wastewater

Industrial activities are a significant source of heavy metals (Santos and Judd, 2010). Mining operations and ore processing, textiles, metallurgy and electroplating, dyes and pigments, paper mills, tannery, and petroleum refining are the main sources of heavy metals in wastewater. Mine wastewater are often acidic and contain dissolved heavy metals, such as Cd, Cr, Zn, and Mn (Hedrich and Johnson, 2014), which mainly come from mineral processing and washing (Zaranyika et al., 2017). Plating rinsing process often introduces Cr, Cu, and Hg into metallurgy and electroplating wastewater (Hideyuki et al., 2003). Metallized complexed azo dye production contributes the presence of most of the heavy metals in textiles wastewater (Edwards and Freeman, 2010). Heavy metals in dyes and pigments mainly are derived from textile warp size, dye, and surface-active agents (Halimoon and Gohsoo Yin, 2010). Paper mills contribute to a high concentration of Hg in wastewater (El-Shafey, 2010). Cr is the dominant heavy metal in tannery wastewater (Mella et al., 2015). Heavy metals from petroleum refining are derived from waste oil-refining catalyst, which mainly include Ni, Hg, Pb, Cr, and Cd. In addition, some heavy metals, such as Cd, Cr, Hg, Ni, Pd, and As have also been recognized in municipal wastewater treatment plants.

2.1.2 Occurrence and concentrations of heavy metals in wastewater

Converging from different sources, heavy metals, for example, Cd, Cr, Ni, Hg, Pb, Mn, Cu, and Zn, have been frequently detected in municipal and industrial wastewater. For example, Ustun (2009) investigated nine metals (Al, Cd, Cr, Cu, Fe, Mn, Ni, Pb, and Zn) in wastewater treatment plants (WWTPs) in Bursa (Turkey) for 23 months in 2002 and 2007 and found that all the metals were detectable. Teijon et al. (2010) assessed the contaminants level of four heavy metals (Cd, Ni, Hg, and Pb) in WWTPs and only Ag was occasionally under the detection limit. Despite of the wide distribution of different heavy metals in WWTPs in geologically different regions, the concentration levels of heavy metals show great variation. Tables 2.1—2.6 list the main heavy metals and their concentrations in both municipal wastewater and industrial wastewater located in different countries.

As listed in Table 2.1, Cd and Pb show high detection frequency in the WWTPs of 13 countries. Concentrations of Cd range from not detected (ND) to 220 μg/L, with the highest concentration in the WWTPs of Turkey. Meanwhile, concentrations of Pb range from ND to 6860 μg/L, with the highest concentration in the WWTPs of China. In addition, Ni and Cr were found to have high levels in the wastewater from Greece and Italy, respectively.

As listed in Table 2.2, Cd often occurs in the industrial wastewater generated from mining operations and ore processing, petroleum refining, textiles production, metallurgy and electroplating, dyes and pigments, paper mills, and tannery. Concentrations of Cd in industrial wastewater range from ND to 200 mg/L, and metallurgy and electroplating wastewater often contains high concentrations of Cd (up to 200 mg/L).

As shown in Table 2.3, Cr often occurs in the industrial wastewater generated from mining operations, ore processing, petroleum refining, textiles production, metallurgy and electroplating, paper mills, and tannery. Concentrations of Cr in industrial wastewaters vary greatly, ranging from 12 μg/L to over 1700 mg/L. Among the different types of industrial wastewater, tannery wastewater often contains the highest concentrations of Cr (sometimes over 500 mg/L).

Hg often occurs in the industrial wastewater generated from mining operations, ore processing, metallurgy, and electroplating. Concentrations of Hg in industrial wastewater are not very high, usually below 1 mg/L (Rahman et al., 2017; Chojnacka et al., 2004).

As shown in Table 2.4, Ni often occurs in the industrial wastewater generated from mining operations, ore processing, petroleum refining, textiles production, metallurgy and electroplating, paper mills, and tannery. Concentrations of Ni in industrial wastewater range from ND to over 50 mg/L, and metallurgy and electroplating wastewater often contain high concentrations of Ni.

Table 2.1 Concentrations of heavy metals in municipal wastewater from different countries.

Country	Cd (μg/L)	Cr (μg/L)	Hg (μg/L)	Ni (μg/L)	Pb (μg/L)	As (μg/L)	Reference
Poland	7–70	<20	n.d.	10–20	10–150	–[a]	Chipasa (2003); Kulbat et al. (2003)
Greece	0.8–3	20–275	n.d.	11–770	37–103	–[a]	Karvelas et al. (2003); Spanos et al. (2016)
Turkey	20–220	20–212	<0.1	20–202	6–358	–[a]	Arslan-Alaton et al. (2007; Ustun (2009)
Italy	0.2 –45.04	0.5 –117.81	0.2–147	1–64.17	1–64	0.3–31	Busetti et al. (2005); Carletti et al. (2008); Hu et al. (2016)
Spain	5	–[a]	0. 4	47.5	5.2	–[a]	Teijon et al. (2010)
USA	0.423	8.69	n.d.	7.39	7.05	5.05	Shafer et al. (1998); Raheem et al. (2017)
China	3.45 –110	37.4–280	0.2 –0.06	7.0–86.2	1.9–6860	3.0	Cheng, 2003; (Feng et al., 2018)
Brazil	0.06 –1.19	2.11 –20.73	<0.05	–[a]	9.66 –334.00	–[a]	da Silva Oliveira et al. (2007)
France	0.39 –0.67	1.5–13.5	–[a]	2.4–15.1	2.2–20.2	–[a]	Buzier et al. (2006)
Portugal	<10	10–30	0.05 –0.96	n.d.	<10	2.00 –4.90	Varela et al. (2014)
South Africa	30–476	–[a]	–[a]	–[a]	2–72	–[a]	Edokpayi et al. (2015)
Kuwait	<10	<10	<10	<10–50	<10–120	–[a]	Al Enezi et al. (2004)
Thailand	<17.16	2.87 –72.54	–[a]	4.88 –116.60	2.93 –79.33	–[a]	Chanpiwat et al. (2010)

[a] -: *no data in the reference.*

Table 2.2 Concentrations of Cd in industrial wastewater from different countries.

Source	Code[a]	Country	Concentrations(mg/L)	Reference
Mining operations and ore processing	CA	Czech Republic	0.227	Doušová et al. (2005)
		Sweden	1.0	Hedrich and Johnson (2014)
		Egypt	0–1	Nosier (2003)
		China	0.002–0.022	Chongyu Lan (1992); Hu et al. (2014)
Petroleum refining	MA	Tunisia	0.6	Ben Hariz et al. (2013)
Textiles production	HA	Pakistan	0.07–0.51	Ali et al (2009); (Noreen et al., 2017)
		Malaysia	0.409	Al-Zoubi et al. (2015)
		Iran	0.28	Ghorbani and Eisazadeh (2013)
		Rwanda	0.05–0.14	Sekomo et al. (2012)
		Egypt	40.65	Mahmoued et al., 2010
		Pakistan	0.004–0.51	Ali et al. (2009); Mahmoued et al. (2010)
Metallurgy and electroplating	SA	Poland	3.81	Chojnacka et al. (2004)
		Egypt	100–200	Nosier (2003)
Dyes and pigments	IA	Rwanda	0.05–0.14	Sekomo et al. (2012)
Paper mills	LA	India	0.01–0.78	Arivoli et al. (2015); Chandra et al. (2017); Kumar et al. (2016)
Tannery	JA	Pakistan	0.02–0.59	Ali et al. (2009); Tariq et al. (2006, 2009)
		Mexico	22	Contreras-Ramos et al. (2004)
		India	1.22	Chandra et al. (2009)

[a] *Categorizing according to ISO 22447.*

As shown in Table 2.5, Pb often occurs in the industrial wastewater generated from mining operations, ore processing, petroleum refining, textiles production, metallurgy and electroplating, paper mills, and tannery. Concentrations of Pb in industrial wastewater range from ND to over 60 mg/L, and metallurgy and electroplating wastewater contains high concentrations of Pd.

As often occurs in the industrial wastewater generated from mining operations, ore processing, and textiles production. Concentrations of As in industrial wastewaters range from ND to 54 mg/L (Doušová et al., 2005; Sekomo et al., 2012; Hedrich and Johnson, 2014; Lim et al., 2010).

2.1.3 Migration and transformation of heavy metals in wastewater

When heavy metals are discharged into wastewater, the distribution of metals between the aqueous and the solid phase of wastewater occurs inevitably (Karvelas et al., 2003; Shafer et al., 1998). Karvelas et al. (2003) investigated distribution of metals between the aqueous and the solid phase of wastewater, revealing high exponential correlation between the metal partition coefficient (logKp) and the suspended solids concentration. In addition, different transformation reactions can occur in wastewater treatment processes. Methylation is a common transformation reaction of heavy metals by microorganisms. For example, Mao et al. (2016) investigated the fate of Hg in WWTPs and found influent MeHg mass was degraded, indicating that WWTPs are an important sink for sewage-borne Hg. In addition, chelation reaction is another common reaction in WWTPs. Heavy metals in wastewater may occur as attached to suspended solids via surface bound organic ligands or adsorbed on to a major insoluble matrix component (e.g., iron(III) oxide, aluminum hydroxide etc.), insoluble salts, inorganic complex solids, or as free or organically bound soluble forms. Their speciation may depend on the influent metal concentration, influent chemical oxygen demand, hardness, alkalinity, and pH of the wastewater (Santos and Judd, 2010). Wang et al. (2003) investigated the interactions of silver in wastewater constituents and found that chloride, sludge particulates, and dissolved organic matter (DOM) had an interaction with silver and the volume of adsorption to DOM substantially affected by the value of pH.

Due to the toxicity effect of heavy metals toward ecosystem, the removal of heavy metals in WWTPs is needed (Qureshi et al., 2016). Generally, heavy metals are resistant to biodegrade in WWTPs, and some physicochemical treatment methods rather than traditional biological treatment have been verified to be effective in removal of heavy metals from wastewater (Karvelas et al., 2003). For example, Fu and Wang (2011) evaluated the current methods that have been used to treat heavy metal wastewater including chemical precipitation, ion-exchange, adsorption, membrane filtration, coagulation-flocculation, flotation, and electrochemical methods. They found that adsorption and membrane filtration are the most frequently studied processes for the treatment of heavy metals in wastewater. In addition, biosorption is another promising method to eliminate heavy metals in wastewater (Veglio and Beolchini, 1997).

2.2 Persistent organic pollutants

POPs refer to one group of organic compounds characterized with high toxicity, environmental persistence, transporting over a long distance through various environmental medium, and accumulating in biological bodies easily (Tieyu et al., 2005). The widely known adverse effects of POPs toward human beings are carcinogenesis,

Table 2.3 Concentrations of Cr in industrial wastewater from different countries.

Source	Code [a]	Country	Concentration (mg/L)	Reference
Mining operations and ore processing	CA	South Africa	0.047–0.059	Shabalala et al. (2017)
		India	1.3–2.0	Mishra et al. (2008); Saha et al. (2017)
		Spain	0.2	Jiménez-Rodríguez et al. (2009)
		China	0.6–2.08	Ma et al. (2015)
Petroleum refining	MA	Tunisia	3.3	Ben Hariz et al. (2013); Hu et al. (2014)
Textiles production	HA	Pakistan	0.023–1.67	Ali et al. (2009); Manzoor et al. (2006)
		India	0.0152–7.854	Sanmuga Priya and Senthamil Selvan (2017)
		Turkey	45.7–66.6	Ertuğrul et al. (2009)
		Malaysia	0.056	Al-Zoubi et al. (2015)
		Rwanda	1.27–2.83	Sekomo et al. (2012)
Metallurgy and electroplating	SA	Saudi Arabia	93.2	Al-Shannag et al. (2015)
		Argentina	0.012–1.45	Maine et al. (2017)
		China	1.6–16.7	Wu et al. (2017)
Paper mills	LA	India	0.11	Chandra et al. (2017)
Tannery	JA	Pakistan	0.037–592.2	Ali et al. (2009); Tariq et al. (2006, 2009); Tariq et al. (2005)
		Mexico	1475	Contreras-Ramos et al. (2004)
		India	9.38	Chandra et al. (2009)
		Northern Ireland	49	Boshoff et al. (2004)
		USA	1.7–55	Polat and Erdogan (2007)

[a] *Categorizing according to ISO 22447.*

tetratogenesis, and mutagenesis effects (Dietz et al., 2018; Raffetti et al., 2018). Generally, POPs include polychlorinated biphenyls (PCBs), polycyclic aromatic hydrocarbons (PAHs), perfluorooctanoic acids (PFOA) and perfluorooctane sulfonate (PFOS), short-chain chlorinated paraffins (SCCPs), and organic pesticides (OCPs) (Bao et al., 2012; Gallistl et al., 2017; Mrema et al., 2013). Furthermore, 28 substances or substance groups of POPs listed in the Stockholm convention deserve more attention (Fernandes et al., 2019; Magulova and Priceputu, 2016; Wei et al., 2007; Zapata et al., 2018). In this section, the source, occurrence and contamination level, and fate

Table 2.4 Concentrations of Ni in industrial wastewater from different countries.

Sources	Code [a]	Country	Concentration (mg/L)	Reference
Mining operations and ore processing	CA	South Africa	0.6–1.3	Shabalala et al. (2017)
		Sweden	0.3	Hedrich and Johnson (2014)
		India	<0.001	Saha et al. (2017)
		China	1.03	Hu et al. (2014)
Petroleum refining	MA	Tunisia	4	Ben Hariz et al. (2013)
Textiles production	HA	Pakistan	0.67	Ali et al. (2009)
		Malaysia	<0.001	Lim et al. (2010)
		India	0.5–3	Sanmuga Priya and Senthamil Selvan (2017)
		Turkey	38.1–47.9	Ertuğrul et al. (2009)
		Poland	0.0693–0.11	Al-Zoubi et al. (2015, Cempel and Nikel (2005)
Metallurgy and electroplating	SA	Saudi Arabia	57.6	Al-Shannag et al. (2015)
		Argentina	0.004–0.101	Maine et al. (2017)
		Poland	0.218	Chojnacka et al. (2004)
		India	6.704	Radhakrishnan et al. (2014)
Paper mills	LA	India	0.06–3.30	Arivoli et al. (2015); Chandra et al. (2017); Kumar et al. (2016)
Tannery	JA	Pakistan	0.116–0.671	Tariq et al. (2006, 2009); Tariq et al. (2005)
		India	0.44	Chandra et al. (2009)

[a] *Categorizing according to ISO 22447.*

of POPs in wastewater, which will provide a clear understanding on ecological risk of POPs in wastewater to readers, are summarized as follows.

2.2.1 Sources of POPs in wastewater

The sources of different categories of POPs are generally different. The presence of pesticides in WWTPs is mainly due to nonagricultural usages, such as grass-management (golf courses, educational facilities, parks, and cemeteries), industrial vegetation control (industrial facilities, electric utilities, roadways, railroads, pipelines), public health (mosquito-abatement districts, rodent-control areas, and aquatic areas), and nonagricultural crops such as commercial forestry and horticulture and plant nurseries (Kock-Schulmeyer et al., 2013).

Table 2.5 Concentrations of Pb in industrial wastewater from different countries.

Source	Code [a]	Country	Concentration (mg/L)	Reference
Mining operations and ore processing	CA	South Africa	<0.03	Shabalala et al. (2017)
		Ghana	0.140	Acheampong et al. (2010)
		China	0.087–1.57	Chongyu Lan (1992); Ma et al. (2015)
Petroleum refining	MA	Tunisia	5	Ben Hariz et al. (2013)
Textiles production	HA	Pakistan	0.1–0.49	Ali et al. (2009); Noreen et al. (2017)
		Malaysia	0.08–0.643	Al-Zoubi et al. (2015); Lim et al. (2010)
		Rwanda	0.25–0.67	Sekomo et al. (2012)
Metallurgy and electroplating	SA	Poland	1.52×10^{-3}	Chojnacka et al. (2004)
		China	67.8	Wu et al. (2017)
Paper mills	LA	India	0.22–1.05	Chandra et al. (2017); Kumar et al. (2016)
Tannery	JA	Pakistan	0.196–0.872	Tariq et al. (2006, 2009); Tariq et al. (2005)
		Mexico	9.1	Contreras-Ramos et al. (2004)
		India	0.34	Chandra et al. (2009)

[a] *Categorizing according to ISO 22447.*

The wide use of PCBs as paints, inks, lubricants, and other additives, as well as insulation fluids in transformers and capacitors during 1950–1983 is the dominant source of PCBs into various environmental medium, especially in wastewater (Rodenburg et al., 2011). It has been indicated that old WWTPs remain a dominant source of PCBs to the environment although the production has been banned for 4 decades (Needham and Ghosh, 2019).

The sources of PAHs are complex and can be divided into natural and anthropogenic ones. Generally, PAHs are formed by the incomplete combustion of coal, oil, tar, gas, wood, garbage, and charbroiled meat during natural or anthropogenic processes. Other sources of PAHs include heavy petroleum, diesel fuels, kerosene, aviation fuel, heavy home-heating oils, oils, waste oil, and many lubricants (Khadhar et al., 2010). In a present study, coke production has been identified as one of the major sources of China, which contributes 16% of the total PAHs in China (Zhang et al., 2012).

PFOA and PFOS are widely used in textile, fire-fighting foam, leather, food packaging, carpet, floor grinding, and shampoo. The production, marketing, and use of

Table 2.6 Concentrations of POPs in wastewater from different countries.

Category	Compound	Country	Wastewater type	Influent concentration (ng/L)	Effluent concentration (ng/L)	Reference
OCPs	Aldrin	Greece	Municipal wastewater	10	–[a]	Katsoyiannis and Samara (2004)
		Spain	Municipal wastewater	n.d.	120	Barco-Bonilla et al. (2013)
	Dieldrin	Greece	Municipal wastewater	27	8.9	Katsoyiannis and Samara (2004)
		Cyprus	Municipal wastewater	12	n.d.	Fatta et al. (2007)
	p,p'-DDD	China	Municipal wastewater	2.42	n.d.	Li et al. (2008)
	p,p'-DDT	China	Municipal wastewater	13.5	5.23	Li et al. (2008)
	Diuron	Spain	Municipal wastewater	n.d.	250	Barco-Bonilla et al. (2013)
		Spain	Municipal wastewater	93	127	Kock-Schulmeyer et al. (2013)
PAHs	Naphtalene	Spain	Municipal wastewater	4500	3490	Sanchez-Avila et al. (2009)
	Acenaphtyl	Spain	Municipal wastewater	30	n.d.	Sanchez-Avila et al. (2009)
	Fluorene	Spain	Municipal wastewater	n.d.	250	Barco-Bonilla et al. (2013)
		Spain	Municipal wastewater	1150	200	Sanchez-Avila et al. (2009)
		China	Coking wastewater	700,000	350,000	Zhang et al. (2012)
		China	Petrochemical wastewater	n.d.	1080	Wang et al. (2007)

PCBs	Pentachlorobiphenyl	USA	Municipal wastewater	0.595	0.397	Pham and Proulx (1997)
	Tetrachlorobiphenyl	USA	Municipal wastewater	0.630	0.229	Pham and Proulx (1997)
	Naphthalene	China	Municipal wastewater	205.6	56.7	Tian et al. (2012)
	Acenaphthene	China	Municipal wastewater	5.3	0.4	Tian et al. (2012)
	Fluorene	China	Municipal wastewater	230	28.9	Tian et al. (2012)
PFOA		Singapore	Municipal wastewater	16.3	5	Yu et al. (2009)
		China	Municipal wastewater	46,000	2,200,000	Chen et al. (2012)
		Denmark	Municipal wastewater	18.6	2	Bossi et al. (2008)
PFOS		Singapore	Municipal wastewater	13.9	7.3	Yu et al. (2009)
		China	Municipal wastewater	88.9	37.9	Chen et al. (2012)
		Denmark	Municipal wastewater	3.3	1.5	Bossi et al. (2008)

[a] *-: no data in the reference.*

commercial products containing PFOA and PFOS materials, as well as the use of products containing PFOA and PFOS will cause these chemials occur in wastewater treatment system through surface runoff and sewer networks (Guo et al., 2010).

SCCPs are a class of highly complex technical mixtures, and contain a great variety of isomers, diastereomers, and enantiomers. Generally, SCCPs are produced in poly vinyl chloride, flame retardant, and metal cutting fluid industries (Wei et al., 2016). Discharge of SCCPs into wastewater can occur during production, usage, and disposal or recycling of these products (Zeng et al., 2012).

2.2.2 Occurrence and concentrations of POPs in wastewater

The concentrations of POPs in different WWTPs vary greatly. Table 2.6 summarizes the information regarding the occurrence and concentrations of PPCPs in wastewater in different countries.

The concentrations of most of POPs such as OCPs, PAHs, PCBs, and PFOS listed in Table 2.6 range from ng/L to μg/L. Only PFOA reported by Chen et al. (2012) in municipal wastewater in China is up to the concentration of 2.2 mg/L. Comparing the concentration of POPs in municipal wastewater with them in industrial wastewater, the markedly higher concentration detected in industrial wastewater can be found. For example, the concentration of PAHs in municipal wastewater ranges from ND to 1150 ng/L. However, the concentration of PAHs in coking wastewater is 700,000 ng/L and 35,000 ng/L for influent and effluent, respectively. In petrochemical wastewater, the concentration of PAHs in effluent is up to 1080 ng/L. Moreover, the concentrations of POPs in wastewater are geographically different. For example, dieldrin has the concentrations of 27 and 12 ng/L in the influent of municipal wastewater in Greece and Cyprus, respectively, while PFOS has the concentrations of 7.3, 37.9, and 1.5 ng/L in effluent in Singapore, China, and Denmark, respectively. Higher levels of PFOS can be detected in WWTPs of China than in the other countries, because industrial wastewater is often discharged into WWTPs after pretreatment in China. It is worth noted that some POPs, such as diuron, aldrin, fluorene, and PFOA, have higher concentrations in effluent than those in influent, which could be explained by that POPs are stored in the activated sludge of WWTPs and then released into the aqueous phase during wastewater treatment process (Chen et al., 2012).

2.2.3 Migration and transformation of POPs in wastewater

In wastewater, hydrolyzation, volatilization to gas phase, photolysis, adsorption to activated sludge or suspended particles, and biodegradation process are the major migration and transformation routes for POPs (Kwon et al., 2014). Depending on the chemical properties of POPs, different types of POPs undergo different migration and transformation processes.

Biodegradation and adsorption are the main removal routes for OCPs in wastewater. For example, Kwon et al. (2014) investigated the fate of lindane in the WWTPs and found that sorption on primary sludge solids was the major removal mechanism. A significant linear correlation between the compound's partition coefficient and the organic fraction of primary sludge was found. Stasinakis et al. (2009) investigated diuron biodegradation in activated sludge batch reactors under aerobic and anoxic conditions. The results indicated that almost 60% of diuron is biodegraded under aerobic conditions, and the existence of anoxic conditions increased diuron biodegradation to more than 95%.

PCBs can be adsorbed, biodegraded, and bioaccumulated during wastewater treatment processes. Because of their high lipid solubility, PCBs have much higher concentrations in sediments and suspended substances than in wastewater (Balasubramani et al., 2014). According to the Kow, (the value is generally larger when PCBs is with a high number of chlorine atoms) PCBs with a low number of chlorine atoms presented lower adsorption strength in sediments compared with a high number of chlorine atoms. It is noteworthy that PCBs in sediments can be re-released into water bodies through the method of biodegradation (Balasubramani et al., 2014). Plenty of studies have investigated the biodegradation of PCBs by aerobic microorganisms in activated sludge, and the results indicated that PCBs with one, two, three, and four chlorine atoms have good biodegradation performance, while other counterpart PCBs are resistant to be biodegraded (Bergqvist et al., 2006). The counterparts with high content chlorine atom can be degraded in anaerobic conditions through reduced dechlorination, which is beneficial to the formation of low chlorinated biphenyls. Furthermore, low chlorinated biphenyls will be degraded by aerobic degradation (Balasubramani et al., 2014). Due to the high lipid solubility of PCBs, they tend to accumulate in fatty tissues. The accumulation of PCBs in fatty tissue is influenced by exposure level, exposure time, molecular structure of the compound, and type of substitution. Generally, PCBs with more chlorine atoms are conductive to the accumulation (Dinn et al., 2012).

PAHs generally are removed partially in wastewater treatment system. The main mechanisms for PAHs removal in these processes include adsorption onto suspended solids and sludge, biotransformation (biodegradation), and volatilization. Other abiotic degradation routes, such as photolysis and hydrolyzation, may also occur naturally. In conventional activated sludge system, significant amounts of PAHs removal rate have been observed (Cao et al., 2018). Adsorption of PAHs to bacterial cells, including active surface reaction and nonspecific partitioning phenomena, has been considered as the most likely removal process (Zhang et al., 2019). Biotransformation of PAHs mainly includes metabolic reaction on mixed substrate and cometabolic reaction. In metabolic reaction, microorganisms use target pollutants together with other compounds as energy and/or carbon source for cell maintenance, growth, and reproduction. Microorganisms utilize other substrates as their energy and/or carbon source for cometabolic reaction. Furthermore, unspecific enzymes, such as monooxygenases, dioxygenases, and hydrolases, or cofactors, produced during transformation of growth substrate could degrade the micropollutants by side reactions. Generally, the mechanism of aerobic biodegradation of PAHs is that dioxygenase enzymes initially oxidize the benzene rings of PAHs for the formation of cis-dihydrodiols, and then bioconverted to dehydroxylated intermediates by dehydrogenation and ended with CO_2 and H_2O productions (Gong et al., 2017).

According to the previous studies, PFOA and PFOS seem not to be consistently removed during secondary biological treatment due to the hydrophobic and oleophobic (Arvaniti et al., 2012). In some cases, the concentrations of specific PFOA and PFOS in treated wastewater are higher compared to raw sewage (Loganathan et al., 2007), indicating their formation via biotransformation of precursor compounds. In a previous study, Key et al. (1998) reported that PFOS is microbiologically inert under aerobic conditions by a pure bacterial culture. Recently, Kwon et al. (2014) found that PFOS can be decomposed up to 67% by a specific microorganism (*Pseudomonas aeruginosa*), but the formation of fluoride ion from PFOS degradation was not observed. However, PFBS and PFHxS were detected as minor products. The biodegradation and transformation of PFOA have been examined by Liou et al. (2010) employing five different anaerobic microbial communities, which originated from WWTPs, industrial sediment, agricultural soil, and soils of two fire training areas, separately. Moreover, the results indicated that PFOA is biologically inactive under all the examined conditions. Sorption could be another important mechanism for PFOAs removal during wastewater treatment. Generally, distribution coefficients (K_d) and organic carbon distribution coefficients (K_{oc}) for PFOS and PFOA are recognized as important factors on adsorption volume (Sinclair and Kannan, 2006).

2.3 Pharmaceutical and personal care products

PPCPs contain a wide range of organic compounds, including pharmaceuticals such as antibiotics, hormones, analgesics, and antiinflammatory drugs, antiepileptic drugs, blood lipid regulators, β-blockers, contrast media, cytostatic drugs, licit stimulants, and illicit drugs. Moreover, personal care products mainly consist of antimicrobial agents/disinfectants, synthetic musks/fragrances, insect repellants, preservatives, and sunscreen UV filters (Liu and Wong, 2013). They have raised significant concerns in recent years for their persistent input and potential threat to ecological environment and human health (Dai et al., 2014; Wu et al., 2019). This section introduces the source, occurrence, concentrations, migration, and transformation of these PPCPs in wastewater, which will provide a clear understanding on ecological risk of PPCPs in wastewater to readers.

2.3.1 Sources of PPCPs in wastewater

As people pay more attention to their own health, pharmaceuticals are produced in large quantities. Generally, the pharmaceutical factory is an important source for pharmaceutical into environment water bodies (Lei et al., 2010). In addition, a large volume of pharmaceuticals is used as veterinary drug, which are often overused to keep poultry healthy. Therefore, some proportion of pharmaceuticals are excreted in poultry droppings and then released into the surrounding water (Jiang et al., 2013).

In analogy, for human beings, many drugs are excreted without metabolism and consequently enter wastewater either in their parent or metabolized form (Fatta-Kassinos et al., 2011). Vieno and Sillanpaa (2014) pointed that only 6%–7% of diclofenac is absorbed while the rest is washed off the skin or glued to the clothing. Furthermore, 65%–67% of diclofenac is excreted by urine, and 20%–30% is present in feces with a parent drug or metabolite forms.

For personal care products, different used personal care products come from different sources generally. Antimicrobials/disinfectants are generally used for the therapeutic treatment of bacterial diseases for human beings, and some are also used for animals, such as cattle, fish, and poultry for growth promotion and for disease therapy and treatment (Miao et al., 2004). Therefore, the source of antimicrobials/disinfectants is basically the same as the source of pharmaceuticals in wastewater. For synthetic musks/fragrances, the low-cost substitutes for natural musks have been largely used as fragrances in most of cleaning agents and cosmetic products (Liu et al., 2014). For preservatives, they generally are used in a wide range of toothpastes, cosmetics, hair styling products, and sunscreens (Kimura et al., 2014). Most of PPCPs are released into the wastewater through sewer lines system.

2.3.2 Occurrence and concentrations of PPCPs in wastewater

As presented in Table 2.7, the concentration of PPCPs in wastewater ranged between ng/L and μg/L level. Generally, the concentration of PPCPs in hospital wastewater or pharmaceutical wastewater is markedly higher than that it in municipal wastewater. Similar trend for the concentration of PPCPs, such as ciprofloxacin, diclofenac, and naproxen, in municipal wastewater, hospital wastewater, and pharmaceutical wastewater can also be found. In different countries, the concentration of different PPCPs in wastewater is different. For example, the concentration of sulfamethoxazole in wastewater in Singapore and China was with significant difference. The concentration of ciprofloxacin in wastewater in Korea and Croatia is at the same level. However, the concentration of atenolol in the United Kingdom is obviously higher than that in Germany. Similar trend can be found for bezafibrate in China and Germany. The concentration of PPCPs decreases during wastewater treatment process, which demonstrates that wastewater treatment plants play a considerable role for the removal of PPCPs.

2.3.3 Migration and transformation of PPCPs in wastewater

PPCPs can be removed or be retained in WWTPs after entering the wastewater. Generally, biodegradation, adsorption, volatilization, hydrolyzation, and photolysis are the likely transformation routes in wastewater treatment process (Li and Zhang, 2010). For most of PPCPs, removal caused by volatilization is insignificant attributed to the relatively greater molecular weight (Yang et al., 2017a). Similarly, photolysis generally cannot account for a mentionable proportion of removal for PPCPs in WWTPs. This can be explained by the turbidity in WWTPs that is unbeneficial to

Table 2.7 Concentration of PPCPs in wastewater in different countries.

Category	Compound	Country	Wastewater type	Influent concentration (ng/L)	Effluent concentration (ng/L)	Reference
Antibiotics	Sulfamethoxazole	Singapore	Municipal wastewater	1172	311.3	Tran et al. (2016)
		China	Municipal wastewater	340.7	64.1	Ben et al. (2018)
		France	Hospital wastewater	–[a]	2100	Dinh et al. (2017)
		China	Breeding wastewater	–[a]	40	Zhi et al. (2018)
		Republic of Korea	Pharmaceutical wastewater	166,000	137,000	Sim et al. (2011)
	Sulfamethazine	Singapore	Municipal wastewater	802.8	135.9	Tran et al. (2016)
	Ciprofloxacin	Republic of Korea	Hospital wastewater	1980	3080	Sim et al. (2011)
		Republic of Korea	Pharmaceutical wastewater	8710	1860	Sim et al. (2011)
		Croatia	Municipal wastewater	2610	–[a]	Senta et al. (2013)
Nonsteroidal antiinflammatory drugs	Diclofenac	China	Municipal wastewater	290	190	Sui et al. (2011)
		USA	Municipal wastewater	280	12	Yu et al. (2013)
		Canada		216,000	214,000	Lajeunesse and Gagnon (2007)
	Naproxen	UK	Municipal wastewater	838	370	Kasprzyk-Hordern et al. (2009)
		Finland	Municipal wastewater	4900	840	Lindqvist et al. (2005)
		Republic of Korea	Pharmaceutical wastewater	59,700	13,300	Sim et al. (2011)

Antiepileptic drugs	Carbamazepine	Germany	Municipal wastewater	660	740	Wick et al. (2009)
		UK	Municipal wastewater	950	826	Kasprzyk-Hordern et al. (2009)
β-blockers	Atenolol	Germany	Municipal wastewater	540	300	Wick et al. (2009)
		UK	Municipal wastewater	12,913	2870	Kasprzyk-Hordern et al. (2009)
Contrast media	Iohexol	China	Municipal wastewater	8.2	8.16	Yang et al. (2017b)
Licit stimulants	Caffeine	China	Municipal wastewater	108	3.65	Yang et al. (2017b)
		India	Municipal wastewater	759	13	Mohapatra et al. (2016)
Lipid regulators	Bezafibrate	China	Municipal wastewater	72.6	11.4	Mohapatra et al. (2016)
		Germany	Municipal wastewater	1500	500	Gurke et al. (2015)

[a] –, no data in the reference.

phototransformation. Biodegradation generally plays a crucial role in degrading PPCPs. For example, biodegradation of diclofenac accounted for about 80% for the total removal volume in activated sludge (Wu et al., 2019). For other PPCPs such as antibiotic, the removal attributed to adsorption, hydrolyzation, volatilization, and biodegradation process has been explored by Li and Zhang (2010) and the results demonstrated that other than biodegradation, other removal routes played a limited role (less 20%) for the removal of antibiotic in wastewater. It is noted that PPCPs generally cannot be eliminated completely by biological treatment methods (Tran et al. 2017). Some advanced treatment methods such as advanced oxidation processes (Zhang et al., 2016b; Fu et al., 2019a) and disinfection (Zhang et al., 2015b) are promising approaches to effectively remove PPCPs in wastewater effluents.

Due to PPCPs cannot be metabolized completely in wastewater treatment system, some intermediate products are generated in wastewater treatment process (Gulde et al., 2016; Helbling et al., 2010). With the development of technology in identifying transformation products, microbial mediated transformation products of PPCPs have increasingly received attention. Some transformation products have been detected and quantified in WWTPs concurrent with parent PPCPs (Beretsou et al., 2016). In addition, some transformation products have been found to be more toxicant than parent compounds (Gao et al., 2018; Escher and Fenner, 2011). Therefore, transformation products and parent compounds should receive equivalent attention.

2.4 Endocrine-disrupting chemicals

Endocrine-disrupting compounds refer to "exogenous substances that alter function(s) of the endocrine system and consequently cause adverse health effects in an intact organism, or its progeny, or (sub)-populations" (WHO/IPCS, 2002). Therefore, any compounds that can trigger reproductive change can be regarded as EDCs. Generally, EDCs consist of estrogens and some plant metabolites, some pesticides such as dichlorodiphenyltrichloroethane dieldrin, and lindane, industrial chemical products such as dioxins and dioxin-like compounds, the formerly widely used ship antifouling agent tributyltin, organotin compounds used in polyvinyl chloride water pipe manufacture (Auriol et al., 2006). In this section, the main sources, concentration, and fate of EDCs in wastewater are introduced.

2.4.1 Sources of EDCs in wastewater

The source of EDCs categorized as pesticide is generally agricultural areas and surface runoff. In addition, in nonagricultural crops such as commercial forestry and horticulture and plant nurseries, pesticide can also be widely used and discharged into wastewater (Kock-Schulmeyer et al., 2013). For EDCs like hormones, pharmaceutical factories are an important point source. In addition, hospital wastewater and human and animal excretion are also important sources (Hamid and Eskicioglu, 2012). For EDCs like industrial chemical products, the major source for them is industrial manufacture (Jackson and Sutton, 2008). For bisphenol A, it usually is used as industrial raw materials; therefore, the source of it is similar as the industrial chemical products (Huang et al., 2012).

2.4.2 Occurrence and concentrations of EDCs in wastewater

As listed in Table 2.8, the concentration for EDCs in wastewater ranges between ng/L and μg/L level. For most EDCs, the concentration is almost under μg/L except for androsterone and bisphenol A in influent of municipal wastewater. Generally, the removal of EDCs in wastewater treatment process is efficient. Therefore, the concentration of EDCs in effluent of municipal wastewater is too low even to be detected for some EDCs. In different regions, although the difference exists for the influent concentration for some EDCs such as androsterone and bisphenol A, the effluent concentration for these EDCs is similar. Compared with municipal wastewater, the concentration of EDCs in hospital wastewater is at the similar level as it in municipal wastewater. It is worth noted that the concentration of androsterone in China and the concentration of Bisphenol A in Spain is especially high. This may be due to the volume of use for EDCs is high for the locals or the municipal wastewater was mixed with industrial wastewater.

2.4.3 Migration and transformation of EDCs in wastewater

In wastewater treatment process, the following process can occur for EDCs: hydrolyzation, volatilization, photolysis, adsorption, and biodegradation. According to the concentration shown in Table 2.8 for the influent and effluent, an efficient removal can be found. Generally, EDCs has a large molecular weight due to its ring structure, which is not conducive to volatilization removal. The adsorption removal of EDCs in wastewater depends on its octanol water distribution coefficient. According to previous reports, removal of EDCs in wastewater attributed by adsorption is limited. In wastewater treatment system, light transmission is generally poor. This is not conducive to the removal of EDCs attributed by photolysis. Biodegradation is the major removal route for EDCs in wastewater (Yu et al., 2018). In previous studies, Manickum and John (2014) demonstrated that the biodegradation make the major contribution for the removal of natural and synthetic estrogen. In addition, the high removal efficiency for most EDCs listed in Table 2.8 has also verified this point.

2.5 Other HRPs

Except for the HRPs mentioned earlier, some other contaminants, such as disinfection by-products (DBPs), microplastic, and nanomaterials, present in wastewater have also raised people's concern. DBPs refer to a series of by-products produced by disinfectant and some organic and inorganic substances in water during disinfection with disinfectant (Krasner et al., 2009; Yang and Zhang, 2016). Microplastics,

Table 2.8 Concentration of EDCs in wastewater in different countries.

Category	Compound	Country	Wastewater type	Influent concentration (ng/L)	Effluent concentration (ng/L)	Reference
	Estrone	Korea	Municipal wastewater	47	6	Behera et al. (2011)
		USA	Municipal wastewater	49.8	9.0	Esperanza et al. (2007)
		Italy	Municipal wastewater	132	8.2	Baronti et al. (2000)
		Belgium	Hospital wastewater	58.3	4	Pauwels et al. (2008)
		China	Municipal wastewater	180	–[a]	Zhou et al. (2012)
	Estradiol	USA	Municipal wastewater	44.6	1.0	Esperanza et al. (2007)
		Korea	Municipal wastewater	4	0	Behera et al. (2011)
		China	Municipal wastewater	20	n.d.	Zhou et al. (2012)
		Brazil	Municipal wastewater	143	n.d.	Pessoa et al. (2014)
	Androstenedione	China	Municipal wastewater	232	3.2	Liu et al. (2011)
		China	Municipal wastewater	177	4.5	Chang et al. (2011)
	Androsterone	China	Municipal wastewater	305	n.d.	Liu et al. (2011)
		China	Municipal wastewater	2766	n.d.	Liu et al. (2011)

	Progesterone		Municipal wastewater	342	8	Manickum and John (2014)
		South Africa	Municipal wastewater	34	5	Fan et al. (2011)
	Dydrogesterone	China	Municipal wastewater	45	1	Yu et al. (2019)
		China	Municipal wastewater	18.5	1.6	Pauwels et al. (2008)
	Bisphenol A	Belgium	Hospital wastewater	100	n.d.	Cesen et al. (2018)
		Slovenia	Municipal wastewater	90	42.3	Xue and Kannan (2019)
		USA	Municipal wastewater	2400	n.d.	Sanchez-Avila et al. (2009)

[a] -: no data in the reference.

often defined as plastic particles <5 mm, have aroused increasing concerns as they pose threat to aquatic species as well as human beings. The rapid growth of nanotechnology has resulted in various implementations of nanomaterials in advantageous products or as process enhancers in manufacturing, and nanoparticles have been largely detected in wastewater treatment system. Researchers found that once released into the environment, nanomaterials might pose the potential risks to human health and microorganisms (Zheng et al., 2011). Therefore, enough attention should be paid on these materials.

2.5.1 Sources of other HRPs in wastewater

Other HRPs, such as artificial sweeteners (Ren et al., 2016; Fu et al., 2019b), DBPs (Zhang et al., 2015a), microplastics (Sun et al., 2019), and nanomaterials (Gottschalk et al., 2013), have received greater concern because of their prevalent and wide occurrence in the environment. DBPs are mainly formed in the process of water and wastewater disinfection; therefore, more attention has been paid in water and wastewater treatment process. In addition, in some countries such as the USA, chlorine or chloramine disinfection has been practiced in wastewater treatment system to guarantee the reclaimed wastewater (Krasner et al., 2009). The rapid growth of nanotechnology has resulted in various implementations of nanomaterials in advantageous products or as process enhancers in manufacturing. For example, carbon nanotubes are used as absorption material due to their excellent physical and chemical properties (Li et al., 2005). Iron oxide nanomaterials have been widely used in water treatment field (Xu et al., 2012). With their extensive use, they inevitably enter the environmental system, and finally enter the sewage treatment system with the sewage pipe network.

Microplastics can be directly manufactured, known as primary microplastics, and be used accompany with many PPCPs. Furthermore, they can be formed by erosion of large plastic debris via exposure to environmental stressors such as water, wind, and sunlight, which are defined as secondary microplastics. The massive usage of plastic products and poor management of plastic waste disposal leads to microplastics being ubiquitously found in aquatic water bodies, including rivers, lakes, estuaries, coastlines, and marine ecosystems (Sun et al., 2019).

2.5.2 Occurrence and concentrations of other HRPs in wastewater

At present, chlorine disinfection is rarely used in sewage treatment, so there are few reports on disinfection by-products in sewage. In a study conducted by Krasner et al. (2009), the well-nitrified effluents had 57 μg/L of trihalomethanes (THMs) and 3 ng/L of N-nitrosodimethyamine (NDMA), while the poorly nitrified effluents had 2 μg/L of THMs and 11 ng/L of NDMA. In addition, the formation of precursor of DBPs has been widely reported (Zhang et al., 2015a). For example, Tang et al. (2012) demonstrated that removal efficiency of organic carbon and nitrogen compound in wastewater treatment plants were positively correlated with the formation of DBPs. Krasner et al. (2009) confirmed that whether or not the well-nitrified effluent effect the formation relative volume of DBPs and NDMA.

Regarding the nanomaterial, different use and different use frequency cause the different exposure level. Gottschalk et al. (2013) reviewed six engineered nanomaterials (TiO_2, ZnO, Ag, fullerenes, CNT, and CeO_2) in wastewater. They found a concentration of 1.58 ± 0.59 mg/L of nano-ZnO particles in their effluent water sample. A high consistency was observed for nano-TiO_2; several studies suggest concentrations of about 5 mg/L.

In a previous review involved in the occurrence of microplastics in wastewater, microplastics in the influent were only measured in a few WWTPs, with the particle concentrations reported varying from 1 to 10,044 particle/L, and the measured microplastics concentrations in the effluent of WWTPs were much lower, which were in the range of 0—447 particle/L (Sun et al., 2019).

2.5.3 Migration and transformation of other HRPs in wastewater

DBPs are usually produced during the final process of disinfection of the water treatment, so the disinfection by-products generated are no longer discharged directly into the downstream water body through the sewage treatment system.

Generally, due to the non-biodegradability, nano-material cannot be degraded in sewage treatment system based biological treatment technology. Absorption is the main removal route for nano-material in wastewater treatment system. Kiser et al. (2009) investigated the removal and release of titanium in WWTPs and found that the sample containing 2250 mg/L TSS had approximately 85% for Ti.

Regarding microplastics, the overall microplastics removal efficiencies of WWTPs without tertiary treatment were above 88% and the number increased to over 97% in the WWTPs with tertiary treatment. The relatively high removal efficiency of microplastics by WWTPs indicated that most microplastics were retained in the sewage sludge (Sun et al., 2019).

2.6 Summary

This chapter provides a comprehensive introduction on chemical HRPs including heavy metals, POPs, PPCPs, and EDCs in wastewater. The source, contamination levels, and transfer characteristics of typical HRPs are summarized in detail. The sources and contamination levels of different HRPs in wastewater are different generally. The concentrations of HRPs range from ppt to ppb level in municipal wastewater and ppb to ppm level in industrial wastewater. The likely migration and transformation routes of HRPs in wastewater treatment processes are absorption, hydrolyzation, volatilization, biodegradation, and bioconcentration. Among these routes, absorption and biodegradation play a dominate role for the removal of most organic HRPs in wastewater. Due to the wide existence and potential risk of chemical HRPs, further study should focus on the complex interaction of chemical HRPs with other substances in wastewater and aquatic environment. Furthermore, the potential adverse effect of these chemical HRPs on ecological systems and human beings deserves great attention. Furthermore, the control of chemical HRPs in wastewater should also be explored.

References

Acheampong, M.A., Meulepas, R.J.W., Lens, P.N.L., 2010. Removal of heavy metals and cyanide from gold mine wastewater. Journal of Chemical Technology & Biotechnology 85, 590–613.

Al Enezi, G., Hamoda, M.F., Fawzi, N., 2004. Heavy metals content of municipal wastewater and sludges in Kuwait. Journal of Environmental Health Science 39, 397–407.

Al-Shannag, M., Al-Qodah, Z., Bani-Melhem, K., Qtaishat, M.R., Alkasrawi, M., 2015. Heavy metal ions removal from metal plating wastewater using electrocoagulation: kinetic study and process performance. Chemical Engineering Journal 260, 749–756.

Al-Zoubi, H., Ibrahim, K.A., Abu-Sbeih, K.A., 2015. Removal of heavy metals from wastewater by economical polymeric collectors using dissolved air flotation process. Journal of Water Process Engineering 8, 19–27.

Ali, N., Hameed, A., Ahmed, S., 2009. Physicochemical characterization and bioremediation perspective of textile effluent, dyes and metals by indigenous bacteria. Journal of Hazardous Materials 164, 322–328.

Arivoli, A., Mohanraj, R., Seenivasan, R., 2015. Application of vertical flow constructed wetland in treatment of heavy metals from pulp and paper industry wastewater. Environmental Science and Pollution Research International 22, 13336–13343.

Arslan-Alaton, I., Tanik, A., Ovez, S., Iskender, G., Gurel, M., Orhon, D., 2007. Reuse potential of urban wastewater treatment plant effluents in Turkey: a case study on selected plants. Desalination 215, 159–165.

Arvaniti, O.S., Ventouri, E.I., Stasinakis, A.S., Thomaidis, N.S., 2012. Occurrence of different classes of perfluorinated compounds in Greek wastewater treatment plants and determination of their solid-water distribution coefficients. Journal of Hazardous Materials 239–240, 24–31.

Auriol, M., Filali-Meknassi, Y., Tyagi, R.D., Adams, C.D., Surampalli, R.Y., 2006. Endocrine disrupting compounds removal from wastewater, a new challenge. Process Biochemistry 41, 525–539.

Balasubramani, A., Howell, N.L., Rifai, H.S., 2014. Polychlorinated biphenyls (PCBs) in industrial and municipal effluents: concentrations, congener profiles, and partitioning onto particulates and organic carbon. The Science of the Total Environment 473–474, 702–713.

Bao, L.J., Maruya, K.A., Snyder, S.A., Zeng, E.Y., 2012. China's water pollution by persistent organic pollutants. Environment and Pollution 163, 100–108.

Barco-Bonilla, N., Romero-Gonzalez, R., Plaza-Bolanos, P., Martinez Vidal, J.L., Garrido Frenich, A., 2013. Systematic study of the contamination of wastewater treatment plant effluents by organic priority compounds in Almeria province (SE Spain). The Science of the Total Environment 447, 381–389.

Baronti, C., Curini, R., D'Ascenzo, G., Di Corcia, A., Gentili, A., Samperi, R., 2000. Monitoring natural and synthetic estrogens at activated sludge sewage treatment plants and in a receiving river water. Environmental Science and Technology 34, 5059–5066.

Behera, S.K., Kim, H.W., Oh, J.E., Park, H.S., 2011. Occurrence and removal of antibiotics, hormones and several other pharmaceuticals in wastewater treatment plants of the largest industrial city of Korea. The Science of the Total Environment 409, 4351–4360.

Ben, W., Zhu, B., Yuan, X., Zhang, Y., Yang, M., Qiang, Z., 2018. Occurrence, removal and risk of organic micropollutants in wastewater treatment plants across China: comparison of wastewater treatment processes. Water Research 130, 38–46.

Ben Hariz, I., Halleb, A., Adhoum, N., Monser, L., 2013. Treatment of petroleum refinery sulfidic spent caustic wastes by electrocoagulation. Separation and Purification Technology 107, 150–157.

Beretsou, V.G., Psoma, A.K., Gago-Ferrero, P., Aalizadeh, R., Fenner, K., Thomaidis, N.S., 2016. Identification of biotransformation products of citalopram formed in activated sludge. Water Research 103, 205–214.

Bergqvist, P.A., Augulytė, L., Jurjonienė, V., 2006. PAH and PCB removal efficiencies in Umeå (Sweden) and Šiauliai (Lithuania) municipal wastewater treatment plants. Water, Air, & Soil Pollution 175, 291–303.

Boshoff, G., Duncan, J., Rose, P.D., 2004. Tannery effluent as a carbon source for biological sulphate reduction. Water Research 38, 2651–2658.

Bossi, R., Strand, J., Sortkjaer, O., Larsen, M.M., 2008. Perfluoroalkyl compounds in Danish wastewater treatment plants and aquatic environments. Environment International 34, 443–450.

Busetti, F., Badoer, S., Cuomo, M., Rubino, B., Traverso, P., 2005. Occurrence and removal of potentially toxic metals and heavy metals in the wastewater treatment plant of fusina (Venice, Italy). Industrial & Engineering Chemistry Research 44, 9264–9272.

Buzier, R., Tusseau-Vuillemin, M.H., dit Meriadec, C.M., Rousselot, O., Mouchel, J.M., 2006. Trace metal speciation and fluxes within a major French wastewater treatment plant: impact of the successive treatments stages. Chemosphere 65, 2419–2426.

Cao, W., Qiao, M., Liu, B., Zhao, X., 2018. Occurrence of parent and substituted polycyclic aromatic hydrocarbons in typical wastewater treatment plants and effluent receiving rivers of Beijing, and risk assessment. Toxic/Hazardous Substances and Environmental Engineering 53, 992–999. Journal of Environmental Science and Health, Part A.

Carletti, G., Fatone, F., Bolzonella, D., Cecchi, F., Carletti, G., 2008. Occurrence and fate of heavy metals in large wastewater treatment plants treating municipal and industrial wastewaters. Water Science and Technology 57, 1329–1336.

Cempel, M., Nikel, G., 2005. Nickel: a review of its sources and environmental toxicology. Polish Journal of Environmental Studies 15, 375–382.

Cesen, M., Lenarcic, K., Mislej, V., Levstek, M., Kovacic, A., Cimrmancic, B., Uranjek, N., Kosjek, T., Heath, D., Dolenc, M.S., Heath, E., 2018. The occurrence and source identification of bisphenol compounds in wastewaters. The Science of the Total Environment 616–617, 744–752.

Chandra, R., Bharagava, R.N., Yadav, S., Mohan, D., 2009. Accumulation and distribution of toxic metals in wheat (*Triticum aestivum* L.) and Indian mustard (*Brassica campestris* L.) irrigated with distillery and tannery effluents. Journal of Hazardous Materials 162, 1514–1521.

Chandra, R., Yadav, S., Yadav, S., 2017. Phytoextraction potential of heavy metals by native wetland plants growing on chlorolignin containing sludge of pulp and paper industry. Ecological Engineering 98, 134–145.

Chang, H., Wan, Y., Wu, S., Fan, Z., Hu, J., 2011. Occurrence of androgens and progestogens in wastewater treatment plants and receiving river waters: comparison to estrogens. Water Research 45, 732–740.

Chanpiwat, P., Sthiannopkao, S., Kim, K.-W., 2010. Metal content variation in wastewater and biosludge from Bangkok's central wastewater treatment plants. Microchemical Journal 95, 326–332.

Chen, H., Zhang, C., Han, J., Yu, Y., Zhang, P., 2012. PFOS and PFOA in influents, effluents, and biosolids of Chinese wastewater treatment plants and effluent-receiving marine environments. Environment and Pollution 170, 26–31.

Cheng, S., 2003. Heavy metal pollution in China: Origin, pattern and control. Environmental Science and Pollution Research 10, 192–198.

Chipasa, K.B., 2003. Accumulation and fate of selected heavy metals in a biological wastewater treatment system. Waste Management 23, 135–143.

Chojnacka, K., Chojnacki, A., Górecka, H., 2004. Trace element removal by *Spirulina* sp. from copper smelter and refinery effluents. Hydrometallurgy 73, 147–153.

Chongyu Lan, G.C., 1992. Use of cattails in treating wastewater from a Pb/Zn mine. Environmental Management 16, 75–80.

Contreras-Ramos, S.M., Alvarez-Bernal, D., Trujillo-Tapia, N., Dendooven, L., 2004. Composting of tannery effluent with cow manure and wheat straw. Bioresource Technology 94, 223–228.

da Silva Oliveira, A., Bocio, A., Beltramini Trevilato, T.M., Magosso Takayanagui, A.M., Domingo, J.L., Segura-Muñoz, S.I., 2007. Heavy metals in untreated/treated urban effluent and sludge from a biological wastewater treatment plant. Environmental Science and Pollution Research International 14, 483–489.

Dai, G., Huang, J., Chen, W., Wang, B., Yu, G., Deng, S., 2014. Major pharmaceuticals and personal care products (PPCPs) in wastewater treatment plant and receiving water in Beijing, China, and associated ecological risks. Bulletin of Environmental Contamination and Toxicology 92, 655–661.

Dietz, R., Desforges, J.P., Gustavson, K., Riget, F.F., Born, E.W., Letcher, R.J., Sonne, C., 2018. Immunologic, reproductive, and carcinogenic risk assessment from POP exposure in East Greenland polar bears (*Ursus maritimus*) during 1983–2013. Environment International 118, 169–178.

Dinh, Q., Moreau-Guigon, E., Labadie, P., Alliot, F., Teil, M.J., Blanchard, M., Eurin, J., Chevreuil, M., 2017. Fate of antibiotics from hospital and domestic sources in a sewage network. The Science of the Total Environment 575, 758–766.

Dinn, P.M., Johannessen, S.C., Ross, P.S., Macdonald, R.W., Whiticar, M.J., Lowe, C.J., van Roodselaar, A., 2012. PBDE and PCB accumulation in benthos near marine wastewater outfalls: the role of sediment organic carbon. Environment and Pollution 171, 241–248.

Doušová, B., Koloušek, D., Kovanda, F., Machovič, V., Novotná, M., 2005. Removal of As(V) species from extremely contaminated mining water. Applied Clay Science 28, 31–42.

Edokpayi, J.N., Odiyo, J.O., Msagati, T.A., Popoola, E.O., 2015. Removal efficiency of faecal indicator organisms, nutrients and heavy metals from a peri-urban wastewater treatment plant in Thohoyandou, Limpopo Province, South Africa. International Journal of Environmental Research and Public Health 12, 7300–7320.

Edwards, L.C., Freeman, H.S., 2010. Synthetic dyes based on environmental considerations. Part 3: Aquatic toxicity of iron-complexed azo dyes. Coioration Technology 121, 265–270.

El-Shafey, E.I., 2010. Removal of Zn(II) and Hg(II) from aqueous solution on a carbonaceous sorbent chemically prepared from rice husk. Journal of Hazardous Materials 175, 319–327.

Ertuğrul, S., San, N.O., Dönmez, G., 2009. Treatment of dye (Remazol Blue) and heavy metals using yeast cells with the purpose of managing polluted textile wastewaters. Ecological Engineering 35, 128–134.

Escher, B.I., Fenner, K., 2011. Recent advances in environmental risk assessment of transformation products. Environmental Science and Technology 45, 3835–3847.

Esperanza, M., Suidan, M.T., Marfil-Vega, R., Gonzalez, C., Sorial, G.A., McCauley, P., Brenner, R., 2007. Fate of sex hormones in two pilot-scale municipal wastewater treatment plants: conventional treatment. Chemosphere 66, 1535–1544.

Fan, Z., Wu, S., Chang, H., Hu, J., 2011. Behaviors of glucocorticoids, androgens and progestogens in a municipal sewage treatment plant: comparison to estrogens. Environmental Science and Technology 45, 2725–2733.

Fatta, D., Canna-Michaelidou, S., Michael, C., Demetriou Georgiou, E., Christodoulidou, M., Achilleos, A., Vasquez, M., 2007. Organochlorine and organophosphoric insecticides, herbicides and heavy metals residue in industrial wastewaters in Cyprus. Journal of Hazardous Materials 145, 169–179.

Fatta-Kassinos, D., Meric, S., Nikolaou, A., 2011. Pharmaceutical residues in environmental waters and wastewater: current state of knowledge and future research. Analytical and Bioanalytical Chemistry 399, 251–275.

Feng, J., Chen, X., Jia, L., Liu, Q., Chen, X., Han, D., Cheng, J., 2018. Effluent concentration and removal efficiency of nine heavy metals in secondary treatment plants in Shanghai, China. Environmental Science and Pollution Research International 25, 17058–17065.

Fernandes, A.R., Mortimer, D., Rose, M., Smith, F., Steel, Z., Panton, S., 2019. Recently listed Stockholm convention POPs: analytical methodology, occurrence in food and dietary exposure. The Science of the Total Environment 678, 793–800.

Fu, F., Wang, Q., 2011. Removal of heavy metal ions from wastewaters: a review. Journal of Environmental Management 92, 407–418.

Fu, Y.Y., Gao, X.S., Geng, J.J., Li, S.L., Wu, G., Ren, H.Q., 2019a. Degradation of three nonsteroidal anti-inflammatory drugs by UV/persulfate: degradation mechanisms, efficiency in effluents disposal. Chemical Engineering Journal 356, 1032–1041.

Fu, Y.Y., Wu, G., Geng, J.J., Li, J.C., Li, S.N., Ren, H., 2019b. Kinetics and modeling of artificial sweeteners degradation in wastewater by the UV/persulfate process. Water Research 150, 12–20.

Gallistl, C., Lok, B., Schlienz, A., Vetter, W., 2017. Polyhalogenated compounds (chlorinated paraffins, novel and classic flame retardants, POPs) in dishcloths after their regular use in households. The Science of the Total Environment 595, 303–314.

Gao, X.S., Geng, J.J., Du, Y.R., Li, S.L., Wu, G., Fu, Y.Y., Ren, H.Q., 2018. Comparative study of the toxicity between three non-steroidal antiinfammatory drugs and their UV/ $Na_2S_2O_8$ degradation products on *Cyprinus carpio*. Scientific Reports 8, 13512.

Geneja, M., Haustein, E., 2003. Heavy metals removal in the mechanical-biological wastewater treatment plant Wschód" in Gdańsk, Polish Journal of Environmental Studies 12, 635–641.

Ghorbani, M., Eisazadeh, H., 2013. Removal of COD, color, anions and heavy metals from cotton textile wastewater by using,polyaniline and polypyrrole nanocomposites coated on rice husk ash. Composites Part B: Engineering 45, 1–7.

Gong, C., Shen, G., Huang, H., He, P., Zhang, Z., Ma, B., 2017. Removal and transformation of polycyclic aromatic hydrocarbons during electrocoagulation treatment of an industrial wastewater. Chemosphere 168, 58–64.

Gottschalk, F., Sun, T., Nowack, B., 2013. Environmental concentrations of engineered nanomaterials: review of modeling and analytical studies. Environment and Pollution 181, 287–300.

Gulde, R., Meier, U., Schymanski, E.L., Kohler, H.P., Helbling, D.E., Derrer, S., Rentsch, D., Fenner, K., 2016. Systematic exploration of biotransformation reactions of amine-containing micropollutants in activated sludge. Environmental Science and Technology 50, 2908–2920.

Guo, R., Sim, W.J., Lee, E.S., Lee, J.H., Oh, J.E., 2010. Evaluation of the fate of perfluoroalkyl compounds in wastewater treatment plants. Water Research 44, 3476–3486.

Gurke, R., Rossler, M., Marx, C., Diamond, S., Schubert, S., Oertel, R., Fauler, J., 2015. Occurrence and removal of frequently prescribed pharmaceuticals and corresponding metabolites in wastewater of a sewage treatment plant. The Science of the Total Environment 532, 762–770.

Halimoon, N., Gohsoo Yin, G.S.R.Y., 2010. Removal of heavy metals from textile wastewater using zeolite. Environment 3, 124–130.

Hamid, H., Eskicioglu, C., 2012. Fate of estrogenic hormones in wastewater and sludge treatment: a review of properties and analytical detection techniques in sludge matrix. Water Research 46, 5813–5833.

Hedrich, S., Johnson, D.B., 2014. Remediation and selective recovery of metals from acidic mine waters using novel modular bioreactors. Environmental Science and Technology 48, 12206–12212.

Helbling, D.E., Hollender, J., Kohler, H.P.E., Singer, H., Fenner, K., 2010. High-throughput identification of microbial transformation products of organic micropollutants. Environmental Science and Technology 44, 6621–6627.

Hideyuki, K., Satoshi, K., Kentaro, I., Kumiko, I., Kunihiro, F., Kazuaki, M., Tohru, S., Kiyohisa, O., 2003. Removal of heavy metals in rinsing wastewater from plating factory by adsorption with economical viable materials. Journal of Environmental Management 187–191.

Hu, X.-F., Jiang, Y., Shu, Y., Hu, X., Liu, L., Luo, F., 2014. Effects of mining wastewater discharges on heavy metal pollution and soil enzyme activity of the paddy fields. Journal of Geochemical Exploration 147, 139–150.

Hu, Q., Zhang, X.X., Jia, S., Huang, K., Tang, J., Shi, P., Ye, L., Ren, H., 2016. Metagenomic insights into ultraviolet disinfection effects on antibiotic resistome in biologically treated wastewater. Water Research 101, 309–317.

Huang, Y.Q., Wong, C.K., Zheng, J.S., Bouwman, H., Barra, R., Wahlstrom, B., Neretin, L., Wong, M.H., 2012. Bisphenol A (BPA) in China: a review of sources, environmental levels, and potential human health impacts. Environment International 42, 91–99.

Jackson, J., Sutton, R., 2008. Sources of endocrine-disrupting chemicals in urban wastewater, Oakland, CA. The Science of the Total Environment 405, 153–160.

Jiang, H.Y., Zhang, D.D., Xiao, S.C., Geng, C.N., Zhang, X., 2013. Occurrence and sources of antibiotics and their metabolites in river water, WWTPs, and swine wastewater in Jiulongjiang River basin, south China. Environmental Science & Pollution Research 20, 9075–9083.

Jiménez-Rodríguez, A.M., Durán-Barrantes, M.M., Borja, R., 2009. Heavy metals removal from acid mine drainage water using biogenic hydrogen sulphide and effluent from anaerobic treatment: effect of pH. Journal of Hazardous Materials 165, 759–765.

Karvelas, M., Katsoyiannis, A., Samara, C., 2003. Occurrence and fate of heavy metals in the wastewater treatment process. Chemosphere 53, 1201–1210.

Kasprzyk-Hordern, B., Dinsdale, R.M., Guwy, A.J., 2009. The removal of pharmaceuticals, personal care products, endocrine disruptors and illicit drugs during wastewater treatment and its impact on the quality of receiving waters. Water Research 43, 363–380.

Katsoyiannis, A., Samara, C., 2004. Persistent organic pollutants (POPs) in the sewage treatment plant of Thessaloniki, Northern Greece: occurrence and removal. Water Research 38, 2685–2698.

Key, B.D., Howell, R.D., Criddle, C.S., 1998. Defluorination of organofluorine sulfur compounds by *Pseudomonas* sp. strain D2. Environmental Science and Technology 32, 2283–2287.

Khadhar, S., Higashi, T., Hamdi, H., Matsuyama, S., Charef, A., 2010. Distribution of 16 EPA-priority polycyclic aromatic hydrocarbons (PAHs) in sludges collected from nine Tunisian wastewater treatment plants. Journal of Hazardous Materials 183, 98–102.

Kimura, K., Kameda, Y., Yamamoto, H., Nakada, N., Tamura, I., Miyazaki, M., Masunaga, S., 2014. Occurrence of preservatives and antimicrobials in Japanese rivers. Chemosphere 107, 393–399.

Kiser, M.A., Westerhoff, P., Benn, T., Wang, Y., Perez-Rivera, J., Hristovski, K., 2009. Titanium nanomaterial removal and release from wastewater treatment plants. Environmental Science and Technology 43, 6757–6763.

Krasner, S.W., Westerhoff, P., Chen, B.Y., Rittmann, B.E., Amy, G., 2009. Occurrence of disinfection byproducts in United States wastewater treatment plant effluents. Environmental Science and Technology 43, 8320–8325.

Kulbat, E., Olańczuk-Neyman, K., Quant, B., Geneja, M., Haustein, E., 2003. Heavy metals removal in the mechanical-biological wastewater treatment plant Wschód" in Gdańsk, Polish Journal of Environmental Studies 12, 635–641.

Kumar, V., Chopra, A.K., Benskin, J., 2016. Reduction of pollution load of paper mill effluent by phytoremediation technique using water caltrop (*Trapa natans* L.). Cogent Environmental Science 2 (1), 1153216.

Kwon, B.G., Lim, H.J., Na, S.H., Choi, B.I., Shin, D.S., Chung, S.Y., 2014. Biodegradation of perfluorooctanesulfonate (PFOS) as an emerging contaminant. Chemosphere 109, 221–225.

Lajeunesse, A., Gagnon, C., 2007. Determination of acidic pharmaceutical products and carbamazepine in roughly primary-treated wastewater by solid-phase extraction and gas chromatography–tandem mass spectrometry. International Journal of Environmental Analytical Chemistry 87, 565–578.

Lei, G., Ren, H., Ding, L., Wang, F., Zhang, X., 2010. A full-scale biological treatment system application in the treated wastewater of pharmaceutical industrial park. Bioresource Technology 101, 5852–5861.

Li, D., Suh, S., 2019. Health risks of chemicals in consumer products: a review. Environment International 123, 580–587.

Li, B., Zhang, T., 2010. Biodegradation and adsorption of antibiotics in the activated sludge process. Environmental Science and Technology 44, 3468–3473.

Li, Y.H., Di, Z., Ding, J., Wu, D., Luan, Z., Zhu, Y., 2005. Adsorption thermodynamic, kinetic and desorption studies of Pb^{2+} on carbon nanotubes. Water Research 39, 605–609.

Li, X., Zhang, Q., Dai, J., Gan, Y., Zhou, J., Yang, X., Cao, H., Jiang, G., Xu, M., 2008. Pesticide contamination profiles of water, sediment and aquatic organisms in the effluent of Gaobeidian wastewater treatment plant. Chemosphere 72, 1145–1151.

Lim, S.L., Chu, W.L., Phang, S.M., 2010. Use of *Chlorella vulgaris* for bioremediation of textile wastewater. Bioresource Technology 101, 7314–7322.

Lindqvist, N., Tuhkanen, T., Kronberg, L., 2005. Occurrence of acidic pharmaceuticals in raw and treated sewages and in receiving waters. Water Research 39, 2219–2228.
Liou, J.S., Szostek, B., DeRito, C.M., Madsen, E.L., 2010. Investigating the biodegradability of perfluorooctanoic acid. Chemosphere 80, 176–183.
Liu, J.L., Wong, M.H., 2013. Pharmaceuticals and personal care products (PPCPs): a review on environmental contamination in China. Environment International 59, 208–224.
Liu, S., Ying, G.G., Zhao, J.L., Chen, F., Yang, B., Zhou, L.J., Lai, H.J., 2011. Trace analysis of 28 steroids in surface water, wastewater and sludge samples by rapid resolution liquid chromatography-electrospray ionization tandem mass spectrometry. Journal of Chromatography A 1218, 1367–1378.
Liu, N., Shi, Y., Li, W., Xu, L., Cai, Y., 2014. Concentrations and distribution of synthetic musks and siloxanes in sewage sludge of wastewater treatment plants in China. The Science of the Total Environment 476–477, 65–72.
Loganathan, B.G., Sajwan, K.S., Sinclair, E., Senthil Kumar, K., Kannan, K., 2007. Perfluoroalkyl sulfonates and perfluorocarboxylates in two wastewater treatment facilities in Kentucky and Georgia. Water Research 41, 4611–4620.
Ma, S.C., Zhang, H.B., Ma, S.T., Wang, R., Wang, G.X., Shao, Y., Li, C.X., 2015. Effects of mine wastewater irrigation on activities of soil enzymes and physiological properties, heavy metal uptake and grain yield in winter wheat. Ecotoxicology and Environmental Safety 113, 483–490.
Magulova, K., Priceputu, A., 2016. Global monitoring plan for persistent organic pollutants (POPs) under the Stockholm convention: triggering, streamlining and catalyzing global POPs monitoring. Environment and Pollution 217, 82–84.
Mahmoued, E.K., 2010. Cement kiln dust and coal filters treatment of textile industrial effluents. Desalination 255, 175–178.
Maine, M.A., Hadad, H.R., Sánchez, G.C., Di Luca, G.A., Mufarrege, M.M., Caffaratti, S.E., Pedro, M.C., 2017. Long-term performance of two free-water surface wetlands for metallurgical effluent treatment. Ecological Engineering 98, 372–377.
Manickum, T., John, W., 2014. Occurrence, fate and environmental risk assessment of endocrine disrupting compounds at the wastewater treatment works in Pietermaritzburg (South Africa). The Science of the Total Environment 468–469, 584–597.
Manzoor, S., Shah, M.H., Shaheen, N., Khalique, A., Jaffar, M., 2006. Characterization, distribution and comparison of selected metals in textile effluents, adjoining soil and groundwater. Journal-Chemical Society of Pakistan 28, 10.
Mao, Y., Cheng, L., Ma, B., Cai, Y., 2016. The fate of mercury in municipal wastewater treatment plants in China: significance and implications for environmental cycling. Journal of Hazardous Materials 306, 1–7.
Mella, B., Glanert, A.C., Gutterres, M., 2015. Removal of chromium from tanning wastewater and its reuse. Process Safety and Environmental Protection 95, 195–201.
Miao, X.-S., Bishay, F., Chen, M., Metcalfe, C.D., 2004. Occurrence of antimicrobials in the final effluents of wastewater treatment plants in Canada. Environmental Science and Technology 38, 3533–3541.
Mishra, V.K., Upadhyaya, A.R., Pandey, S.K., 2008. Heavy metal pollution induced due to coal mining effluent on surrounding aquatic ecosystem and its management through naturally occurring aquatic macrophytes. Bioresource Technology 99, 930–936.
Mohapatra, S., Huang, C.H., Mukherji, S., Padhye, L.P., 2016. Occurrence and fate of pharmaceuticals in WWTPs in India and comparison with a similar study in the United States. Chemosphere 159, 526–535.

Mrema, E.J., Rubino, F.M., Brambilla, G., Moretto, A., Tsatsakis, A.M., Colosio, C., 2013. Persistent organochlorinated pesticides and mechanisms of their toxicity. Toxicology 307, 74–88.

Nagajyoti, P.C., Lee, K.D., Sreekanth, T.V.M., 2010. Heavy metals, occurrence and toxicity for plants: a review. Environmental Chemistry Letters 8, 199–216.

Needham, T.P., Ghosh, U., 2019. Four decades since the ban, old urban wastewater treatment plant remains a dominant source of PCBs to the environment. Environment and Pollution 246, 390–397.

Noreen, M., Shahid, M., Iqbal, M., Nisar, J., 2017. Measurement of cytotoxicity and heavy metal load in drains water receiving textile effluents and drinking water in vicinity of drains. Measurement 109, 88–99.

Nosier, S.A., 2003. Removal of cadmium ions from industrial wastewater by cementation. Chemical and Biochemical Engineering Quarterly 17, 219–224.

Pauwels, B., Noppe, H., De Brabander, H., Verstraete, W., 2008. Comparison of steroid hormone concentrations in domestic and hospital wastewater treatment plants. Journal of Environmental Engineering-Asce 134, 933–936.

Pessoa, G.P., de Souza, N.C., Vidal, C.B., Alves, J.A., Firmino, P.I., Nascimento, R.F., dos Santos, A.B., 2014. Occurrence and removal of estrogens in Brazilian wastewater treatment plants. The Science of the Total Environment 490, 288–295.

Pham, T.T., Proulx, S., 1997. PCBs and PAHs in the montreal urban community (Quebec, Canada) wastewater treatment plant and in the effluent plume in the st Lawrence river. Water Research 31, 1887–1896.

Polat, H., Erdogan, D., 2007. Heavy metal removal from waste waters by ion flotation. Journal of Hazardous Materials 148, 267–273.

Qureshi, A.S., Hussain, M.I., Ismail, S., Khan, Q.M., 2016. Evaluating heavy metal accumulation and potential health risks in vegetables irrigated with treated wastewater. Chemosphere 163, 54–61.

Radhakrishnan, K., Sethuraman, L., Panjanathan, R., Natarajan, A., Solaiappan, V., Thilagaraj, W.R., 2014. Biosorption of heavy metals from actual electroplating wastewater using encapsulated *Moringa oleifera* beads in fixed bed column. Desalination and Water Treatment 57, 3572–3587.

Raffetti, E., Donato, F., Speziani, F., Scarcella, C., Gaia, A., Magoni, M., 2018. Polychlorinated biphenyls (PCBs) exposure and cardiovascular, endocrine and metabolic diseases: a population-based cohort study in a North Italian highly polluted area. Environment International 120, 215–222.

Raheem, A., Sikarwar, V.S., He, J., Dastyar, W., Dionysiou, D.D., Wang, W., Zhao, M., 2017. Opportunities and challenges in sustainable treatment and resource reuse of sewage sludge: a review. Chemical Engineering Journal 337, 616–641.

Rahman, M.T., Kameda, T., Kumagai, S., Yoshioka, T., 2017. Effectiveness of Mg–Al-layered double hydroxide for heavy metal removal from mine wastewater and sludge volume reduction. International Journal of Environmental Science and Technology 15, 263–272.

Rasheed, T., Bilal, M., Nabeel, F., Adeel, M., Iqbal, H.M.N., 2019. Environmentally-related contaminants of high concern: Potential sources and analytical modalities for detection, quantification, and treatment. Environment International 122, 52–66.

Ren, Y.H., Geng, J.J., Li, F.C., Ren, H.Q., Ding, L.L., Xu, K., 2016. The oxidative stress in the liver of *Carassius auratus* exposed to acesulfame and its UV irradiance products. The Science of the Total Environment 571, 755–762.

Rodenburg, L.A., Du, S., Xiao, B., Fennell, D.E., 2011. Source apportionment of polychlorinated biphenyls in the New York/New Jersey Harbor. Chemosphere 83, 792–798.

Saha, P., Shinde, O., Sarkar, S., 2017. Phytoremediation of industrial mines wastewater using water hyacinth. International Journal of Phytoremediation 19, 87–96.

Sanchez-Avila, J., Bonet, J., Velasco, G., Lacorte, S., 2009. Determination and occurrence of phthalates, alkylphenols, bisphenol A, PBDEs, PCBs and PAHs in an industrial sewage grid discharging to a municipal wastewater treatment plant. The Science of the Total Environment 407, 4157–4167.

Sanmuga Priya, E., Senthamil Selvan, P., 2017. Water hyacinth (*Eichhornia crassipes*) – an efficient and economic adsorbent for textile effluent treatment – a review. Arabian Journal of Chemistry 10, S3548–S3558.

Santos, A., Judd, S., 2010. The fate of metals in wastewater treated by the activated sludge process and membrane bioreactors: a brief review. Journal of Environmental Monitoring 12, 110–118.

Sekomo, C.B., Rousseau, D.P.L., Saleh, S.A., Lens, P.N.L., 2012. Heavy metal removal in duckweed and algae ponds as a polishing step for textile wastewater treatment. Ecological Engineering 44, 102–110.

Senta, I., Terzic, S., Ahel, M., 2013. Occurrence and fate of dissolved and particulate antimicrobials in municipal wastewater treatment. Water Research 47, 705–714.

Shabalala, A.N., Ekolu, S.O., Diop, S., Solomon, F., 2017. Pervious concrete reactive barrier for removal of heavy metals from acid mine drainage - column study. Journal of Hazardous Materials 323 (Pt B), 641–653.

Shafer, M.M., Overdier, J.T., Armstong, D.E., 1998. Removal, partitioning, and fate of silver and other metals in wastewater treatment plants and effluent-receiving streams. Environmental Toxicology & Chemistry 17, 630–641.

Sim, W.J., Lee, J.W., Lee, E.S., Shin, S.K., Hwang, S.R., Oh, J.E., 2011. Occurrence and distribution of pharmaceuticals in wastewater from households, livestock farms, hospitals and pharmaceutical manufactures. Chemosphere 82 (2), 179–186.

Sinclair, E., Kannan, K., 2006. Mass loading and fate of perfluoroalkyl surfactants in wastewater treatment plants. Environmental Science and Technology 40, 1408–1414.

Spanos, T., Ene, A., Styliani Patronidou, C., Xatzixristou, C., 2016. Temporal variability of sewage sludge heavy metal content from Greek wastewater treatment plants. Ecological Chemistry and Engineering S 23, 271–283.

Stasinakis, A.S., Kotsifa, S., Gatidou, G., Mamais, D., 2009. Diuron biodegradation in activated sludge batch reactors under aerobic and anoxic conditions. Water Research 43, 1471–1479.

Sui, Q., Huang, J., Deng, S., Chen, W., Yu, G., 2011. Seasonal variation in the occurrence and removal of pharmaceuticals and personal care products in different biological wastewater treatment processes. Environmental Science and Technology 45, 3341–3348.

Sun, J., Dai, X., Wang, Q., van Loosdrecht, M.C.M., Ni, B.J., 2019. Microplastics in wastewater treatment plants: detection, occurrence and removal. Water Research 152, 21–37.

Tang, H.L., Chen, Y.C., Regan, J.M., Xie, Y.F., 2012. Disinfection by-product formation potentials in wastewater effluents and their reductions in a wastewater treatment plant. Journal of Environmental Monitoring 14, 1515–1522.

Tariq, S.R., Shah, M.H., Shaheen, N., Khalique, A., Manzoor, S., Jaffar, M., 2005. Multivariate analysis of selected metals in tannery effluents and related soil. Journal of Hazardous Materials 122, 17–22.

Tariq, S.R., Shah, M.H., Shaheen, N., Khalique, A., Manzoor, S., Jaffar, M., 2006. Multivariate analysis of trace metal levels in tannery effluents in relation to soil and water: a case study from Peshawar, Pakistan. Journal of Environmental Management 79, 20–29.

Tariq, S.R., Shah, M.H., Shaheen, N., 2009. Comparative statistical analysis of chrome and vegetable tanning effluents and their effects on related soil. Journal of Hazardous Materials 169, 285–290.

Teijon, G., Candela, L., Tamoh, K., Molina-Diaz, A., Fernandez-Alba, A.R., 2010. Occurrence of emerging contaminants, priority substances (2008/105/CE) and heavy metals in treated wastewater and groundwater at Depurbaix facility (Barcelona, Spain). The Science of the Total Environment 408, 3584–3595.

Tian, W., Bai, J., Liu, K., Sun, H., Zhao, Y., 2012. Occurrence and removal of polycyclic aromatic hydrocarbons in the wastewater treatment process. Ecotoxicology and Environmental Safety 82, 1–7.

Tieyu, W., Yonglong, L., Hong, Z., Yajuan, S., 2005. Contamination of persistent organic pollutants (POPs) and relevant management in China. Environment International 31, 813–821.

Tran, N.H., Chen, H., Reinhard, M., Mao, F., Gin, K.Y., 2016. Occurrence and removal of multiple classes of antibiotics and antimicrobial agents in biological wastewater treatment processes. Water Research 104, 461–472.

Tran, N.H., Reinhard, M., Gin, K.Y., 2017. Occurrence and fate of emerging contaminants in municipal wastewater treatment plants from different geographical regions-a review. Water Ressearch 133, 182–207.

Ustun, G.E., 2009. Occurrence and removal of metals in urban wastewater treatment plants. Journal of Hazardous Materials 172, 833–838.

Varela, A.R., Andre, S., Nunes, O.C., Manaia, C.M., 2014. Insights into the relationship between antimicrobial residues and bacterial populations in a hospital-urban wastewater treatment plant system. Water Research 54, 327–336.

Veglio, F., Beolchini, F., 1997. Removal of metals by biosorption: a review. Hydrometallurgy 44, 301–316.

Vieno, N., Sillanpaa, M., 2014. Fate of diclofenac in municipal wastewater treatment plant – a review. Environment International 69, 28–39.

Wang, J., Huang, C.P., Pirestani, D., 2003. Interactions of silver with wastewater constituents. Water Research 37, 4444–4452.

Wang, L.C., Wang, I.C., Chang, J.E., Lai, S.O., Chang-Chien, G.P., 2007. Emission of polycyclic aromatic hydrocarbons (PAHs) from the liquid injection incineration of petrochemical industrial wastewater. Journal of Hazardous Materials 148, 296–302.

Wei, D., Kameya, T., Urano, K., 2007. Environmental management of pesticidal POPs in China: past, present and future. Environment International 33, 894–902.

Wei, G.L., Liang, X.L., Li, D.Q., Zhuo, M.N., Zhang, S.Y., Huang, Q.X., Liao, Y.S., Xie, Z.Y., Guo, T.L., Yuan, Z.J., 2016. Occurrence, fate and ecological risk of chlorinated paraffins in Asia: a review. Environment International 92–93, 373–387.

Whaley, P., Halsall, C., Agerstrand, M., Aiassa, E., Benford, D., Bilotta, G., Coggon, D., Collins, C., Dempsey, C., Duarte-Davidson, R., FitzGerald, R., Galay-Burgos, M., Gee, D., Hoffmann, S., Lam, J., Lasserson, T., Levy, L., Lipworth, S., Ross, S.M., Martin, O., Meads, C., Meyer-Baron, M., Miller, J., Pease, C., Rooney, A., Sapiets, A., Stewart, G., Taylor, D., 2016. Implementing systematic review techniques in chemical risk assessment: challenges, opportunities and recommendations. Environment International 92–93, 556–564.

WHO (World Health Organization)/International Programme on Chemical Safety (IPCS), 2002. Global Assessment of the State of-the-science of Endocrine Disruptors. http://www.who.int/ipcs/publications/new_issues/endocrine_disruptors/en/. (Accessed 4 October 2002).

Wick, A., Fink, G., Joss, A., Siegrist, H., Ternes, T.A., 2009. Fate of beta blockers and psycho-active drugs in conventional wastewater treatment. Water Research 43, 1060–1074.

Wu, P., Jiang, L.Y., He, Z., Song, Y., 2017. Treatment of metallurgical industry wastewater for organic contaminant removal in China: status, challenges, and perspectives. Environmental Sciences: Water Research and Technology 3, 1015–1031.

Wu, G., Geng, J., Li, S., Li, J., Fu, Y., Xu, K., Ren, H., Zhang, X., 2019. Abiotic and biotic processes of diclofenac in enriched nitrifying sludge: kinetics, transformation products and reactions. The Science of the Total Environment 683, 80–88.

Xu, P., Zeng, G.M., Huang, D.L., Feng, C.L., Hu, S., Zhao, M.H., Lai, C., Wei, Z., Huang, C., Xie, G.X., Liu, Z.F., 2012. Use of iron oxide nanomaterials in wastewater treatment: a review. The Science of the Total Environment 424, 1–10.

Xue, J., Kannan, K., 2019. Mass flows and removal of eight bisphenol analogs, bisphenol A diglycidyl ether and its derivatives in two wastewater treatment plants in New York State, USA. The Science of the Total Environment 648, 442–449.

Yang, M., Zhang, X., 2016. Current trends in the analysis and identification of emerging disinfection byproducts. Trends in Environmental Analytical Chemistry 10, 24–34.

Yang, Y., Ok, Y.S., Kim, K.H., Kwon, E.E., Tsang, Y.F., 2017a. Occurrences and removal of pharmaceuticals and personal care products (PPCPs) in drinking water and water/sewage treatment plants: a review. The Science of the Total Environment 596–597, 303–320.

Yang, Y.Y., Liu, W.R., Liu, Y.S., Zhao, J.L., Zhang, Q.Q., Zhang, M., Zhang, J.N., Jiang, Y.X., Zhang, L.J., Ying, G.G., 2017b. Suitability of pharmaceuticals and personal care products (PPCPs) and artificial sweeteners (ASs) as wastewater indicators in the Pearl River Delta, South China. The Science of the Total Environment 590–591, 611–619.

Yu, J., Hu, J., Tanaka, S., Fujii, S., 2009. Perfluorooctane sulfonate (PFOS) and perfluorooctanoic acid (PFOA) in sewage treatment plants. Water Research 43, 2399–2408.

Yu, Y., Wu, L., Chang, A.C., 2013. Seasonal variation of endocrine disrupting compounds, pharmaceuticals and personal care products in wastewater treatment plants. The Science of the Total Environment 442, 310–316.

Yu, Q., Geng, J., Huo, H., Xu, K., Huang, H., Hu, H., Ren, H., 2018. Bioaugmentated activated sludge degradation of progesterone: kinetics and Mechanism. Chemical Engineering Journal 352, 214–224.

Yu, Q., Geng, J., Zong, X., Zhang, Y., Xu, K., Hu, H., Deng, Y., Zhao, F., Ren, H., 2019. Occurrence and removal of progestagens in municipal wastewater treatment plants from different regions in China. The Science of the Total Environment 668, 1191–1199.

Zapata, P., Ballesteros-Cano, R., Colomer, P., Bertolero, A., Viana, P., Lacorte, S., Santos, F.J., 2018. Presence and impact of Stockholm Convention POPs in gull eggs from Spanish and Portuguese natural and national parks. The Science of the Total Environment 633, 704–715.

Zaranyika, M.F., Nyati, W., 2017. Uptake of heavy metals by Typha capensis from wetland sites polluted by effluent from mineral processing plants: implications of metal-metal interactions. Biotechnology 7, 286.

Zeng, L., Wang, T., Ruan, T., Liu, Q., Wang, Y., Jiang, G., 2012. Levels and distribution patterns of short chain chlorinated paraffins in sewage sludge of wastewater treatment plants in China. Environment and Pollution 160, 88–94.

Zhang, W., Wei, C., Chai, X., He, J., Cai, Y., Ren, M., Yan, B., Peng, P., Fu, J., 2012. The behaviors and fate of polycyclic aromatic hydrocarbons (PAHs) in a coking wastewater treatment plant. Chemosphere 88, 174–182.

Zhang, Y.Y., Zhuang, Y., Geng, J.J., Ren, H.Q., 2015a. Inactivation of antibiotic resistance genes in municipal wastewater effluent by chlorination and sequential UV/chlorination disinfection. The Science of the Total Environment 512–513, 125–132.

Zhang, B., Xian, Q., Zhu, J., Li, A., Gong, T., 2015b. Characterization, DBPs formation, and mutagenicity of soluble microbial products (SMPs) in wastewater under simulated stressful conditions. Chemical Engineering Journal 279, 258–263.

Zhang, Y.Y., Geng, J.J., Ma, H.J., Ren, H.Q., Xu, K., Ding, L.L., 2016b. Characterization of microbial community and antibiotic resistance genes in activated sludge under tetracycline and sulfamethoxazole selection pressure. The Science of the Total Environment 571, 479–486.

Zhang, X., Yu, T., Li, X., Yao, J., Liu, W., Chang, S., Chen, Y., 2019. The fate and enhanced removal of polycyclic aromatic hydrocarbons in wastewater and sludge treatment system: a review. Critical Reviews in Environmental Science and Technology 1–51.

Zheng, X., Wu, R., Chen, Y., 2011. Effects of ZnO nanoparticles on wastewater biological nitrogen and phosphorus removal. Environmental Science and Technology 45, 2826–2832.

Zhi, S., Zhou, J., Yang, F., Tian, L., Zhang, K., 2018. Systematic analysis of occurrence and variation tendency about 58 typical veterinary antibiotics during animal wastewater disposal processes in Tianjin, China. Ecotoxicology and Environmental Safety 165, 376–385.

Zhou, Y., Zha, J., Xu, Y., Lei, B., Wang, Z., 2012. Occurrences of six steroid estrogens from different effluents in Beijing, China. Environmental Monitoring and Assessment 184, 1719–1729.

Zhou, S., Di Paolo, C., Wu, X., Shao, Y., Seiler, T.B., Hollert, H., 2019. Optimization of screening-level risk assessment and priority selection of emerging pollutants - the case of pharmaceuticals in European surface waters. Environment International 128, 1–10.

Further Reading

Martin, M., Shafer, J.T.O., Armstong, D.E., 1998. Removal, partitioning, and fate of silver and other metals in wastewater treatment plants and effluent-receiving streams.pdf. Environmental Toxicology & Chemistry 17, 630–641.

CHAPTER

Biological HRPs in wastewater

3

Shuyu Jia, Xuxiang Zhang, PhD

State Key Laboratory of Pollution Control and Resource Reuse, School of the Environment, Nanjing University, Nanjing, China

Chapter outline

High-Risk Pollutants in Wastewater. https://doi.org/10.1016/B978-0-12-816448-8.00003-4

Due to extensive industrialization and increase in population worldwide, biological high-risk pollutants (HRPs) are becoming an increasing cause for concern. Biological HRPs are the substances in the environment that come from living organisms and can lead to adverse effects on both human health and the environment safety, which usually include bacteria, viruses, protozoa, helminth, biotoxins, antibiotic resistance genes (ARGs), and antibiotic resistant bacteria (ARB). There are many sources of these biological HRPs, such as soil, air, sediment, surface water, wastewater, and fecal wastes. Notably, wastewater is the worldwide hotspot for biological HRPs, and also is an important reservoir of biological HRPs, which can introduce these biological HRP into receiving surface waterbody along with wastewater discharge and can induce the microbial community shift in these receiving media. Some of these biological HRPs are more resistant to the wastewater treatment process and more infectious. For instance, bacteria, having the capacity to cause a wide range of several water-related infections and diseases, are the most common biological HRPs in wastewater, and are frequently detected in the receiving water environment. The major pathogenic protozoans are also more prevalent in wastewater than in any other environmental sources. Thus, the occurrence of these biological HRPs in wastewater is not only conservative but also persistent and accumulative.

3.1 Bacteria

3.1.1 Classification of bacteria in wastewater

Bacteria are single-celled prokaryotes, which have several shapes (including spheres, rods, and spirals) according to their morphology. They are usually in a few micrometers in length and exist together. Although bacterial cells are much smaller and simpler in structure (Todar, 2013), the sizes of most bacteria range from 0.3 to 3 μm (Gerardi, 2006). There are many kinds of bacteria, which are widely distributed in nature and closely related to human beings. Some beneficial bacteria have been discovered by human beings. However, there are also countless harmful bacteria in the living environment of human beings, which pollute the air, water, soil, and food and pose a great threat to human health. There are a large number of bacteria in wastewater, and various bacteria and even human pathogenic bacteria can spread by water pollution caused by wastewater discharge, which poses risks to human health and ecological security.

Bacteria are an exceedingly diverse group of organisms that differ in shape, size, habitat, metabolism, and other features. A few different criteria are used to classify bacteria, based on the differences in shape, metabolism, cell walls, and genetic makeup. One way of classifying them is based on shape, and there are three basic shapes (Dusenbery, 2009): (1) spherical: the bacterial shape likes a ball and a single bacterium is a coccus; (2) rod-shaped: these bacteria are called bacilli (singular bacillus) and part of the rod-shaped bacteria are curved, which are called as vibrio; (3) spiral: these bacteria are known as spirilla (singular spirillus), and they are known as spirochetes if their coil is very tight. Furthermore, a small number of other unusual shapes have been described, such as icosahedron, cube, or star-shaped (Fritz et al., 2004; Yang et al., 2016). The shapes and structures of bacteria are often reflected in their names (Table 3.1).

The other way of classifying them is according to the metabolism style, and most bacteria are divided into autotrophic and heterotrophic bacteria (Foster and Slonczewski, 2017; Hellingwerf et al., 1994; Nealson, 1999; Zillig, 1991). Autotrophic bacteria, which only need carbon dioxide as their carbon source, make their food by (1) photosynthesis using sunlight, carbon dioxide, and water or (2) chemosynthesis using carbon dioxide, water, and some chemicals (such as ammonia, nitrogen, and sulfur). Heterotrophic bacteria, relying on organic matters as carbon source, get their energy by consuming organic carbon outside the cells. According to their reaction to oxygen, most bacteria can be divided into the following three categories (Joubert and Britz, 1987): (1) aerobic bacteria that can only grow in the presence of oxygen; (2) anaerobic bacteria that can only grow in the absence of oxygen; and (3) facultative anaerobic bacteria that can grow regardless of aerobic and oxygen free.

Table 3.1 Bacteria groups and their corresponding shapes.

Shape	Examples
Spherical	*Streptococcus* group, *Aerococcus*, and *Pediococcus*
Rod-shaped	*Bacillus anthracis*, *Bacillus cereus*, and *Coxiella burnetii*
Spiral	*Leptospira interrogans*, *Treponema pallidum*, and *Borrelia recurrentis*

The structure of bacteria with no membrane structure of the organelles, such as mitochondria and chloroplasts, is very simple. However, the cell wall of most bacteria is divided into two different types based on the reaction of cells to the Gram stain, which classifies bacteria into Gram-positive bacteria and Gram-negative bacteria (Beveridge, 2001). Gram-positive bacteria have a thick cell wall with peptidoglycan and teichoic acids. On the contrary, a relatively thin cell wall consisting of several layers of peptidoglycan that are surrounded by a second lipid membrane with lipopolysaccharides and lipoproteins is possessed by Gram-negative bacteria (Hugenholtz, 2002). Furthermore, some bacteria have cell wall structures that differ from those possessed by Gram-positive and Gram-negative bacteria. For instance, *Mycobacteria*, a clinically important bacterium, have the thick peptidoglycan cell wall and also the second outer layer of lipids (Alderwick et al., 2015).

3.1.2 Typical bacteria in wastewater

Bacteria also can be classified by the roles that they perform in the wastewater based on different condition (Gerardi, 2006). Table 3.2 summarizes the typical groups of bacteria and their corresponding roles in the wastewater according to the study by Gerardi (2006).

3.1.3 Pathogenic bacteria and their typical characteristics and hazards

Pathogenic bacteria are bacteria that can cause disease and are one of the major threats to public health in the world. Pathogenic bacteria are infections and able to contribute to globally important diseases, such as pneumonia, foodborne illnesses, tetanus, typhoid fever, diphtheria, syphilis, and leprosy (Chan et al., 2013; El-Lathy et al., 2009). Wastewater may contain millions of bacteria per milliliter and is also a potential source of various pathogenic bacteria. Recently, a wide range of pathogenic bacteria, such as *Escherichia coli*, *Clostridium perfringens*, *Mycobacterium tuberculosis*, *Legionella pneumophila*, *Pseudomonas aeruginosa*, *Shigella flexneri*, *Salmonella enterica*, *Vibrio cholera*, and *Yersinia enterocolitica*, have been detected in the wastewater (Cai and Zhang, 2013; Dudley et al., 1980; Stevik et al., 2004). Several reports have summarized pathogenic bacteria commonly detected in wastewater (Table 3.3) (Gerardi, 2006; Gerardi and Zimmerman, 2004).

3.1.3.1 Pathogenic Escherichia coli

Escherichia coli, commonly known as nonpathogenic bacteria, are long-lived bacterium found in the intestines of humans and animals. The bacterial antigen type is "O" type, the flagellar antigen type is "H" type, and the surface antigen type is "K" type. According to the different antigen structures, the serotypes of the bacteria can be divided into more than 180 kinds. However, some strains are highly pathogenic and can cause diarrhea and other diseases, which are often called pathogenic *Escherichia coli*. The pathogenic *Escherichia coli* can produce or contain

Table 3.2 Typical bacteria groups in wastewater.

Typical groups	Examples
Acetongenic bacteria	*Acetobacter*, *Syntrobacter*, and *Syntrophomonas*
Coliforms	*Escherichia*, *Citrobacter*, *Enterobacter*, *Hafnia*, *Klebsiella*, *Serratia*, and *Yersinia*
Cyanobacteria	*Anabaena*, *Chlorella*, *Euglena*, and *Oscillatoria*
Denitrifying bacteria	*Alcaligenes*, *Bacillus*, and *Pseudomonas*
Fecal coliforms	*Escherichia*
Fermentative bacteria	*Bacteroides*, *Bifidobacteria*, *Clostridium*, *Escherichia*, *Lactobacillus*, and *Proteus*
Floc-forming bacteria	*Achromobacter*, *Aerobacter*, *Citromonas*, *Flavobacterium*, *Pseudomonas*, and *Zoogloea*
Gliding bacteria	*Beggiatoa*, *Flexibacter*, and *Thiothrix*
Gram-negative aerobic cocci and rods	*Acetobacter*, *Acinetobacter*, *Alcaligens*, *Nitrobacter*, *Nitrosomonas*, *Pseudomonas*, and *Zoogloea*
Gram-negative facultative anaerobic rods	*Aeromonas*, *Escherichia*, *Flavobacterium*, *Klebsiella*, *Proteus*, and *Salmonella*
Hydrolytic bacteria	*Bacteroides*, *Bifidobacteria*, and *Clostridium*
Methane-forming bacteria	*Methanobacterium*, *Methanococcus*, *Methanomonas*, and *Methanosarcinia*
Nitrifying bacteria	*Nitrosomonas*, *Nitrosospira*, *Nitrobacter*, and *Nitrospira*
Pathogenic bacteria	*Campylobacter jejuni*, and *Leptospira interrogans*
Poly-P bacteria	*Acinetobacter*, *Aerobacter*, *Beggiatoa*, *Enterobacter*, *Klebsiella*, and *Proteobacter*
Saprophytic bacteria	*Achromobacter*, *Alcaligenes*, *Bacillus*, *Flavobacterium*, *Micrococcus*, and *Pseudomonas*
Sulfur-reducing bacteria	*Desulfovibrio* and *Desulfotomaculum*

colonization factors, enterotoxin, K antigen, and related substances, and also has the ability of endotoxin secretion. Pathogenic *Escherichia coli* cause disease outbreaks through the contamination of drinking water, food, and other ways.

Pathogenic *Escherichia coli* are mainly responsible for three types of infections in humans: (1) neonatal meningitis, (2) urinary tract infections, and (3) intestinal diseases. Pathogenic *Escherichia coli* can be divided into several categories according to its serological characteristics and virulence properties, mainly consisting of enterotoxigenic *Escherichia coli*, enterohemorrhagic *Escherichia coli*, enteropathogenic *Escherichia coli*, enteroaggregative *Escherichia coli*, and enteroinvasive *Escherichia coli* (Kaper et al., 2004; Todar, 2008). The most infamous member of enterohemorrhagic *Escherichia coli* is the strain O157:H7 that can cause bloody diarrhea and fever, and it is prominent and important in North America, the United Kingdom, and Japan (Kaper et al., 2004).

Table 3.3 Pathogenic bacteria commonly detected in wastewater.

Bacterium/Bacteria	Disease
Actinomyces israelii	Actinomycosis
Bacillus anthracis	Anthrax
Brucella spp.	Brucellosis (Malta fever)
Campylobacter jejuni	Gastroenteritis
Clostridium perfringens	Gangrene (gas gangrene)
Clostridium tetani	Tetanus
Clostridium spp.	Gas gangrene
Enteroinvasive *Escherichia coli*	Gastroenteritis
Enteropathogenic *Escherichia coli*	Gastroenteritis
Enterotoxigenic *Escherichia coli*	Gastroenteritis
Enterohemorrhagic *Escherichia coli* 0157: H7	Gastroenteritis and hemolytic uremic syndrome
Francisella tularensis	Tularemia
Leptospira interrogans	Leptospirosis
Mycobacterium tuberculosis	Tuberculosis
Nocardia spp.	Nocardiosis
Salmonella paratyphi	Paratyphoid fever
Salmonella spp.	Salmonellosis
Salmonella typhi	Typhoid fever
Shigella spp.	Shigellosis
Vibrio cholerae	Cholera (Asiatic cholera)
Vibrio parahaemolyticus	Gastroenteritis
Yersinia enterocolitica	Yersiniosis (bloody diarrhea)

Pathogenic *Escherichia coli* are reported to be usually detected in wastewater. Some pathogenic *Escherichia coli* strains survive during the treatment stages of sewage treatment plants (STPs) and in the surrounding environmental waterbodies of STPs (Anastasi et al., 2012). Strains of *Escherichia coli* O157:H7 have been not only commonly isolated from urban sewage and animal wastewater in Spain but also are present in human and animal wastewaters with other Shiga toxin-producing *Escherichia coli* (Garcia-Aljaro et al., 2005). The level of *Escherichia coli* O157:H7 is about $10-10^2$ CFU/100 mL for municipal sewage and 10^2-10^3 CFU/100 mL for animal wastewater from slaughterhouses (Garcia-Aljaro et al., 2005). Shannon et al. (2007) detected that the level of *Escherichia coli* in raw wastewater was about 1.51×10^7 gene copy number per 100 mL, and had a reduction of 3.52–3.98 orders of magnitude after final treatment while *Escherichia coli* O157:H7 was not present or was below the detection limit in all treatment stages of the investigative STP.

3.1.3.2 *Salmonella enterica* serovar Typhi

Typhoid is caused by a highly virulent and aggressive intestinal bacterium called *Salmonella enterica* serovar Typhi. This bacterium infects only humans and is usually acquired by ingestion of food or water contaminated by the feces of patients with typhoid or asymptomatic carriers (Dougan and Baker, 2014). There are three strains of *Salmonella enterica* serovar Typhi including *Salmonella typhi*, *Salmonella paratyphi*, and *Salmonella schottmuelleri*. They are Gram-negative, facultative anaerobes, and rod shaped, which do not have spores and capsules. *Salmonella* pathogenicity islands (SPIs), large genomic regions of 10–134 kb, are responsible for most of the virulence factors. Most of the effector molecules associated with complex pathogenesis are encoded by SPIs (Hensel, 2004). *Salmonella typhimurium* is also a potential source of human illness, having the ability to transfer from irrigation water to the edible parts of the plants (Lapidot and Yaron, 2009). They are common in wastewater and can be induced into the viable but nonculturable state after typical wastewater disinfection (Oliver et al., 2005). Because the infectious dose of them is only as low as 20 cells per mL, the residual level of them in wastewater also has potential health risks (Oliver et al., 2005). In addition, strains of *Salmonella* with a greater pathogenic potential have also been isolated from wastewater and activated sludge, and the most frequent serotypes are *Salmonella hadar* (38.1%), followed by *Salmonella enteritidis* (23.8%), *Salmonella london* (14.3%), and *Salmonella anatum* (9.5%) in raw and treated wastewater (Espigares et al., 2006).

3.1.3.3 *Shigella dysenteriae*

Shigella dysenteriae belongs to the genus of enterobacteriaceae *Shigella*, and it is a Gram-negative bacterium without spore and flagellates. It can grow on common medium and is cold resistant and facultatively anaerobic. Furthermore, *Shigella dysenteriae* is one of the most common pathogenic bacteria leading to dysentery in human and primate, and typical observed symptoms caused by it are diarrhea, abdominal pain, and fever after infection.

The pathogenic mechanism of *Shigella dysenteriae* is summarized as follows (Athman et al., 2005; Jennison and Verma, 2004; Schroeder and Hilbi, 2008): (1) *Shigella dysenteria* upregulates the acidic gene, so that it is possible to survive in the stomach of the host; (2) *Shigella dysenteriae* invades colonic epithelial cells and is tightly linked to the associated proteins to replicate virulence factors; (3) *Shigella dysenteriae* leads to the apoptosis of macrophages and induces the release of interleukin IL-21, resulting in the accumulation of inflammatory cells and polymorphonuclear leukocytes. The accumulated polymorphonuclear leukocytes can pass through the intestinal epithelial cells and destroy the connections between the epithelial cells, allowing more *Shigella dysenteriae* to reach the submucosal layer through the crack; (4) *Shigella dysenteriae* further infects adjacent cells, causing an inflammatory reaction when the number of infected cells reaches a certain level, thereby resulting in typical bacterial dysentery symptoms such as congestion, hemorrhage, and edema of the intestinal mucosa.

Notably, *Shigella dysenteriae*, considered as the most virulent bacterium, is widely distributed in the wastewater, which is considered as a reservoir. For example, the occurrence rates of *Shigella dysenteriae* are up to 40–60% in wastewater effluents and the receiving waterbodies in South Africa (Teklehaimanot et al., 2014, 2015). *Shigella dysenteriae* has also been detected in 35 sewage samples collected from hospital and residential areas (Peng et al., 2002). Furthermore, *Shigella dysenteriae* was isolated from water and riverbed sediment of the Apies River, South Africa (Ekwanzala et al., 2017)

3.1.3.4 Vibrio cholerae

Vibrio cholerae, which is a member of the family Vibrionaceae and is capable of respiratory and fermentative metabolism, is a Gram-negative, comma-shaped, facultative anaerobic, and nonspore-forming bacterium (Morris Jr and Acheson, 2003; WHO, 2004). *Vibrio cholerae* is suitable for survival in salt-containing water and is dependent on aquatic organisms and plankton (Colwell, 1996). It can use chitin on the surface of zooplankton as the carbon source and nitrogen source for growth. Meanwhile, chitin can also induce horizontal gene transfer of *Vibrio cholera* (Kirn et al., 2005).

Some strains of *Vibrio cholerae* can cause the disease of cholera, an acute intestinal infection. *Vibrio cholerae* can be divided into more than 200 serotypes, according to the O antigens located on the lipopolysaccharide of the cell membrane (Longini et al., 2002). After being ingested by humans, *Vibrio cholerae* enters the small intestine through the oral cavity and gastric juice and then colonizes through the mucus layer. Once entering the small intestine, virulence genes of *Vibrio cholerae* are induced and expressed by the host conditions, and the virulence coregulating pili and cholera toxin are important pathogenic factors leading to the virulence of *Vibrio cholera* (Wang, 2015).

Due to the contribution of wastewater to the spread of cholera, the fate of *Vibrio cholerae* in wastewater and wastewater treatment systems has been investigated. The prevalence of *Vibrio cholerae* has been observed in four wastewater treatment plants (WWTPs) located in Gauteng Province, South Africa (Dungeni et al., 2010). Pathogenic bacteria strains of non-O1 *Vibrio cholerae* have been detected in domestic wastewater with an average abundance between $2.5\times10^{1-}$ – 1.7×10^{3} MPN mL^{-1} (Mezrioui and Oufdou, 1996). The high incidence of *Vibrio cholerae* in wastewater is further validated by other *Vibrio* pathogens to some extent. For example, Nongogo and Okoh (2014) also have found the occurrence of other *Vibrio* pathogens in the final effluents of five WWTPs located in South Africa over 12 months.

3.1.3.5 Legionella

Legionella, a pathogenic group of Gram-negative bacteria, is an intracellular pathogen that is widespread in natural waterbodies and can cause two forms of *Legionella* diseases (including *Legionella* pneumonia and Pontiac fever) (Bartram et al., 2007). *Legionella* survives and proliferates in host cells by the phagocytosis of free-living single-celled protozoa (Kim et al., 2004). *Legionella* can also form a biofilm with other aquatic microorganisms (Smid et al., 1996). The extracellular polymeric substances in biofilm not only help bacteria to capture and concentrate environmental nutrients, but also protect bacteria against environmental pressures such as antibiotics, disinfectants, dryness, and high temperature (Cooper and Hanlon, 2010; Kim et al., 2002; Wright et al., 1991).

Notably, the occurrence of *Legionella* in wastewater has been reported worldwide in recent years. For example, *Legionella* has been detected in 10 of the 17 investigated WWTPs (58.8%) and also has been observed in hospital, industrial, and domestic wastewater systems in Taiwan (China), respectively (Huang et al., 2009). *Legionella* spp. have been detected at varying concentrations from 4.8 to 5.6 log GU/mL in activated sludge tanks in three WWTPs in Germany (Caicedo et al., 2016). In total, Caicedo et al. (2019) have summarized the occurrence of *Legionella* in several municipal wastewater treatment plants (MWTPs) based on the activated sludge process in France, Norway, USA, China, and other countries (Brissaud et al., 2008; Cai and Zhang, 2013; Kulkarni et al., 2018; Lund et al., 2014).

3.2 Viruses

3.2.1 Definition, morphology, and composition of viruses

Viruses are in the noncellular form, which are composed of the protein and nucleic acid molecule (DNA or RNA), and they are organic species that are parasitic in the living and nonliving body. They are neither biological nor abiotic, and are not attributed to the five kingdoms (including prokaryotes, protists, fungi, plants, and animals). Viruses vary in shape, ranging from simple spirals, regular icosahedral to other complex structures, and their particle size is about 1% of bacteria. Viruses are ubiquitous and are the most abundant biological entities on earth (Bergh et al., 1989; Fuhrman, 2009).

Viruses consist of two or three components: (1) Viruses contain genetic material; (2) All viruses also have capsids made of protein to encapsulate and protect the genetic material; and (3) Some viruses can form a lipid envelope around the cell when they reach the cell surface. Furthermore, viruses can self-replicate using the cellular system of hosts, but cannot grow and replicate independently. Furthermore, viruses can infect almost any living organisms with the cellular structure. However, not all the viruses can cause diseases, because the replication of many viruses cannot cause apparent damage to infected organs. Some viruses, such as human immunodeficiency virus (HIV), can coexist with the human body for a long time and remain infectious without being affected by the immune system of their hosts. In general, viral infections can trigger immune responses that destroy invading viruses. Vaccination can contribute to these immune responses, so that the vaccinated person or animal can immunize the corresponding virus for life. Microorganisms such as bacteria also have effective mechanisms to protect themselves against viral infections (Wilson and Murray, 1991).

3.2.2 Types of viruses

Viruses are mainly classified by phenotypic characteristics, such as nucleic acid type, replication mode, host organism, morphology, and disease types that they cause. Among these classifications, the formal taxonomic classification of viruses is based on the current classification system developed by the international committee on taxonomy of viruses (ICTV). By 2018, ICTV have defined 1 single phylum, 2 subphyla, 6 classes, 14 orders, 5 suborders, 143 families, 64 subfamilies, 846 genera, and 4958 species of viruses (Lefkowitz et al., 2017).

In addition, the Baltimore classification of viruses is based on the mechanism of mRNA production, because viruses must generate mRNAs to produce proteins and replicate themselves. However, this process is different in each virus family. Viruses can be placed in one of the seven following groups (Web book 2008): (1) double-stranded DNA viruses; (2) single-stranded DNA viruses; (3) double-stranded RNA viruses; (4) single-stranded RNA viruses—positive sense; (5) single-stranded RNA viruses—negative sense; (6) positive-sense single-stranded RNA viruses that replicate through a DNA intermediate; and (7) double-stranded DNA viruses that replicate through a single-stranded RNA intermediate.

3.2.3 Common viruses in wastewater

A variety of viruses that are directly harmful to human health are present in wastewater. The implementation of current wastewater treatment processes significantly reduced the levels of virus contamination in wastewater. However, viruses are still widely spread and are disseminated in the environment by discharging untreated or treated wastewater (Bofill-Mas et al., 2006, 2010) to the receiving water environments and food productions (Cantalupo et al., 2011). So the viruses, including enterovirus, hepatitis virus, rotavirus, coronavirus, adenovirus, parvovirus, reovirus, and astrovirus, that are present in wastewater, may deserve special attention.

3.2.3.1 Enterovirus

Enterovirus is a type of single-stranded RNA virus and is associated with several human diseases that mainly infects intestine. Enteroviruses are highly contagious and are transmitted mainly through two routes, including the consumption of contaminated food, water, saliva, etc. and respiratory tract (droplets, cough, etc.).

Enteroviruses are globally distributed, and humans are the only hosts. Some enteroviruses are more resistant to natural conditions and disinfectants than common bacteria (Hu et al., 2011). Moreover, enterovirus can survive in wastewater for several months. For instance, a previous study has also revealed the occurrence and distribution of cultivable enteroviruses in wastewater and surface waterbody of north-eastern Spain (Costan-Longares et al., 2008).

Enterovirus is one of the most common viruses in the water environment, and also one of the most studied viruses in water virology. In the study of the virus safety of water, enterovirus is often used as a representative, because the discharge of the virus is large and detoxification often needs a long time. The incomplete removal of enterovirus by the wastewater treatment also highlights the resistance and potential health risk for the public after discharging into the natural environment (Battistone et al., 2014).

3.2.3.2 Hepatitis virus

Hepatitis virus is the pathogen causing viral hepatitis, which has been considered as a major public health problem affecting human beings in the world and has considerable morbidity and mortality in the human. Hepatitis viruses are highly resistant to common chemical disinfectants and can survive for months or years in dry or frozen environments.

Human hepatitis viruses include hepatitis A virus (HAV), hepatitis B virus (HBV), hepatitis C virus (HCV), hepatitis D virus (HDV), and hepatitis E virus (HEV). Among them, HAV and HEV are spread through intestinal infection, and other viruses are transmitted by blood and other body fluid. HAV is a small RNA virus that is distinct from other members of picornavirus family in morphology, and it causes infectious or epidemic hepatitis by the route from feces to oral. For HBV, it is a double-stranded DNA virus and belongs to the hepadnavirus group, which is endemic in the human population and hyperendemic in the world. The resistance of HBV to environmental factors is relatively high in vitro that it cannot be inactivated by ultraviolet, heating and chemical disinfectants (such as phenol and thiomersal) in general concentration. Furthermore, they are also widely distributed in wastewater.

3.2.3.3 Rotavirus

Rotavirus is a nonencapsulated and icosahedral symmetrical virus, affiliated to the family of *Reoviridae* and the genus of *Rotavirus*. Rotavirus, with highly contagious, can survive for several weeks on the surface of water and has strong adaptability to change in physical and chemical factors. It is resistant to some organic reagents (such as chloroform ether), and still has an activity after repeated freezing and thawing as well as ultrasonic treatment. Furthermore, this virus remains infectious under the pH ranging from 3.5 to 10.0. However, ozone, iodine, chlorine, phenol, and other reagents can make it inactive.

Moreover, 285 sewage samples from four Italian cities were tested for rotavirus VP7 and VP4 genes, which revealed that rotavirus was detected in 172 (60.4%) samples and 26 samples contained multiple rotavirus G genotypes (Ruggeri et al., 2015). In China, 46 sewage samples were monthly collected for the detection of rotavirus, and 93.5% of samples were found to be positive (Zhou et al., 2016).

Rotavirus is the primary cause of acute diarrhea in children worldwide, and the peak of disease outbreak is mainly in autumn and winter. It is estimated that more than 140 million children worldwide suffer from rotavirus gastroenteritis every year. Rotavirus is mainly transmitted through the oral—fecal route and can be transmitted by consuming contaminated drinking water, eating contaminated food, or contacting contaminated objects. At present, effective treatment of humans or animals infected with rotavirus is still not available, and its viral harm cannot be fundamentally eliminated. On this account, the social impact and the economic burden of the family caused by rotavirus are enormous.

3.2.3.4 Coronavirus

Coronavirus (CoV), belonging to the family of Coronaviridae and the order of *Nidovirales*, consists of four genera (including α-CoV, β-CoV, γ-CoV, and δ-CoV) based on phylogeny. Among them, α-CoV and β-CoV mainly infect mammals; γ-CoV mainly infects birds; and δ-CoV can infect both birds and mammals. CoV is an enveloped virus with a diameter of about 60–200 nm, and its average diameter is approximately 100 nm. The length of the single-stranded linear genome is about 12–30 kb, so CoV is the longest positive single strand RNA virus in the genome. The composition and expression of genomes of all the CoVs are similar. Nucleocapsid protein is the main structural protein of CoV and also the most abundant protein in CoV coding protein, exhibiting multiple functions in replication and immune regulation. Since the emergence of severe acute respiratory syndrome coronavirus (SARS-CoV) in 2003, the research on the CoV is becoming a global concern (Zhong et al., 2003). Recently, with the change of environment conditions, the influence of immune pressure or other factors, new mutant strains of CoV have emerged and been prevalent. Novel CoVs have been successive identified, such as Middle East respiratory syndrome coronavirus and Porcine deltacoronavirus, and the emergence and reemergence of these novel CoVs cause high morbidity and mortality in human and animal and pose heavily public health threat and heavily economic losses (Dong et al., 2016; Li et al., 2012b; Zaki et al., 2012; Zhao et al., 2013).

Among the human respiratory viruses, human CoVs are the most difficult to detect in the laboratory and are not detected in most routine diagnostic virology laboratories (Mackie, 2003). In addition, CoVs were reported to die off rapidly in wastewater (Gundy et al., 2009). A previous study had demonstrated that the RNA of SARS-CoV are detectable in the hospital sewage concentrates before disinfection and occasionally after disinfection though there was no live SARS-CoV (Wang et al., 2005).

3.2.3.5 Adenovirus

In 1954, adenovirus was first isolated from tonsil tissue and had been discovered for more than 60 years (Hllleman and Werner, 1954). Adenovirus is a virus without an envelope and is one of the largest and most complex viruses among the viruses that have been discovered so far. Its genome consists of linear double-stranded DNA molecules with the length ranging from 25 to 45 kb (Smith et al., 2008). Compared with CoVs, many members of adenovirus are readily propagated in routine laboratory cell culture systems (Mackie, 2003).

Adenovirus can cause respiratory disease in a small group of normal people, but it can cause 5%–10% of respiratory diseases in children and people with low immunity. Furthermore, human adenoviruses can lead to respiratory infections, conjunctivitis, gastroenteritis, hepatitis, nervous system disorders, and other diseases. For human adenoviruses, the most common infections are in the upper respiratory tract and the epithelial cells of eyes. Adenovirus can be transmitted through a variety of routes, including humans, water, and equipment. Generally, adenovirus is more susceptible to people with low immunity, such as children, HIV-infected patients, patients with immune genetic defects, etc. For human adenovirus, belonging to the family of Adenoviridae, there are 7 subgroups and 84 kinds of adenoviruses have been reported so far (Kaján et al., 2017). Adenovirus was also frequently detected in wastewater samples. For instance, adenoviruses were detectable in wastewater, and combined sewer overflows discharged samples from a WWTP in Michigan between August 2005 and August 2006 (Fong et al., 2010).

3.2.3.6 Norovirus

Norovirus is a positive-stranded RNA virus of the Caliciviridae family, which has 5 genera (including *Vesivirus*, *Lagovirus*, *Nebovirus*, *Sapovirus*, and *Norovirus*). The norovirus genus contains at least 5 genotypes (GI-V). Among the 5 genotypes, GI, GII, and GIV can infect human beings and cause acute gastroenteritis, and GII, GIII, and GV have also been isolated from pigs, cattle and sheep, and mice, respectively (Thorne and Goodfellow, 2014).

At present, norovirus is considered to be the main cause of gastroenteritis in developed and developing countries (Glass et al., 2009; Hall et al., 2013). This virus is transmitted through the fecal–oral route and usually causes acute self-limiting gastrointestinal infections. Norovirus infection is also associated with many clinical symptoms, such as necrotizing enterocolitis, infantile epileptic encephalopathy, emphysema enteropathy, and sporadic intravascular coagulation (Chan et al., 2010; Centers for disease control and prevention 2002; Medici et al., 2010; Turcios-Ruiz et al., 2008). In developing countries, norovirus infection causes approximately 200,000 deaths of children fewer than 5 years old each year (Hall et al., 2013). Recent reports indicate that norovirus is the second leading cause of gastroenteritis-related deaths in the United States, resulting in approximately 797 deaths each year (Hall et al., 2013; Papafragkou et al., 2013). A previous study quantified norovirus genogroups I (GI) and II (GII) in wastewater in France and explored their removals by treatment systems of waste stabilization pond, activated sludge, and submerged membrane bioreactor treatments, which found that all the treatment systems efficiently decreased the level of norovirus contamination in receiving waterbodies (da Silva et al., 2007).

3.3 Protozoa

3.3.1 Biological characteristics of protozoa

Protozoa are the same as multicellular animals, with physiological functions including metabolism, exercise, reproduction, reaction to an external stimulus, and adaptability to the environment. Some common protozoa are distinguished from sizes, and protozoa individuals range in size from as little as 1 μm to several millimeters, or more (Singleton and Sainsbury, 2003). Moreover, all the protozoa have cell membranes, and the cell membranes of most protozoa are strong and elastic, so that protozoa could remain a certain shape. Generally, protozoa have one or more nuclei, which are various in shapes. However, there are also some protozoa, such as ameba, which have only one layer of the very thin plasma membrane and cannot maintain a fixed shape. Protozoa produce morphological differentiation in their cells and form organelles capable of performing various life activities and physiological functions. In terms of movement organelles, there are flagella, pseudopods, and cilia. Furthermore, some types of protozoa have myofilaments distributed in the cell membrane, which has the function of contraction and deformation.

There are three types of protozoa that are common in water treatment systems: (1) sarcopods, whose cytoplasm is flexible enough to form a pseudopod acting as an organelle for exercise and feeding; (2) flagellates, which have one or more flagella; (3) infusorians, which have cilia on the body or part of its surface acting as a tool for action or feeding. Some of the protozoa have two stages of the life cycle, alternating between proliferative stage (trophozoite) and dormant stage (cyst). When protozoa are in the stage of trophozoite, they can actively feed. As cysts, protozoa can survive in the harsh condition including extreme temperatures, harmful chemicals, and fewer nutrients, water or oxygen for a long time. The conversion from trophozoite to cyst is known as encystation, while the process of transforming from cyst to trophozoite is known as excystation.

3.3.2 Hazards and risks of protozoa

The outbreak of diseases associated with exposure to low levels of waterborne protozoa has received more concern in the world, so the occurrence, hazards, and risks of protozoa are frequently investigated. Some protozoa have a hazardous effect on human health, which can cause clinical diseases and also are responsible for outbreaks of waterborne diseases. The top 12 diseases caused by some protozoa include malaria, amebiasis, trypanosomiasis, chagas disease, lambliasis, babesiosis, cryptosporidiosis, sappinia amebic encephalitis, blastocystosis, trichomoniasis, toxoplasmosis, and schistosomiasis (Kirsten, 2018). These diseases are found in very different parts of human beings and have been documented around the world. There were 214 million cases of symptomatic malaria reported in 2015 (Saheb, 2018). For toxoplasmosis caused by the infection of *Toxoplasma gondii*, approximately 14% of the individuals in the United States are seropositive to it by the age of 40 years (Saheb, 2018). At the same time, the prevalence of *Toxoplasma gondii* infection also has been found in some American countries (Cong et al., 2015).

Moreover, risk assessment methods for these pathogenic protozoa have been explored over past several years, and a semiquantitative risk assessment is possible for most protozoa disseminated in the environment including *Cryptosporidium*, *Giardia*, *Toxoplasma*, and *Entamoeba* (Filipkowska, 2003). Furthermore, the risk assessment for *Cryptosporidium* and *Giardia* has been established to highlight the full control and adequate prevention of protozoa infections to protect human health

(Gibson et al., 1999). For *Cryptosporidium*, source water used by drinking treatment plants with an average concentration of oocysts of less than 2 oocysts per 100 L would meet the acceptable low-risk level (Haas et al., 1996). For *Giardia*, source water used by drinking treatment plants with geometric mean concentration of less than 0 ± 7 cysts per 100 L would result in acceptably low-risk level (Regli et al., 1991).

3.3.3 Common pathogenic protozoa in wastewater

Compared with waterborne epidemics caused by inorganic poisons, organic pesticides, and bacteria, diseases caused by pathogenic protozoa such as *Giardia* and *Cryptosporidium* have characteristics including high outbreak rate, high patient populations, and poor effect of treatment. It can be seen that pathogenic protozoa are the main causes of waterborne diseases among various pathogenic microorganisms. Table 3.4 summaries the common pathogenic protozoa lived in different types of wastewater (Cacciò et al., 2003; Ferrari et al., 2006; Hench et al., 2003; King et al., 2017; Sukprasert et al., 2008).

3.3.3.1 Cryptosporidium

Cryptosporidiosis is one of the most common zoonotic infectious diseases. The main symptom is diarrhea, and the severity of diarrhea depends on the characteristics of the host and the parasite (Cama et al., 2008). Most people can recover from the disease in a short period of time, but immunocompromised people may face prolonged infection and fatal danger (Chen et al., 2002). The outbreak of cryptosporidiosis is often associated with drinking water contaminated with oocysts as well as fresh food contacted by infected hosts. *Cryptosporidium* is the causative pathogen of cryptosporidiosis, which is a kind of intracellular parasitic eukaryotic single-celled organism. Many studies have shown that *Cryptosporidium* can infect humans and more than 170 kinds of animals, and mammalian *Cryptosporidium* infection plays an important role in the transmission of human cryptosporidiosis (Zhang and Jiang, 2001; Zhang et al., 2013b). *Cryptosporidium hominis* and *Cryptosporidium parvum* are two of the most common species infecting humans (Molloy et al., 2010). After entering the human body and animal body via fecal—oral route, they mainly parasitize on the intestinal epithelial cells as well as organs (including stomach, respiratory tract, lung, liver, tonsil, pancreas, gallbladder, and others) (Ortega and Sanchez, 2010; Ryan et al., 2014; Xiao and Feng, 2008).

Table 3.4 Common pathogenic protozoa in different types of wastewater.

Protozoa	Different types of wastewater
Cryptosporidium	Municipal wastewater, industrial wastewater
Giardia	Municipal wastewater, industrial wastewater
Entamoeba histolytica	Municipal wastewater

Moreover, *Cryptosporidium* is seriously harmful to the safety of drinking water. Compared to *Giardia*, *Ameba*, *Toxoplasma*, *Neisseria*, and *Cyclospora*, *Cryptosporidium* survives longer in the environment and has the strongest resistance to chemical disinfection. Furthermore, it is difficult to be removed by filtration and chemical disinfection. Therefore, *Cryptosporidium* is considered as an indicator of pathogenic protozoan parasites in the public water supply system (WHO, 2009).

3.3.3.2 Giardia

Giardia is an anaerobic flagellated protozoan parasite and colonizes and reproduces in the small intestines of several vertebrates, which can cause abdominal pain, diarrhea, and indigestion as well as giardiasis. Moreover, *Giardia* has two different developmental stages in the life cycle, including swimming trophozoite and infective cyst. *Giardia*, which can be discharged into the water via human or animal feces, is more resistant to environmental factors such as water temperature and other chemicals and can persist in a water environment (King and Monis, 2007). Waterborne transmission is the main route of human infection with *Giardia*. People can infect *Giardia* by directly contacting with contaminated water (diving, swimming, bathing, etc.) or eating contaminated food (Baldursson and Karanis, 2011; Feng and Xiao, 2011; Santín and Fayer, 2011).

3.3.3.3 Entamoeba histolytica

Entamoeba histolytica is an anaerobic parasitic amoebozoa (Ryan and Ray, 2004). The life cycle of *Entamoeba histolytica* includes trophozoite and cyst. The trophozoite is an active, feeding, and proliferating stage, and it is divided into small trophozoite and large trophozoite. Small trophozoite can develop to the stage of the cyst, which is nonactive and nonfeeding. *Entamoeba histolytica* can cause tissue destruction that leads to clinical disease, and the tissue damage is induced by direct host cell death, inflammation, and parasite invasion (Ghosh et al., 2019). It is estimated that *Entamoeba histolytica* infection kill more than 55,000 people each year (Shirley et al., 2018). The source of infection is food, drinking water, or utensils contaminated with feces containing mature cysts.

3.3.4 Roles of protozoa in wastewater treatment

Protozoa play an important ecological role not only in the matter cycling and self-purification in a natural ecosystem but also in the artificial system of WWTPs (Madoni, 2011; Pauli et al., 2001). Activated sludge, a widely used wastewater treatment all over the world, has a flocculent structure with protozoa living or crawling around it. There are about 230 species of protozoa observable in the activated sludge system, including Mastigophora (such as *Peranema deflexum* and *Anisonema acinus*), Sarcodina (such as *Mayorella penardi* and *Arcella hemisphaerica*), ciliated protozoa, and others. Among them, ciliates account for about 70% of the total protozoa in the sludge. During the wastewater purification process, protozoa are responsible for improving the quality and safety of the effluent by keeping the density of dispersed bacteria (Madoni, 2011). The roles of protozoa in wastewater treatment process are (1) direct use of organic matter in wastewater; (2) promotion of flocculation process; and (3) capable of swallowing bacteria and other microorganisms. Moreover, protozoa have a faster rate of bacterial predation and shorter generation time. However, their feeding range is narrow, and they mainly feed on free bacteria, so they are not suitable to be the main predator in the sewage treatment system. Furthermore, other physiological characteristics of bacteria in the sludge, such as athletic ability, morphology, and surface characteristics, can affect the predation of protozoa.

3.4 Helminths

3.4.1 Biological characteristics of helminths

Helminths are oligochaete annelids and worm-like parasite that can generally be seen with the naked eye in their adult stages, including both free-living and parasitic worm in nature. Helminths are the most common parasites in the world, which have three main life-cycle stages, including eggs, larvae, and adults. Different from other pathogens (such as viruses, bacteria, protozoa, and fungi), helminths cannot multiply within their hosts when they are in adult form (Castro, 1996). Compared to other infectious pathogens, helminths develop slowly, so any resultant diseases are slow in onset.

Helminths are invertebrates characterized by elongated, flat, or round bodies. Helminths can be transmitted to humans via water, food, and soil as well as vectors of arthropod and molluscan. It is noteworthy that helminths can infect many organs and are prevalent in intestines. So the most common parasitic infections in humans is helminth infection in the world. The highest prevalence of helminth infections occurs in tropical countries where food supplies are inadequate, and parasite eggs, insects, and other invertebrate vectors are abundant. Moreover, most helminth infections can be prevented by avoiding contacting with vectors, developing sanitation, and avoiding consuming foods and water that might be contaminated with helminths.

3.4.2 Classification of helminths

The definitive classification of helminths is based on the external and internal morphology of the adult stage, larval, and egg (Castro, 1996). Recently, most parasitic helminths are in three major assemblages: flukes (trematodes), tapeworms (cestodes), and roundworms (nematodes) (Castro, 1996). The further subdivision is based on the host organs. Table 3.5 is adapted from previous studies (Garcia, 1997; Cohen et al., 2017) and summarizes the major helminths in the environment.

Flukes (trematodes) are leaf-shaped flatworms, which vary in length from a few millimeters to 8 cm. Flukes are hermaphroditic except for blood flukes, which are bisexual. Flukes go through several larval stages before reaching adulthood. Tapeworms (cestodes) are hermaphroditic and vary in length from 2 mm to 10m,

Table 3.5 Major helminths in the environment (Garcia, 1997; Cohen et al., 2017).

Classification of helminths	Host organs	Major helminths
Flukes (trematodes)	Intestinal	*Fasciolopsis buski*
		Echinostoma ilocanum
		Heterophyes heterophyes
		Metagonimus yokogawai
	Liver/lung	*Clonorchis (opisthorchis) sinensis*
		Opisthorchis viverrini
		Fasciola hepatica
		Paragonimus westermani
		Paragonimus mexicanus
		Paragonimus heterotremus
		Paragonimus skrjabini
		Paragonimus spp.
	Blood	*Schistosoma mansoni*
		Schistosoma haematobium
		Schistosoma japonicum
		Schistosoma intercalatum
		Schistosoma mekongi
Tapeworms (cestodes)	Intestinal	*Diphyllobothrium latum*
		Dipylidium caninum
		Hymenolepis nana
		Hymenolepis diminuta
		Taenia solium
		Taenia saginata
	Tissue	*Taenia solium*
		Echinococcus granulosus
		Echinococcus multilocularis
		Multiceps multiceps
		Taenia multiceps
		Spirometra mansonoides
		Diphyllobothrium spp.
Roundworms (nematodes)	Intestinal	*Ascaris lumbricoides*
		Enterobius vermicularis
		Ancylostoma duodenale
		Necator americanus
		Strongyloides stercoralis
		Trichostrongylus spp.
		Trichuris trichiura
		Capillaria philippinensis
	Tissue	*Trichinella spiralis*

Table 3.5 Major helminths in the environment (Garcia, 1997; Cohen et al., 2017).—*cont'd*

Classification of helminths	Host organs	Major helminths
		Visceral larva migrans (*Toxocara canis* or *Toxocara cati*)
		Ocular larva migrans (*Toxocara canis* or *Toxocara cati*)
		Dracunculus medinensis
		Neural larva migrans (*Baylisascaris procyonis*)
		Angiostrongylus cantonensis
		Angiostrongylus costaricensis
		Gnathostoma spinigerum
		Anisakis spp. (larvae from saltwater fish)
		Phocanema spp. (larvae from saltwater fish)
		Contracaecum spp. (larvae from saltwater fish)
		Capillaria hepatica
		Thelazia spp.
	Blood and tissues	*Wuchereria bancrofti*
		Brugia malayi
		Brugia timori
		Onchocerca volvulus
		Mansonella ozzardi
		Mansonella streptocerca
		Mansonella perstans
		Dirofilaria spp. (may be found in subcutaneous nodules)

colonizing the human intestinal lumen. Adult tapeworms are elongated, flattened, and segmented. Adult roundworms (nematodes), which are cylindrical in structure, are usually bisexual (Castro, 1996). Most nematodes inhabit in the intestine and extraintestinal sites. The main developmental stages in nematodes include egg, larval, and adult stage. The eggs lay by most nematodes that are parasitic in humans may include either the zygote, blastomere, or formed larva (Castro, 1996).

The concentration of helminth eggs was high in wastewater and sludge, especially in developing countries (Amoah et al., 2018). The soil-transmitted helminth eggs relate to infection risk through different exposure routes. For instance, wastewater irrigation to the soil, which transmitted helminth infections were found among vegetable farmers in Ghana (Amoah et al., 2016). The raw and treated wastewater samples collected from 8 WWTPs in Tehran and 2 WWTPs in Isfahan were explored for the presence of helminth eggs during 2002–03, revealing that level of eggs was high in influent and it significantly reduced after treatment (Mahvi and Kia, 2006). Moreover, helminth ova are considered as one of the main target pollutants and should be removed from reused wastewater for agriculture and aquaculture, and their fates should be further investigated during different wastewater treatment processes (Jimenez, 2007).

3.5 Biotoxins

3.5.1 Characteristics and classification of biotoxins

Biotoxins mainly refer to toxic substances that can be produced by animals, plants, or microorganisms in certain conditions. Biotoxins, which are difficult to achieve by chemical synthesis, are with high toxicity and have numerous kinds. Biotoxins can be extremely hazardous even in minute quantities and are threats to human health, which are classified as biological hazards. They are fairly stable in undiluted forms, but, usually, do not persist for long periods in certain environments. Table 3.6 summarizes some biotoxins and their corresponding LD50 (μg/kg) (Institutional biosafety committee et al., 2014; OHSU research integrity Office, 2014).

In addition, many biotoxins may be further classified according to the effects on the human body, such as cytotoxins and neurotoxins. According to the source of biotoxins, it also can be divided into plant toxins, animal toxins, bacterial toxins, mycotoxins, and marine biotoxins.

3.5.2 Bacterial toxins

Among different biotoxins, bacterial toxins are mostly secondary metabolites during their growth and reproduction, which can disable the immune system and directly damage the tissue to contribute to infection and disease. Bacterial toxins include exotoxins and endotoxins. Exotoxins are highly toxic and relatively unstable, and they are cellular products excreted from certain viable Gram-positive and -negative bacteria (Chatterjee and Raval, 2019), such as *Corynebacterium diphtheriae*, *Clostridium tetani*, *Clostridium botulinum*, and *Staphylococcus aureus*. Endotoxins are moderately toxic and relatively stable, and they are lipopolysaccharide complexes derived from the cell membrane of Gram-negative bacteria (Galanos, 1998; Todar, 2018). The properties of endotoxins and classic exotoxins have been summarized by the microbiologist Kenneth Todar of the University of Wisconsin, who found that exotoxins have relatively higher potency (1 μg) than that of endotoxins (>100 μg) as well as higher specificity than that of endotoxins (Todar, 2018). Furthermore, exotoxins have enzymatic activity while endotoxins have no enzymatic activity (Todar, 2018).

Endotoxin is a kind of proinflammatory factors and pyrogen materials. Endotoxin exposure can cause a variety of symptoms, including fever, diarrhea, vomiting, wheezing, dyspnea, shock, and intravascular coagulation (Anderson et al., 2002; Basinas et al., 2015; Liebers et al., 2008). Furthermore, endotoxin may enhance

Table 3.6 Some biotoxins and their corresponding LD50.

Toxin	LD50 (μg/kg)
Abrin	0.7
Aerolysin	7
Botulinum toxin A	0.0012
Botulinum toxin B	0.0012
Botulinum toxin C1	0.0011
Botulinum toxin C2	0.0012
Botulinum toxin D	0.0004
Botulinum toxin E	0.0011
Botulinum toxin F	0.0025
β-bungarotoxin	14
Caeruleotoxin	53
Cereolysin	40–80
Cholera toxin	250
Clostridium difficile enterotoxin A	0.5
Clostridium perfringens delta toxin	5
Clostridium perfringens epsilon toxin	0.1
Clostridium perfringens lecithinase	3
Clostridium perfringens perfringolysin O	13–16
Conotoxins	12–30
Crotoxin	12–30
Diacetoxyscirpenol	1000–10,000
Diphtheria toxin	0.1
HT-2 toxin	5–10
Leucocidin	50
Listeria listeriolysin or hemolysin	3–10
Listeriolysin	3–12
Modeccin	1–10
Nematocyst toxins	33–70
Notexin	25
Pertussis toxin	15
Pneumolysin	1.5
Pseudomonas aeruginosa exotoxin A	3
Ricin	2.7
Saxitoxin	8
Shigella dysenteriae neurotoxin	1.3
Staphylococcus enterotoxin B	25
Staphylococcus enterotoxin F	2–10
Staphylococcus enterotoxins A, C, D, and E	20 (A); <50 (C)
Streptolysin O	8
Streptolysin S	25

Continued

Table 3.6 Some biotoxins and their corresponding LD50.—*cont'd*

Toxin	LD50 (μg/kg)
T-2 toxin	5–10
Taipoxin	2
Tetanus toxin	0.001
Tetrodotoxin	8
Viscum Album lectin 1	2.4–80
Volkensin	1.4
Yersinia pestis murine toxin	10

the toxic effects of other toxic substances, such as algal toxins (Best et al., 2002; Roth et al., 1997). Therefore, some control standards of endotoxin have proposed that the endotoxin content of water for injection should be less than 0.25 EU/mL (Anderson et al., 2002). For the workers in the wastewater treatment plant, endotoxin exposure ranged from 0.6 to 2093 EU/m^3, and the geometric mean exposure was low (27 EU/m^3) (Smit et al., 2005). Some symptoms appeared to be more prevalent in workers when they were exposed to endotoxin with a level higher than 50 EU/m^3 (Smit et al., 2005). Endotoxin activity has been assessed in several WWTPs using the samples collected from influent, effluent, return sludge, and advanced treatment effluent in Sapporo and Japan, which have revealed that active endotoxin materials occurred in wastewater and endotoxin activity was high in wastewater (Guizani et al., 2009). Moreover, the discharge of effluent of treatment plants has increased the endotoxicity in the receiving river water (Ohkouchi et al., 2007). Guizani (2010) have reported that biological treatment cannot control endotoxicity and can produce organic matters with endotoxicity during wastewater reclamation.

3.6 Antibiotic resistance

3.6.1 ARGs and ARB

Recently, extensive use and abuse of antibiotics may induce the development of ARGs and ARB in the environment, which has conferred enormous and complicated impacts on human health and environmental safety (Blaser, 2011; He et al., 2016; Rizzo et al., 2013). As early as in 2004, American scholars Rysz and Alvarez (2004) considered ARGs as a new type of environmental pollutants. In 2006, Pruden et al. (2006) put forward ARGs as a new type of environmental pollutants, which has drawn more and more attention in the research field of environmental sciences. ARGs can spread and transfer between different bacteria and have more adverse effects on the environment than the ARGs themselves, which is one of the reasons for the growing pollution in the environment (Dodd, 2012; Jiao et al., 2017). In the meantime, increasing ARB has been detected in the environment. It has been reported that multiresistant New Delhi Metallo-β-lactamase-1 (NDM-1) emerges in wastewater of two STPs in Haihe River basin of China, and the ARGs of NDM-1 can transfer to the indigenous bacteria in the receiving river providing source water for millions of people nearby, which poses huge health risk (Luo et al., 2013). The purpose of this section is to thoroughly summarize the pollution status of ARGs and ARB in the wastewater and propose future research directions.

3.6.2 Mechanisms of antibiotic resistance in bacteria

Bacterial antibiotic resistance, especially for the multiple-antibiotic resistance, has emerged as both medical and social problems in the world, which poses a significant threat against antiinfective therapy in the environment. The antibiotic resistance in bacteria can be mediated via several mechanisms, which fall into four main groups (Blair et al., 2015; Munita and Arias, 2016; Ramirez and Tolmasky, 2010; Zhang et al., 2009):

3.6.2.1 Enzyme-catalyzed inactivation of antibiotics

The inactivation or passivation enzyme can destruct antibiotic molecules by hydrolysis or modification, which induces antibiotic resistance in bacteria. For example, β-lactamase can destroy the amide bond of the β-lactam ring of penicillin and cephalosporin antibiotics, rendering the antimicrobial ineffective. Aminoglycoside modifying enzymes inactivate some active groups in the molecules of aminoglycoside and quinolones so that their binding ability to the target is reduced.

3.6.2.2 Changes in antibiotic targets

Some antibiotics are specifically combined with bacterial target sites and affect their normal physiological functions, leading to bacterial death. Changes of the genes coding the corresponding targets will lead to the change of targets structure that prevents efficient antibiotic binding and target recognition. For example, methylase synthesized by the bacteria resistant to macrolides, leading to methylation of 23S rRNA adenine in the 50S subunit of the ribosome, which prevents the binding of antibiotics to binding sites and exhibits resistance to macrolides.

3.6.2.3 Bacterial efflux pumps

Bacterial efflux pumps are capable to actively extrude many antibiotics out of the cell, and can also result in antimicrobial resistance. The efflux pump systems are common in all kinds of bacteria and are major contributors to the intrinsic resistance to many drugs. For example, tetracycline efflux pump genes can encode related membrane proteins to extrude the tetracycline out of the cell, which reduces the concentration of intracellular tetracycline.

3.6.2.4 Changes in the permeability of bacterial cell walls or cell membranes

Changes in the outer membrane proteins of bacteria can reduce the permeability of the outer membrane, and limit antibiotic entry into the target sites of the bacterial cell, leading to cross-resistance to different types of antibiotics, especially for β-lactam and quinolone antibiotics.

For intrinsic resistance, bacteria also acquire or develop resistance to antibiotics via mutations and by horizontal gene transfer (Munita and Arias, 2016). Acquisition of external DNA through horizontal gene transfer by three main strategies (transformation, transduction, and conjugation) is one of the most important drivers of bacterial evolution, and it is frequently responsible for the development and dissemination of resistance to many frequently used antibiotics.

3.6.3 Fates of ARGs and ARB in wastewater

As a reservoir for ARGs and ARB, wastewater is one of the important contamination sources for the antibiotic resistance dissemination in the environment. Recently, a variety of ARG types (tetracycline, sulfonamide, multidrug, aminoglycoside, bacitracin, chloramphenicol, β-lactam, quinolone, trimethoprim, polymyxin, and vancomycin as well as other types) have been found in medical wastewater, pharmaceutical wastewater, domestic wastewater and wastewater from aquaculture systems, and livestock breeding (Guo et al., 2018; Jia et al., 2017; Tong et al., 2019; Zhao et al., 2018). Furthermore, ARB and multiple ARB, such as *Pseudomonas aeruginosa*, *Escherichia coli*, *Acinetobacter* spp., and Enterobacteriaceae are observed in wastewater (Huang et al., 2018; Kümmerer, 2009). The main sources and distribution of ARGs in wastewater were reviewed:

3.6.3.1 Medical and pharmaceutical wastewater

Medical and pharmaceutical wastewater contains a large amount of ARGs and ARB, which is the main source of ARGs and ARB in the water environment. Rodriguez-Mozaz et al. (2015) have detected ARGs in hospital sewage, urban sewage treatment system, and the receiving river in Spain, and found that the abundances of bla_{TEM}, *qnr*S, *sul*I, and *tet*(W) in hospital sewage were higher than those in the urban sewage treatment system and the receiving river. ARGs within β-lactam (bla_{VIM} and bla_{SHV}), aminoglycosides (*aac*C2), chloramphenicol (*cat*A1 and *flo*R), macrolide–lincosamide–streptogramins (*emr*A and *mef*A), sulfonamide (*sul*I and *sul*II), and tetracycline (*tet*(A), *tet*(B), *tet*(C), *tet*(O), and *tet*(W)) associated with transposons have been found in Romanian hospital sewage (Szekeres et al., 2017). In addition, sulfonamide (*sul*I and *sul*II)), tetracycline (*tet*(O), *tet*(T), *tet*(M), *tet*(Q), and *tet*(W), β-lactam (bla_{OXA-1}, bla_{OXA-2}, and bla_{OXA-10}), and macrolide (*erm*B) resistance genes were detected based on quantitative PCR analysis in typical pharmaceutical wastewater treatment systems (Zhai et al., 2016). The maximum concentrations of ARGs detected in the final effluents of pharmaceutical WWTPs were up to 3.68×10^6 copies/mL by Wang et al. (2015) and 2.36×10^7 copies/mL by Zhai et al. (2016), respectively, which were much higher than the concentration in MWTPs as revealed by Mao et al. (2015).

In hospital sewage in New Delhi, Lamba et al. (2017) have detected 748 extended-spectrum β-lactam resistant bacterial strains (including *Escherichia coli*,

Klebsiella, and *Pseudomonas putida*) and 953 carbapenem-resistant Enterobacteriaceae strains (including *Klebsiella pneumoniae*, *Pseudomonas putida*, and *Klebsiella pneumonia* subsp. Pneumonia). ARB resistant to carbapenem also existed in hospital sewage in China, Croatia, and other countries in the world (Zhang et al., 2013a; Hrenovic et al., 2015).

3.6.3.2 The domestic wastewater treatment system

Domestic wastewater treatment system is the node where different types of wastewater converge, containing a large number of exogenous ARGs and ARB. At the same time, the activated sludge in the domestic wastewater treatment plant is an ideal habitat for bacteria due to its rich nutrition and large amount of aeration, which also makes the domestic wastewater treatment system to become an important pollution source of ARGs and ARB in the water environment (Donlan, 2002; Li et al., 2019; Yang et al., 2013). Hrenovic et al. (2017) have detected carbapenem-resistant bacteria in influent, activated sludge, effluent, digested sludge, and stabilized sludge of the largest Croatian secondary wastewater treatment plant in Zagreb, and found that the number of carbapenem-resistant bacteria in effluent was reduced by more than 99% compared with influent while the relative abundance increased. Hembach et al. (2017) have detected *mcr-1* gene and some clinical-related genes (including $bla_{CTX\text{-}M\text{-}32}$, bla_{TEM}, $bla_{CTX\text{-}M}$, *tet*(M), $bla_{CMY\text{-}2}$, and *erm*B) in the influent and effluent of seven German WWTPs. Luo et al. (2013) have found the NDM-1 gene in disinfected effluent from two STPs in north China with the absolute abundance of 1316 ± 232 and 1431 ± 247 copies/mL, respectively. It is worth noting that ARGs cannot be completely removed by STPs, and ARGs can be transmitted in the receiving river (Rodriguez-Mozaz et al., 2015).

The research group of Professor Zhang from the University of Hong Kong conducted a detailed and deep research on the fate of ARGs in STPs. *Tet*(A), *tet*(C), *tet*(G), *tet*(M), *tet*(S), and *tet*(X) have been detected in all the 15 STPs across China, Canada, the United States, and Singapore (Zhang and Zhang, 2011). Yang et al. (2013) have investigated the variation of ARGs in activated sludge over a 4-year period, which revealed that more than 200 ARG subtypes have been detected in the activated sludge, and aminoglycoside and tetracycline resistance genes have highest abundances followed by sulfonamide, multidrug, and chloramphenicol resistance genes. On this basis, a total of 271 ARG subtypes within 18 ARG types were detectable in the influent, activated sludge, effluent, and anaerobic digestion sludge in a sewage treatment plant in Hong Kong from 2011 to 2012, and the abundance of ARGs was highest in influent followed by effluent, activated sludge, and anaerobic digestion sludge (Yang et al., 2014). Meanwhile, the research group also found 78 persistent ARGs in the process of biological wastewater and sludge treatment, revealing that anaerobic digestion of sludge cannot completely remove ARGs (Zhang et al., 2015). Furthermore, the abundance of ARGs is correlated with antibiotic resistance phenotypes in municipal sewage, and the abundance of ARGs in resistant coliforms is also relatively high and up to 33.8 ± 4.2 copies per cell (Li et al., 2017).

3.6.3.3 Wastewater from aquaculture system and livestock breeding

The use of antibiotics in aquaculture system and livestock breeding could increase the potential risk of antibiotic resistance in products and ecosystems (He et al., 2016; Huang et al., 2015). Increasing studies have shown that aquaculture and animal farms are important sources of ARB and ARGs in the water environment, where ARB and ARGs are abundant and may infect humans through the food chain (Huang et al., 2015; Jia et al., 2017). Currently, antibiotic resistant pathogens such as *Aeromonas* (Penders and Stobberingh, 2008), *Vibrio* (Oh et al., 2011), and *Salmonella* (Budiati et al., 2013) have been found in aquaculture systems. A total of 4767 strains nonsusceptible to sulfamethoxazole/trimethoprim, tetracycline, erythromycin or cefotaxime were isolated from fish intestine, fish surface rinsing water, fish feed, pond mud, and pond water from an American aquaculture farm, and 80% of them exhibited multiple antibiotic resistance (Huang et al., 2015). Furthermore, strains of *Escherichia coli* were isolated in pigs and poultry in China which had high rates of resistance to ampicillin, tetracycline, doxycycline, trimethoprim–sulfamethoxazole, amoxicillin, streptomycin, and chloramphenicol (Jiang et al., 2011). Notably, antibiotic resistance was observed in aquaculture and livestock breeding system as well as clinically isolated strains, indicating that potential antibiotic resistance could be transmitted between food and humans (Brooks et al., 2014; Huang et al., 2015). Xiong et al. (2015) found a total of 15 ARG subtypes (*sul*I, *sul*II, *sul*III, *tet*(M), *tet*(O), *tet*(W), *tet*(S), *tet*(Q), *tet*(X), *tet*(B/P), *qep*A, *oqx*A, *oqx*B, *aac(6′)-Ib*, and *qnr*S) within four ARG types (sulfonamide, tetracycline, aminoglycoside, and quinolone) in fresh water aquaculture environment and the relative abundance was up to 2.8×10^{-2} (ARG copy/16S rRNA copy). Tajbakhsh et al. (2015) collected 150 water samples from aquaculture water of fish fields in different geographical regions in Iran and found 18% of the water samples contained *Escherichia coli* resistant to ampicillin, ciprofloxacin, gentamycin, chloramphenicol, tetracycline, and imipenem. Furthermore, Zhu et al. (2013) comprehensively evaluated the diversity and abundance of ARGs in a pig farm and detected a total of 149 ARG subtypes resistance to aminoglycoside, β-lactam, tetracycline, vancomycin, and other antibiotics. At the same time, their abundances significantly correlated with levels of antibiotics (Zhu et al., 2013). Quinolone-resistant genes *qnr*S and *oqx*A have been detected in pig farm wastewater using real-time quantitative PCR technology, and they have spread to the surrounding environment based on the comparison with the levels of them in the surrounding soil (Li et al., 2012a).

3.7 Summary

Characteristics, classification, fates, functions, and health implications of different biological HRPs in wastewater are summarized in this chapter. A variety of biological HRPs are prevalent in wastewater, including bacteria, viruses, protozoa, helminth, biotoxins, ARGs, and ARB. The biological HRPs in wastewater possibly contact with human bodies along with wastewater discharge into the environment or reuse for different purposes, which can induce serious infection to affect public health. Notably, some wastewater treatment processes have the potential to reduce the level of various biological HRPs and the induced risk on environment and health. However, due to the lack of reliable risk assessment methodology for the biological HRPs in the wastewater, whether the treated wastewater is safe enough for public health currently remains known. Thus, the health risk associated with these biological HRPs in the wastewater deserves greater concerns, and more efforts have to be devoted to extending our knowledge regarding their health hazards and risks and developing effective technologies to prevent, remove, or kill biological HRPs in the wastewater.

References

Alderwick, L.J., Harrison, J., Lloyd, G.S., Birch, H.L., 2015. The *Mycobacterial* cell wall—peptidoglycan and arabinogalactan. Cold Spring Harbor Perspectives in Medicine 5 (8), a021113.

Amoah, I.D., Abubakari, A., Stenström, T.A., Abaidoo, R.C., Seidu, R., 2016. Contribution of wastewater irrigation to soil transmitted helminths infection among vegetable farmers in Kumasi, Ghana. PLoS Neglected Tropical Diseases 10 (12), e0005161.

Amoah, I.D., Adegoke, A.A., Stenström, T.A., 2018. Soil-transmitted helminth infections associated with wastewater and sludge reuse: a review of current evidence. Tropical Medicine and International Health 23 (7), 692—703.

Anastasi, E., Matthews, B., Stratton, H., Katouli, M., 2012. Pathogenic *Escherichia coli* found in sewage treatment plants and environmental waters. Applied and Environmental Microbiology 78 (16), 5536—5541.

Anderson, W.B., Slawson, R.M., Mayfield, C.I., 2002. A review of drinking-water-associated endotoxin, including potential routes of human exposure. Canadian Journal of Microbiology 48 (7), 567—587.

Athman, R., Fernandez, M.I., Gounon, P., Sansonetti, P., Louvard, D., Philpott, D., Robine, S., 2005. *Shigella flexneri* infection is dependent on villin in the mouse intestine and in primary cultures of intestinal epithelial cells. Cellular Microbiology 7 (8), 1109—1116.

Baldursson, S., Karanis, P., 2011. Waterborne transmission of protozoan parasites: review of worldwide outbreaks—an update 2004—2010. Water Research 45 (20), 6603—6614.

Bartram, J., Chartier, Y., Lee, J.V., Pond, K., Surman-Lee, S., 2007. *Legionella* and the Prevention of Legionellosis. World Health Organization. http://www.who.int/water_sanitation_health/emerging/legionella.pdf.

Basinas, I., Sigsgaard, T., Kromhout, H., Heederik, D., Wouters, I.M., Schlünssen, V., 2015. A comprehensive review of levels and determinants of personal exposure to dust and endotoxin in livestock farming. Journal of Exposure Science and Environmental Epidemiology 25 (2), 123—137.

Battistone, A., Buttinelli, G., Bonomo, P., Fiore, S., Amato, C., Mercurio, P., Cicala, A., Simeoni, J., Foppa, A., Triassi, M., 2014. Detection of enteroviruses in influent and effluent flow samples from wastewater treatment plants in Italy. Food and Environmental Virology 6 (1), 13—22.

Bergh, Ø., BØrsheim, K.Y., Bratbak, G., Heldal, M., 1989. High abundance of viruses found in aquatic environments. Nature 340, 467–468.

Best, J.H., Pflugmacher, S., Wiegand, C., Eddy, F.B., Metcalf, J.S., Codd, G.A., 2002. Effects of enteric bacterial and cyanobacterial lipopolysaccharides, and of microcystin-LR, on glutathione S-transferase activities in zebra fish (*Danio rerio*). Aquatic Toxicology 60 (3–4), 223–231.

Beveridge, T.J., 2001. Use of the gram stain in microbiology. Biotechnic and Histochemistry 76 (3), 111–118.

Blair, J.M.A., Webber, M.A., Baylay, A.J., Ogbolu, D.O., Piddock, L.J.V., 2015. Molecular mechanisms of antibiotic resistance. Nature Reviews Microbiology 13, 42–51.

Blaser, M., 2011. Antibiotic overuse: stop the killing of beneficial bacteria. Nature 476, 393–394.

Bofill-Mas, S., Albinana-Gimenez, N., Clemente-Casares, P., Hundesa, A., Rodriguez-Manzano, J., Allard, A., Calvo, M., Girones, R., 2006. Quantification and stability of human adenoviruses and polyomavirus JCPyV in wastewater matrices. Applied and Environmental Microbiology 72 (12), 7894–7896.

Bofill-Mas, S., Rodriguez-Manzano, J., Calgua, B., Carratala, A., Girones, R., 2010. Newly described human polyomaviruses Merkel cell, KI and Wu are present in urban sewage and may represent potential environmental contaminants. Virology Journal 7, 141.

Brissaud, F., Blin, E., Hemous, S., Garrelly, L., 2008. Water reuse for urban landscape irrigation: aspersion and health related regulations. Water Science and Technology 57 (5), 781–787.

Brooks, J.P., Adeli, A., McLaughlin, M.R., 2014. Microbial ecology, bacterial pathogens, and antibiotic resistant genes in swine manure wastewater as influenced by three swine management systems. Water Research 57, 96–103.

Budiati, T., Rusul, G., Wan-Abdullah, W.N., Arip, Y.M., Ahmad, R., Thong, K.L., 2013. Prevalence, antibiotic resistance and plasmid profiling of *Salmonella* in catfish (*Clarias gariepinus*) and tilapia (*Tilapia mossambica*) obtained from wet markets and ponds in Malaysia. Aquaculture 372, 127–132.

Cacciò, S.M., De Giacomo, M., Aulicino, F.A., Pozio, E., 2003. *Giardia* cysts in wastewater treatment plants in Italy. Applied and Environmental Microbiology 69 (6), 3393–3398.

Cai, L., Zhang, T., 2013. Detecting human bacterial pathogens in wastewater treatment plants by a high-throughput shotgun sequencing technique. Environmental Science and Technology 47 (10), 5433–5441.

Caicedo, C., Beutel, S., Scheper, T., Rosenwinkel, K., Nogueira, R., 2016. Occurrence of *Legionella* in wastewater treatment plants linked to wastewater characteristics. Environmental Science and Pollution Research 23 (16), 16873–16881.

Caicedo, C., Rosenwinkel, K.H., Exner, M., Verstraete, W., Suchenwirth, R., Hartemann, P., Nogueira, R., 2019. *Legionella* occurrence in municipal and industrial wastewater treatment plants and risks of reclaimed wastewater reuse: review. Water Research 149, 21–34.

Cama, V.A., Bern, C., Roberts, J., Cabrera, L., Sterling, C.R., Ortega, Y., Gilman, R.H., Xiao, L., 2008. *Cryptosporidium* species and subtypes and clinical manifestations in children, Peru. Emerging Infectious Diseases 14 (10), 1567–1574.

Cantalupo, P.G., Calgua, B., Zhao, G., Hundesa, A., Wier, A.D., Katz, J.P., Grabe, M., Hendrix, R.W., Girones, R., Wang, D., Pipas, J.M., 2011. Raw sewage harbors diverse viral populations. mBio 2 (5) e00180-11.

Castro, G.A., 1996. Helminths: structure, classification, growth, and development. In: Baron, S. (Ed.), Medical Microbiology. University of Texas Medical Branch at Galveston. Galveston, Texas. Chapter 86.

Centers for Disease Control and Prevention, 2002. Outbreak of acute gastroenteritis associated with Norwalk-like viruses among British military personnel — Afghanistan, May 2002. Morbidity and Mortality Weekly Report 51 (22), 477—479.

Chan, W.K., Lee, K.W., Fan, T.W., 2010. Pneumatosis intestinalis in a child with nephrotic syndrome and norovirus gastroenteritis. Pediatric Nephrology 25 (8), 1563—1566.

Chan, G.J., Lee, A.C., Baqui, A.H., Tan, J., Black, R.E., 2013. Risk of early-onset neonatal infection with maternal infection or colonization: a global systematic review and meta-analysis. PLoS Medicine 10 (8), e1001502.

Chatterjee, S., Raval, I.H., 2019. Pathogenic microbial genetic diversity with reference to health. In: Das, S., Dash, H.R. (Eds.), Microbial Diversity in the Genomic Era. Academic Press, Bhavnagar, Gujarat, India, pp. 550—577.

Chen, X.-M., Keithly, J.S., Paya, C.V., LaRusso, N.F., 2002. Cryptosporidiosis. New England Journal of Medicine 346 (22), 1723—1731.

Cohen, J., Powderly, W.G., Opal, S.M., 2017. Infectious Diseases, fourth ed. Elsevier, Amsterdam, Netherlands, pp. 1763—1779.

Colwell, R.R., 1996. Global climate and infectious disease: the cholera paradigm. Science 274, 2025—2031.

Cong, W., Dong, X.-Y., Meng, Q.-F., Zhou, N., Wang, X.-Y., Huang, S.-Y., Zhu, X.-Q., Qian, A.-D., 2015. *Toxoplasma gondii* infection in pregnant women: a seroprevalence and case-control study in Eastern China. Biomed Research International 2015, 170278.

Cooper, I., Hanlon, G., 2010. Resistance of *Legionella pneumophila* serotype 1 biofilms to chlorine-based disinfection. Journal of Hospital Infection 74 (2), 152—159.

Costan-Longares, A., Moce-Llivina, L., Avellon, A., Jofre, J., Lucena, F., 2008. Occurrence and distribution of culturable enteroviruses in wastewater and surface waters of north-eastern Spain. Journal of Applied Microbiology 105 (6), 1945—1955.

da Silva, A.K., Le Saux, J.-C., Parnaudeau, S., Pommepuy, M., Elimelech, M., Le Guyader, F.S., 2007. Evaluation of removal of noroviruses during wastewater treatment, using real-time reverse transcription-PCR: different behaviors of genogroups I and II. Applied and Environmental Microbiology 73 (24), 7891—7897.

Dodd, M.C., 2012. Potential impacts of disinfection processes on elimination and deactivation of antibiotic resistance genes during water and wastewater treatment. Journal of Environmental Monitoring 14 (7), 1754—1771.

Dong, N., Fang, L., Yang, H., Liu, H., Du, T., Fang, P., Wang, D., Chen, H., Xiao, S., 2016. Isolation, genomic characterization, and pathogenicity of a Chinese porcine deltacoronavirus strain CHN-HN-2014. Veterinary Microbiology 196, 98—106.

Donlan, R.M., 2002. Biofilms: microbial life on surfaces. Emerging Infectious Diseases 8 (9), 881—890.

Dougan, G., Baker, S., 2014. *Salmonella enterica Serovar Typhi* and the pathogenesis of typhoid fever. Annual Review of Microbiology 68, 317—336.

Dudley, D.J., Guentzel, M.N., Ibarra, M., Moore, B., Sagik, B., 1980. Enumeration of potentially pathogenic bacteria from sewage sludges. Applied and Environmental Microbiology 39 (1), 118—126.

Dungeni, M., van Der Merwe, R., Momba, M., 2010. Abundance of pathogenic bacteria and viral indicators in chlorinated effluents produced by four wastewater treatment plants in the Gauteng Province, South Africa. Water SA 36 (5), 607—614.

Dusenbery, D.B., 2009. Living at the Micro Scale: the Unexpected Physics of Being Small. Harvard University Press, Cambridge, MA.

Ekwanzala, M.D., Abia, A.L.K., Keshri, J., Momba, M.N.B., 2017. Genetic characterization of *Salmonella* and *Shigella* spp. isolates recovered from water and riverbed sediment of the Apies River, South Africa. Water SA 43 (3), 387–397.

El-Lathy, A., El-Taweel, G.E., El-Sonosy, M., Samhan, F., Moussa, T.A., 2009. Determination of pathogenic bacteria in wastewater using conventional and PCR techniques. Environmental Biotechnology 5, 73–80.

Espigares, E., Bueno, A., Espigares, M., Gálvez, R., 2006. Isolation of *Salmonella* serotypes in wastewater and effluent: effect of treatment and potential risk. International Journal of Hygiene and Environmental Health 209 (1), 103–107.

Feng, Y., Xiao, L., 2011. Zoonotic potential and molecular epidemiology of *Giardia* species and giardiasis. Clinical Microbiology Reviews 24 (1), 110–140.

Ferrari, B.C., Stoner, K., Bergquist, P.L., 2006. Applying fluorescence based technology to the recovery and isolation of *Cryptosporidium* and *Giardia* from industrial wastewater streams. Water Research 40 (3), 541–548.

Filipkowska, Z., 2003. Sanitary and bacteriological aspects of sewage treatment. Acta Microbiologica Polonica 52 (Suppl. 1), 57–66.

Fong, T.-T., Phanikumar, M.S., Xagoraraki, I., Rose, J.B., 2010. Quantitative detection of human adenoviruses in wastewater and combined sewer overflows influencing a Michigan river. Applied and Environmental Microbiology 76 (3), 715–723.

Foster, J.W., Slonczewski, J.L., 2017. Microbiology: An Evolving Science, fourth International Student ed. WW Norton & Company, New York.

Fritz, I., Strömpl, C., Abraham, W.-R., 2004. Phylogenetic relationships of the genera *stella, labrys* and *angulomicrobium* within the 'alphaproteobacteria' and description of *Angulomicrobium amanitiforme* sp. nov. International Journal of Systematic and Evolutionary Microbiology 54 (3), 651–657.

Fuhrman, J.A., 2009. Microbial community structure and its functional implications. Nature 459, 193–199.

Galanos, C., 1998. Encyclopedia of Immunology, second ed. Elsevier, Oxford.

Garcia, L.S., 1997. Classification of human parasites. Clinical Infectious Diseases 25 (1), 21–23.

Garcia-Aljaro, C., Muniesa, M., Blanco, J., Blanco, M., Blanco, J., Jofre, J., Blanch, A., 2005. Characterization of *Shiga* toxin-producing *Escherichia coli* isolated from aquatic environments. FEMS Microbiology Letters 246 (1), 55–65.

Gerardi, M.H., 2006. Wastewater Bacteria. Wiley-Interscience, New Jersey.

Gerardi, M.H., Zimmerman, M.C., 2004. Wastewater Pathogens. John Wiley & Sons, New Jersey.

Ghosh, S., Padalia, J., Moonah, S., 2019. Tissue destruction caused by *Entamoeba histolytica* parasite: cell death, inflammation, invasion, and the gut microbiome. Current Clinical Microbiology Reports 6 (1), 51–57.

Gibson, C., Haas, C., Rose, J., 1999. Risk assessment of waterborne protozoa: current status and future trends. Parasitology 117 (7), 205–212.

Glass, R.I., Parashar, U.D., Estes, M.K., 2009. Norovirus gastroenteritis. New England Journal of Medicine 361 (18), 1776–1785.

Guizani, M., 2010. Control of Endotoxins and Their Fate during Wastewater Reclamation (Ph.D. thesis report). Hokkaido University, Japan.

Guizani, M., Dhahbi, M., Funamizu, N., 2009. Assessment of endotoxin activity in wastewater treatment plants. Journal of Environmental Monitoring 11 (7), 1421–1427.

Gundy, P.M., Gerba, C.P., Pepper, I.L., 2009. Survival of coronaviruses in water and wastewater. Food and Environmental Virology 1, 10–14.

Guo, X., Yan, Z., Zhang, Y., Xu, W., Kong, D., Shan, Z., Wang, N., 2018. Behavior of antibiotic resistance genes under extremely high-level antibiotic selection pressures in pharmaceutical wastewater treatment plants. Science of the Total Environment 612, 119–128.

Haas, C.N., Crockett, C.S., Rose, J.B., Gerba, C.P., Fazil, A.M., 1996. Assessing the risk posed by oocysts in drinking water. Journal of the American Water Works Association 88 (9), 131–136.

Hall, A.J., Lopman, B.A., Payne, D.C., Patel, M.M., Gastañaduy, P.A., Vinjé, J., Parashar, U.D., 2013. Norovirus disease in the United States. Emerging Infectious Diseases 19 (8), 1198–1205.

He, L.-Y., Ying, G.-G., Liu, Y.-S., Su, H.-C., Chen, J., Liu, S.-S., Zhao, J.-L., 2016. Discharge of swine wastes risks water quality and food safety: antibiotics and antibiotic resistance genes from swine sources to the receiving environments. Environment International 92, 210–219.

Hellingwerf, K., Crielaard, W., Hoff, W., Matthijs, H., Mur, L., Van Rotterdam, B., 1994. Photobiology of bacteria. Antonie van Leeuwenhoek 65 (4), 331–347.

Hembach, N., Schmid, F., Alexander, J., Hiller, C., Rogall, E.T., Schwartz, T., 2017. Occurrence of the *mcr-1* colistin resistance gene and other clinically relevant antibiotic resistance genes in microbial populations at different municipal wastewater treatment plants in Germany. Frontiers in Microbiology 8, 1282.

Hench, K.R., Bissonnette, G.K., Sexstone, A.J., Coleman, J.G., Garbutt, K., Skousen, J.G., 2003. Fate of physical, chemical, and microbial contaminants in domestic wastewater following treatment by small constructed wetlands. Water Research 37 (4), 921–927.

Hensel, M., 2004. Evolution of pathogenicity islands of *Salmonella enterica*. International Journal of Medical Microbiology 294 (2–3), 95–102.

Hllleman, M., Werner, J.H., 1954. Recovery of new agent from patients with acute respiratory illness. Proceedings of the Society for Experimental Biology and Medicine 85 (1), 183–188.

Hrenovic, J., Goic-Barisic, I., Kazazic, S., Kovacic, A., Ganjto, M., Tonkic, M., 2015. Carbapenem-resistant isolates of *Acinetobacter baumannii* in a municipal wastewater treatment plant, Croatia, 2014. Euro Surveillance 21 (15). https://doi.org/10.2807/1560-7917.ES.2016.21.15.30195.

Hrenovic, J., Ivankovic, T., Ivekovic, D., Repec, S., Stipanicev, D., Ganjto, M., 2017. The fate of carbapenem-resistant bacteria in a wastewater treatment plant. Water Research 126, 232–239.

Hu, H., Wu, Q., Huang, J., Zhao, X., 2011. Safety Assessment and Guarantee Principle of Reclaimed Water Quality. Science Press, Beijing.

Huang, S., Hsu, B., Ma, P., Chien, K., 2009. *Legionella* prevalence in wastewater treatment plants of Taiwan. Water Science and Technology 60 (5), 1303–1310.

Huang, Y., Zhang, L., Tiu, L., Wang, H., 2015. Characterization of antibiotic resistance in commensal bacteria from an aquaculture ecosystem. Frontiers in Microbiology 6, 914.

Huang, K., Mao, Y., Zhao, F., Zhang, X.-X., Ju, F., Ye, L., Wang, Y., Li, B., Ren, H., Zhang, T., 2018. Free-living bacteria and potential bacterial pathogens in sewage treatment plants. Applied and Environmental Microbiology 102 (5), 2455–2464.

Hugenholtz, P., 2002. Exploring prokaryotic diversity in the genomic era. Genome Biology 3 (2) reviews 0003.1–0003.8.

Institutional Biosafety Committee, Office of Research Safety, Office of Biological Safety, University of Chicago, 2014. Biological Toxin. Biohazard Recognition and Control, third ed. Chapter IX. Available from: https://researchsafety.uchicago.edu/sites/researchsafety.uchicago.edu/files/uploads/UC%20Biosafety%20Manual%20Third%20Edition.pdf.

Jennison, A.V., Verma, N.K., 2004. *Shigella flexneri* infection: pathogenesis and vaccine development. FEMS Microbiology Reviews 28 (1), 43–58.

Jia, S., Zhang, X.X., Miao, Y., Zhao, Y., Ye, L., Li, B., Zhang, T., 2017. Fate of antibiotic resistance genes and their associations with bacterial community in livestock breeding wastewater and its receiving river water. Water Research 124, 259–268.

Jiang, H.-X., Lü, D.-H., Chen, Z.-L., Wang, X.-M., Chen, J.-R., Liu, Y.-H., Liao, X.-P., Liu, J.-H., Zeng, Z.-L., 2011. High prevalence and widespread distribution of multi-resistant *Escherichia coli* isolates in pigs and poultry in China. The Veterinary Journal 187 (1), 99–103.

Jiao, Y.N., Chen, H., Gao, R.X., Zhu, Y.G., Rensing, C., 2017. Organic compounds stimulate horizontal transfer of antibiotic resistance genes in mixed wastewater treatment systems. Chemosphere 184, 53–61.

Jimenez, B., 2007. Helminth ova removal from wastewater for agriculture and aquaculture reuse. Water Science and Technology 55 (1–2), 485–493.

Joubert, W., Britz, T., 1987. Characterization of aerobic, facultative anaerobic, and anaerobic bacteria in an acidogenic phase reactor and their metabolite formation. Microbial Ecology 13 (2), 159–168.

Kaján, G.L., Kajon, A.E., Pinto, A.C., Bartha, D., Arnberg, N., 2017. The complete genome sequence of human adenovirus 84, a highly recombinant new human mastadenovirus D type with a unique fiber gene. Virus Research 242, 79–84.

Kaper, J.B., Nataro, J.P., Mobley, H.L.T., 2004. Pathogenic *Escherichia coli*. Nature Reviews Microbiology 2, 123.

Kim, B., Anderson, J., Mueller, S., Gaines, W., Kendall, A., 2002. Literature review—efficacy of various disinfectants against *Legionella* in water systems. Water Research 36 (18), 4433–4444.

Kim, H.-O., Park, S.-W., Park, H.-D., 2004. Inactivation of *Escherichia coli* O157: H7 by cinnamic aldehyde purified from *Cinnamomum cassia* shoot. Food Microbiology 21 (1), 105–110.

King, B., Monis, P., 2007. Critical processes affecting *Cryptosporidium* oocyst survival in the environment. Parasitology 134 (3), 309–323.

King, B., Fanok, S., Phillips, R., Lau, M., van den Akker, B., Monis, P., 2017. *Cryptosporidium* attenuation across the wastewater treatment train: recycled water fit for purpose. Applied and Environmental Microbiology 83 (5) e03068-03016.

Kirn, T.J., Jude, B.A., Taylor, R.K., 2005. A colonization factor links *Vibrio cholerae* environmental survival and human infection. Nature 438, 863–867.

Kirsten, J., 2018. Top 12 Diseases Caused by Protozoa. https://www.bioexplorer.net/diseases-caused-by-protozoa.html/.

Kulkarni, P., Olson, N.D., Paulson, J.N., Pop, M., Maddox, C., Claye, E., Goldstein, R.E.R., Sharma, M., Gibbs, S.G., Mongodin, E.F., 2018. Conventional wastewater treatment and reuse site practices modify bacterial community structure but do not eliminate some opportunistic pathogens in reclaimed water. Science of the Total Environment 639, 1126–1137.

Kümmerer, K., 2009. Antibiotics in the aquatic environment—a review—part II. Chemosphere 75 (4), 435—441.

Lamba, M., Graham, D.W., Ahammad, S.Z., 2017. Hospital wastewater releases of carbapenem-resistance pathogens and genes in urban India. Environmental Science and Technology 51 (23), 13906—13912.

Lapidot, A., Yaron, S., 2009. Transfer of *Salmonella enterica* Serovar Typhimurium from contaminated irrigation water to parsley is dependent on curli and cellulose, the biofilm matrix components. Journal of Food Protection 72 (3), 618—623.

Lefkowitz, E.J., Dempsey, D.M., Hendrickson, R.C., Orton, R.J., Siddell, S.G., Smith, D.B., 2017. Virus taxonomy: the database of the international committee on taxonomy of viruses (ICTV). Nucleic Acids Research 46 (D1), D708—D717.

Li, J., Wang, T., Shao, B., Shen, J., Wang, S., Wu, Y., 2012a. Plasmid-mediated quinolone resistance genes and antibiotic residues in wastewater and soil adjacent to swine feedlots: potential transfer to agricultural lands. Environmental Health Perspectives 120 (8), 1144—1149.

Li, Z.-L., Zhu, L., Ma, J.-Y., Zhou, Q.-F., Song, Y.-H., Sun, B.-L., Chen, R.-A., Xie, Q.-M., Bee, Y.-Z., 2012b. Molecular characterization and phylogenetic analysis of porcine epidemic diarrhea virus (PEDV) field strains in south China. Virus Genes 45 (1), 181—185.

Li, A.-D., Ma, L., Jiang, X.-T., Zhang, T., 2017. Cultivation-dependent and high-throughput sequencing approaches studying the co-occurrence of antibiotic resistance genes in municipal sewage system. Applied Microbiology and Biotechnology 101 (22), 8197—8207.

Li, B., Qiu, Y., Zhang, J., Liang, P., Huang, X., 2019. Conjugative potential of antibiotic resistance plasmids to activated sludge bacteria from wastewater treatment plants. International Biodeterioration and Biodegradation 138, 33—40.

Liebers, V., Raulf-Heimsoth, M., Brüning, T., 2008. Health effects due to endotoxin inhalation. Archives of Toxicology 82 (4), 203—210.

Longini Jr., I.M., Yunus, M., Zaman, K., Siddique, A., Sack, R.B., Nizam, A., 2002. Epidemic and endemic cholera trends over a 33-year period in Bangladesh. The Journal of Infectious Diseases 186 (2), 246—251.

Lund, V., Fonahn, W., Pettersen, J.E., Caugant, D.A., Ask, E., Nysaeter, A., 2014. Detection of *Legionella* by cultivation and quantitative real-time polymerase chain reaction in biological water treatment plants in Norway. Journal of Water and Health 12 (3), 543—554.

Luo, Y., Yang, F., Mathieu, J., Mao, D., Wang, Q., Alvarez, P., 2013. Proliferation of multidrug-resistant New Delhi metallo-β-lactamase genes in municipal wastewater treatment plants in northern China. Environmental Science and Technology Letters 1 (1), 26—30.

Mackie, P.L., 2003. The classification of viruses infecting the respiratory tract. Paediatric Respiratory Reviews 4 (2), 84—90.

Madoni, P., 2011. Protozoa in wastewater treatment processes: a minireview. Italian Journal of Zoology 78 (1), 3—11.

Mahvi, A., Kia, E., 2006. Helminth eggs in raw and treated wastewater in the Islamic Republic of Iran. EMHJ—Eastern Mediterranean Health Journal 12 (1—2), 137—143.

Mao, D., Yu, S., Rysz, M., Luo, Y., Yang, F., Li, F., Hou, J., Mu, Q., Alvarez, P., 2015. Prevalence and proliferation of antibiotic resistance genes in two municipal wastewater treatment plants. Water Research 85, 458—466.

Medici, M.C., Abelli, L.A., Dodi, I., Dettori, G., Chezzi, C., 2010. Norovirus RNA in the blood of a child with gastroenteritis and convulsions—a case report. Journal of Clinical Virology 48 (2), 147–149.

Mezrioui, N., Oufdou, K., 1996. Abundance and antibiotic resistance of non-O1 *Vibrio cholerae* strains in domestic wastewater before and after treatment in stabilization ponds in an arid region (Marrakesh, Morocco). FEMS Microbiology Ecology 21 (4), 277–284.

Molloy, S.F., Smith, H.V., Kirwan, P., Nichols, R.A., Asaolu, S.O., Connelly, L., Holland, C.V., 2010. Identification of a high diversity of *Cryptosporidium* species genotypes and subtypes in a pediatric population in Nigeria. The American Journal of Tropical Medicine and Hygiene 82 (4), 608–613.

Morris Jr., J.G., Acheson, D., 2003. Cholera and other types of vibriosis: a story of human pandemics and oysters on the half shell. Clinical Infectious Diseases 37 (2), 272–280.

Munita, J.M., Arias, C.A., 2016. Mechanisms of antibiotic resistance. Microbiology Spectrum 4 (2). https://doi.org/10.1128/microbiolspec.VMBF-0016-2015.

Nealson, K.H., 1999. Post-viking microbiology: new approaches, new data, new insights. Origins of Life and Evolution of the Biosphere 29 (1), 73–93.

Nongogo, V., Okoh, A.I., 2014. Occurrence of *Vibrio* pathotypes in the final effluents of five wastewater treatment plants in Amathole and Chris Hani District Municipalities in South Africa. International Journal of Environmental Research and Public Health 11 (8), 7755–7766.

Oh, E.-G., Son, K.-T., Yu, H., Lee, T.-S., Lee, H.-J., Shin, S., Kwon, J.-Y., Park, K., Kim, J., 2011. Antimicrobial resistance of *Vibrio parahaemolyticus* and *Vibrio alginolyticus* strains isolated from farmed fish in Korea from 2005 through 2007. Journal of Food Protection 74 (3), 380–386.

Ohkouchi, Y., Ishikawa, S., Takahashi, K., Itoh, S., 2007. Factors associated with endotoxin fluctuation in aquatic environment and characterization of endotoxin removal in water treatment process. Environmental Engineering Research 44, 247–254.

OHSU Research Integrity Office, 2014. OHSU IBC Toxin Fact Sheet. https://ohsu.ellucid.com/documents/view/8064/?security=2257a019d2bddb6b45859db09355427aced526c3.

Oliver, J.D., Dagher, M., Linden, K., 2005. Induction of *Escherichia coli* and *Salmonella typhimurium* into the viable but nonculturable state following chlorination of wastewater. Journal of Water and Health 3 (3), 249–257.

Ortega, Y.R., Sanchez, R., 2010. Update on *Cyclospora cayetanensis*, a food-borne and waterborne parasite. Clinical Microbiology Reviews 23 (1), 218–234.

Papafragkou, E., Hewitt, J., Park, G.W., Greening, G., Vinje, J., 2013. Challenges of culturing human norovirus in three-dimensional organoid intestinal cell culture models. PLoS One 8 (6), e63485.

Pauli, W., Jax, K., Berger, S., Biodegradation and Persistence, 2001. In: Beek, B. (Ed.). Springer, Berlin, Germany, pp. 205–211.

Penders, J., Stobberingh, E.E., 2008. Antibiotic resistance of motile aeromonads in indoor catfish and eel farms in the southern part of the Netherlands. International Journal of Antimicrobial Agents 31 (3), 261–265.

Peng, X., Luo, W., Zhang, J., Wang, S., Lin, S., 2002. Rapid detection of *Shigella* species in environmental sewage by an immunocapture PCR with universal primers. Applied and Environmental Microbiology 68 (5), 2580–2583.

Pruden, A., Pei, R., Storteboom, H., Carlson, K.H., 2006. Antibiotic resistance genes as emerging contaminants: studies in northern Colorado. Environmental Science and Technology 40 (23), 7445–7450.

Ramirez, M.S., Tolmasky, M.E., 2010. Aminoglycoside modifying enzymes. Drug Resistance Updates 13 (6), 151–171.

Regli, S., Rose, J.B., Haas, C.N., Gerba, C.P., 1991. Modeling the risk from *Giardia* and viruses in drinking water. Journal of the American Water Works Association 83 (11), 76–84.

Rizzo, L., Manaia, C., Merlin, C., Schwartz, T., Dagot, C., Ploy, M., Michael, I., Fatta-Kassinos, D., 2013. Urban wastewater treatment plants as hotspots for antibiotic resistant bacteria and genes spread into the environment: a review. Science of the Total Environment 447, 345–360.

Rodriguez-Mozaz, S., Chamorro, S., Marti, E., Huerta, B., Gros, M., Sànchez-Melsió, A., Borrego, C.M., Barceló, D., Balcázar, J.L., 2015. Occurrence of antibiotics and antibiotic resistance genes in hospital and urban wastewaters and their impact on the receiving river. Water Research 69, 234–242.

Roth, R.A., Harkema, J.R., Pestka, J.P., Ganey, P.E., 1997. Is exposure to bacterial endotoxin a determinant of susceptibility to intoxication from xenobiotic agents? Toxicology and Applied Pharmacology 147 (2), 300–311.

Ruggeri, F.M., Bonomo, P., Ianiro, G., Battistone, A., Delogu, R., Germinario, C., Chironna, M., Triassi, M., Campagnuolo, R., Cicala, A., 2015. Rotavirus genotypes in sewage treatment plants and in children hospitalized with acute diarrhea in Italy in 2010 and 2011. Applied and Environmental Microbiology 81 (1), 241–249.

Ryan, K.J., Ray, C.G., 2004. Sherris Medical Microbiology. McGraw Hill, New York.

Ryan, U., Fayer, R., Xiao, L., 2014. *Cryptosporidium* species in humans and animals: current understanding and research needs. Parasitology 141 (13), 1667–1685.

Rysz, M., Alvarez, P.J., 2004. Amplification and attenuation of tetracycline resistance in soil bacteria: aquifer column experiments. Water Research 38 (17), 3705–3712.

Saheb, E.J., 2018. The prevalence of parasitic protozoan diseases in Iraq, 2016. Karbala International Journal of Modern Science 4 (1), 21–25.

Santín, M., Fayer, R., 2011. Microsporidiosis: *Enterocytozoon bieneusi* in domesticated and wild animals. Research in Veterinary Science 90 (3), 363–371.

Schroeder, G.N., Hilbi, H., 2008. Molecular Pathogenesis of *Shigella* spp.: controlling host cell signaling, invasion, and death by type III secretion. Clinical Microbiology Reviews 21 (1), 134–156.

Shannon, K., Lee, D.-Y., Trevors, J., Beaudette, L., 2007. Application of real-time quantitative PCR for the detection of selected bacterial pathogens during municipal wastewater treatment. Science of the Total Environment 382 (1), 121–129.

Shirley, D.-A.T., Farr, L., Watanabe, K., Moonah, S., 2018. A review of the global burden, new diagnostics, and current therapeutics for amebiasis. Open Forum Infectious Diseases 5 (7), ofy161.

Singleton, P., Sainsbury, D., 2003. Dictionary of microbiology and molecular biology, third edition. Virus Research 93 (1), 123.

Smid, E.J., Koeken, J.G.P., Gorris, L.G.M., 1996. Fungicidal and Fungistatic Action of the Secondary Plant Metabolites Cinnarnaldehyde and Carvone. Modern Fungicides and Antimicrobial Compounds, Intercept. Andover, U.S.

Smit, L.A., Spaan, S., Heederik, D., 2005. Endotoxin exposure and symptoms in wastewater treatment workers. American Journal of Industrial Medicine 48 (1), 30–39.

Smith, J.G., Cassany, A., Gerace, L., Ralston, R., Nemerow, G.R., 2008. Neutralizing antibody blocks adenovirus infection by arresting microtubule-dependent cytoplasmic transport. Journal of Virology 82 (13), 6492–6500.

Stevik, T.K., Aa, K., Ausland, G., Hanssen, J.F., 2004. Retention and removal of pathogenic bacteria in wastewater percolating through porous media: a review. Water Research 38 (6), 1355–1367.

Sukprasert, S., Rattaprasert, P., Hamzah, Z., Shipin, O.V., Chavalitshewinkoon-Petmitr, P., 2008. PCR detection of *Entamoeba* spp. from surface and waste water samples using genus-specific primers. Southeast Asian Journal of Tropical Medicine and Public Health 39 (Suppl. 1), 6–9.

Szekeres, E., Baricz, A., Chiriac, C.M., Farkas, A., Opris, O., Soran, M.-L., Andrei, A.-S., Rudi, K., Balcázar, J.L., Dragos, N., Coman, C., 2017. Abundance of antibiotics, antibiotic resistance genes and bacterial community composition in wastewater effluents from different Romanian hospitals. Environmental Pollution 225, 304–315.

Tajbakhsh, E., Khamesipour, F., Ranjbar, R., Ugwu, I.C., 2015. Prevalence of class 1 and 2 integrons in multi-drug resistant *Escherichia coli* isolated from aquaculture water in Chaharmahal Va Bakhtiari province, Iran. Annals of Clinical Microbiology and Antimicrobials 14 (1), 37.

Teklehaimanot, G.Z., Coetzee, M.A., Momba, M.N., 2014. Faecal pollution loads in the wastewater effluents and receiving water bodies: a potential threat to the health of Sedibeng and Soshanguve communities, South Africa. Environmental Science and Pollution Research 21 (16), 9589–9603.

Teklehaimanot, G.Z., Genthe, B., Kamika, I., Momba, M., 2015. Prevalence of enteropathogenic bacteria in treated effluents and receiving water bodies and their potential health risks. Science of the Total Environment 518, 441–449.

Thorne, L.G., Goodfellow, I.G., 2014. Norovirus gene expression and replication. Journal of General Virology 95 (2), 278–291.

Todar, K., 2008. Pathogenic *E. coli*. Online Textbook of Bacteriology. University of Wisconsin—Department of Bacteriology. https://www.textbookofbacteriology.net.

Todar, K., 2013. Structure and function of bacterial cells. Seikagaku the Journal of Japanese Biochemical Society 65 (9), 1174–1179.

Todar, K., 2018. Bacterial Endotoxin, Online Textbook of Bacteriology. University of Wisconsin—Department of Bacteriology. https://www.textbookofbacteriology.net.

Tong, J., Tang, A., Wang, H., Liu, X., Huang, Z., Wang, Z., Zhang, J., Wei, Y., Su, Y., Zhang, Y., 2019. Microbial community evolution and fate of antibiotic resistance genes along six different full-scale municipal wastewater treatment processes. Bioresource Technology 272, 489–500.

Turcios-Ruiz, R.M., Axelrod, P., John, K.S., Bullitt, E., Donahue, J., Robinson, N., Friss, H.E., 2008. Outbreak of necrotizing enterocolitis caused by norovirus in a neonatal intensive care unit. The Journal of Pediatrics 153 (3), 339–344.

Wang, Y., 2015. The Study of Physological Regulation Factors in *Vibrio cholerae* that Affect Colonization in Host. Nanjing Agricultural University, Nanjing, China.

Wang, X., Li, J., Guo, T., Zhen, B., Kong, Q., Yi, B., Li, Z., Song, N., Jin, M., Xiao, W., 2005. Concentration and detection of SARS coronavirus in sewage from Xiao Tang Shan Hospital and the 309th Hospital of the Chinese People's Liberation Army. Water Science and Technology 52 (8), 213–221.

Wang, J., Mao, D., Mu, Q., Luo, Y., 2015. Fate and proliferation of typical antibiotic resistance genes in five full-scale pharmaceutical wastewater treatment plants. Science of the Total Environment 526, 366–373.

Web Books, 2008. Molecular Biology Web Book. http://web-books.com.

WHO, 2004. Guidelines for Drinking-Water Quality, third ed. Geneva, Switzerland.

WHO, 2009. Risk Assessment of *Cryptosporidium* in Drinking Water. In: Water, Sanitation, Hygiene and Health. Public Health and Environment, Geneva, Switzerland.

Wilson, G.G., Murray, N.E., 1991. Restriction and modification systems. Annual Review of Genetics 25 (1), 585–627.

Wright, J., Ruseska, I., Costerton, J., 1991. Decreased biocide susceptibility of adherent *Legionella pneumophila*. Journal of Applied Bacteriology 71 (6), 531–538.

Xiao, L., Feng, Y., 2008. Zoonotic cryptosporidiosis. FEMS Immunology and Medical Microbiology 52 (3), 309–323.

Xiong, W., Sun, Y., Zhang, T., Ding, X., Li, Y., Wang, M., Zeng, Z., 2015. Antibiotics, antibiotic resistance genes, and bacterial community composition in fresh water aquaculture environment in China. Microbial Ecology 70 (2), 425–432.

Yang, Y., Li, B., Ju, F., Zhang, T., 2013. Exploring variation of antibiotic resistance genes in activated sludge over a four-year period through a metagenomic approach. Environmental Science and Technology 47 (18), 10197–10205.

Yang, Y., Li, B., Zou, S., Fang, H.H.P., Zhang, T., 2014. Fate of antibiotic resistance genes in sewage treatment plant revealed by metagenomic approach. Water Research 62, 97–106.

Yang, D.C., Blair, K.M., Salama, N.R., 2016. Staying in shape: the impact of cell shape on bacterial survival in diverse environments. Microbiology and Molecular Biology Reviews 80 (1), 187–203.

Zaki, A.M., Van Boheemen, S., Bestebroer, T.M., Osterhaus, A.D., Fouchier, R.A., 2012. Isolation of a novel coronavirus from a man with pneumonia in Saudi Arabia. New England Journal of Medicine 367 (19), 1814–1820.

Zhai, W., Yang, F., Mao, D., Luo, Y., 2016. Fate and removal of various antibiotic resistance genes in typical pharmaceutical wastewater treatment systems. Environmental Science and Pollution Research 23 (12), 12030–12038.

Zhang, L., Jiang, J., 2001. Research progress of *Cryptosporidium* and *cryptosporidiosis*. Acta Parasetologica et Medica Entomologica Sinica 8 (3), 184–192.

Zhang, X.-X., Zhang, T., 2011. Occurrence, abundance, and diversity of tetracycline resistance genes in 15 sewage treatment plants across China and other global locations. Environmental Science and Technology 45 (7), 2598–2604.

Zhang, X.X., Zhang, T., Fang, H.H., 2009. Antibiotic resistance genes in water environment. Applied Microbiology and Biotechnology 82 (3), 397–414.

Zhang, C., Qiu, S., Wang, Y., Qi, L., Hao, R., Liu, X., Shi, Y., Hu, X., An, D., Li, Z., 2013a. Higher isolation of NDM-1 producing *Acinetobacter baumannii* from the sewage of the hospitals in Beijing. PLoS One 8 (6), e64857.

Zhang, R., Wang, L., Sun, B., 2013b. Current status of *Cryptosporidium* contamination in drinking water and human infection. Journal of Environmental Hygiene 3 (6), 571–574.

Zhang, T., Yang, Y., Pruden, A., 2015. Effect of temperature on removal of antibiotic resistance genes by anaerobic digestion of activated sludge revealed by metagenomic approach. Applied Microbiology and Biotechnology 99 (18), 7771–7779.

Zhao, J., Shi, B.-j., Huang, X.-g., Peng, M.-y., Zhang, X.-m., He, D.-n., Pang, R., Zhou, B., Chen, P.-y., 2013. A multiplex RT-PCR assay for rapid and differential diagnosis of four porcine diarrhea associated viruses in field samples from pig farms in East China from 2010 to 2012. Journal of Virological Methods 194 (1–2), 107–112.

Zhao, Y., Zhang, X.-x., Zhao, Z., Duan, C., Chen, H., Wang, M., Ren, H., Yin, Y., Ye, L., 2018. Metagenomic analysis revealed the prevalence of antibiotic resistance genes in the gut and living environment of freshwater shrimp. Journal of Hazardous Materials 350, 10–18.

Zhong, N., Zheng, B., Li, Y., Poon, L., Xie, Z., Chan, K., Li, P., Tan, S., Chang, Q., Xie, J., 2003. Epidemiology and cause of severe acute respiratory syndrome (SARS) in Guangdong, People's Republic of China, in February, 2003. The Lancet 362 (9393), 1353–1358.

Zhou, N., Lv, D., Wang, S., Lin, X., Bi, Z., Wang, H., Wang, P., Zhang, H., Tao, Z., Hou, P., 2016. Continuous detection and genetic diversity of human rotavirus A in sewage in eastern China, 2013–2014. Virology Journal 13 (1), 153.

Zhu, Y.-G., Johnson, T.A., Su, J.-Q., Qiao, M., Guo, G.-X., Stedtfeld, R.D., Hashsham, S.A., Tiedje, J.M., 2013. Diverse and abundant antibiotic resistance genes in Chinese swine farms. Proceedings of the National Academy of Sciences 110 (9), 3435–3440.

Zillig, W., 1991. Comparative biochemistry of archaea and bacteria. Current Opinion in Genetics and Development 1 (4), 544–551.

CHAPTER

Technologies for detection of HRPs in wastewater

4

Yan Zhang, PhD, Ruxia Qiao, Cheng Sheng, Huajin Zhao

State Key Laboratory of Pollution Control and Resource Reuse, School of the Environment, Nanjing University, Nanjing, China

Chapter outline

High-Risk Pollutants in Wastewater. https://doi.org/10.1016/B978-0-12-816448-8.00004-6

At present, wastewater has been regarded as an important source of pollution, which can cause a series of environmental health risks, due to the presence of many high-risk pollutants (HRPs) in the wastewater. These pollutants mainly include heavy metals, organic pollutants, and biological pollutants. To evaluate the potential health risks of these pollutants, precise detection of the types and concentration of these pollutants in wastewater is necessary. However, these pollutants are diverse and have very low concentrations in the wastewater. Therefore, many highly specific and sensitive detection methods have been developed. For example, atomic absorption spectrometry (AAS), atomic fluorescence spectrometry (AFS), anodic stripping voltammetry (ASV), and inductively coupled plasma mass spectrometry (ICP-MS) have been widely used to detect heavy metals in wastewater. In organic pollutants, gas chromatography (GC), gas chromatography−mass spectrometry (GC−MS), and high-performance liquid chromatography (HPLC) can be used for qualitative and quantitative analysis of various organic pollutants in wastewater. For biological pollutants, such as various pathogenic microorganisms, many molecular biological methods such as gene probe, polymerse chain reaction (PCR), microarray, and high-throughput sequencing technology have been considerably booned. In addition, to deal with the frequent occurrence of environmental emergencies and to improve the precise management of wastewater treatment, new requirements have been put forward for the detection of high-risk pollutants in wastewater, that is, in situ, on-line, and rapid detection. In this chapter, the basic principle, operation process, and advantages and disadvantages of the detection methods for the HRPs in wastewater are introduced, which will provide fundamental knowledge for selecting suitable technology for detection of HRPs in wastewater.

4.1 Detection techniques of heavy metals in wastewater

The most encountered toxic metals in wastewater include cadmium (Cd), arsenic (As), mercury (Hg), lead (Pb), cooper (Cu), and zinc (Zn) (Karvelas et al., 2003). The unproperly treated heavy metals in the wastewater will enter the biological chain through various routes and endanger ecological safety or even human health (Fatima and Ahmad, 2005). For example, ground water contaminated with arsenic have caused arsenic poisoning in the Bay of Bengal, which was the most serious and largest scale poisoning incident in human history. The accumulation of methylmercury in the brains of humans and animals has caused the world-famous public nuisance incident, Minamata disease in Japan. The cadmium pollution caused "bone pain disease" in Toyama Prefecture, Japan.

Although the risk of most toxic metals in the wastewater can be controlled by reducing their concentration, some metals such as As, Hg, and Cd are special because they are toxic for animals at any concentration and should not be taken into the body even in ultratrace levels. Therefore, it is necessary to accurately determine the concentration of heavy metals in wastewater to provide basic data for health risk assessment.

In general, the determination of trace heavy metals in wastewater requires several steps, including sample collection, sample pretreatment, detection method selection, sample detection, and results analysis. Each step is important and a possible source of error if not applied appropriately (Eaton et al., 2005). However, the sample pretreatment is extremely important for wastewater detection, because wastewater samples may contain particulates or organic materials such as dissolved organic matters that could dramatically influence detection results.

Different pretreatment methods are needed for different purposes. For instance, if only the dissolved metals are needed to be detected, filtration of sample and analyzing of filtrate will be enough even though the wastewater samples contain particulates. If only the metals in particulates are also required to be detected, the wastewater samples need to be filtered and then the filter needs to be digested and analyzed.

Moreover, the organic materials in the wastewater samples need to be digested before determination of heavy metals. In general, the procedures of digesting organic materials include three groups: wet digestion by acid mixtures before metal detection, dry ashing followed by acid dissolution of the ash, and microware digestion (Dimpe et al., 2014a). For most digestion methods, nitric acid is widely used and is suitable for flame and electrothermal atomic absorption spectroscopy (AAS) and inductively coupled plasma-mass spectrometry (ICP-MS) (Bhandari and Amarasiriwardena, 2000). If the wastewater samples containing readily oxidizable organic matters, HNO_3-H_2SO_4 or HNO_3-HCl can be used. If the samples have high organic contents, HNO_3-$HClO_4$, HNO_3-H_2O_2, or HNO_3-$HClO_4$-HF can be used and these mixtures can also be used to digest the particulates.

The heavy metal concentration can be detected after the pretreatment of wastewater sample. Currently, a lot of research has been carried out on the detection methods of heavy metals in wastewater and abundant results have been achieved. The main detection methods include AAS, atomic fluorescence spectrometry (AFS), anodic stripping voltammetry (ASV), ICP-MS, inductively coupled plasma-optical emission spectrometry (ICP-OES), and recently laser-induced breakdown spectroscopy (LIBS). The advantages and disadvantages of these techniques will be discussed in this chapter. A comparison of detection techniques of heavy metals in wastewater is given in Table 4.1.

4.1.1 Atomic absorption spectroscopy

AAS is an instrumental analysis method for determining the concentration of the element in the sample based on the absorption intensity of the ground state atom of the element. According to different atomization techniques, AAS can be classified as flame AAS, graphite furnace AAS, hydride AAS, and cold vapor AAS. For flame AAS, the sample should be liquid and the detection limits are around ppm range (Wang et al., 2012). Different from the flame AAS, graphite furnace AAS uses graphite tube, which can stand for 3000 °C atomization, to replace flame and its detection limit (around ppb range) is higher than flame AAS (Gomez-Nieto et al., 2013). Hydride AAS is suitable to detect the metalloid elements such as arsenic

Table 4.1 Comparison of detection techniques of heavy metals in wastewater.

Detection techniques	Metals	Detection limit	References
AAS	Pb, Cd, Cu, Zn,	1.3 ppt-60 ppb	Cadorim et al. (2019) Khayatian et al. (2018a) Islam et al. (2014a) Mirzaei et al. (2011) Vellaichamy and Palanivelu (2011)
AFS	Hg, As, Cd, Pb	0.0008−0.1 ppb	Yuan et al. (2018a) Carneado et al. (2015) Luo (2012a) Liu et al. (2014b) Zhou et al. (2011)
ASV	Hg, As, Cd, Cu, Pb, Zn	0.021−10 ppb	Sonthalia et al. (2004) Vieira dos Santos and Masini (2006) Javanbakht et al. (2009) Kyrisoglou et al. (2012) Allafchian et al. (2017b)
ICP-MS	As, Cd, Pb, Zn, Hg	8 ppt-1000 ppb	Liu et al. (2014a) Raposo et al. (2014) Castillo et al. (2006b) Chen et al. (2007a) Castillo et al. (2008)
ICP-OES	Hg, As, Cd, Cu, Pb, Zn	0.01−8 ppb	Giersz et al. (2017) Peng et al. (2016) Zhang et al. (2015) Dimpe et al. (2014b) Sereshti et al. (2011)
LIBS	Cd, Cu, Pb, Zn	2.59 ppb-5 ppm	Zhao et al. (2019b) Wang et al. (2015a) Gondal and Hussain (2007) Järvinen et al. (2014) Yang et al. (2016)

and lead that are introduced to instrument in gas phase. This method can reduce the detection limit by 10−100 times (Maragou et al., 2017). Cold vapor AAS is generally used to detect mercury because this element has enough vapor pressure at room temperature. However, this technique cannot be used to detect organic mercury compounds, as they cannot be reduced to the element by sodium tetrahydroborate. Therefore, digestion is necessary before the detection using this technique. The detection limit of cold vapor AAS is around ppb range (Adlnasab et al., 2014).

AAS has the advantages of high sensitivity, strong selectivity, wide analysis range, strong antiinterference ability, accurate and reliable results, simple and rapid operation, simple instrument, and automation of the whole operation. Based on these advantages, AAS is unparalleled in the field of heavy metal analysis and detection of water, and is even listed as an arbitration method for multimetal analysis of water. However, the instrument of AAS is expensive and its operating cost is high, which limits its application.

Cadorim et al. (2019) used disposable pipette extraction coupled with high-resolution continuum source graphite furnace atomic absorption spectrometry (HR-CS GF AAS) to detect Pb and Cd in the wastewater and the limit of detection was 0.2 ppb for Pb and 0.1 ppb for Cd, respectively. Khayatian et al. (2018b) used FAAS to detect Cu(II) and Pb(II) in refinery wastewater and the detection limit was 4 ppb for Cu(II) and 11 ppb for Pb(II). Islam et al. (2014b) developed a novel solid-phase extractant for the preconcentration of lead in electroplating wastewater and detected this metal using FAAS.

4.1.2 Atomic fluorescence spectrometry

Atomic fluorescence is a spectroscopic process which is based on absorption of radiation of specific wavelengths by an atomic vapour with subsequent detection of radiationally deactivated states through emission in a direction orthogonal to the excitation source. The absorption and the subsequent atomic emission processes both occur at wavelengths which are characteristic of the atomic species present. The concentration of the element can be detected by measuring the fluorescence intensity. AFS is a relatively mature analytical technique with high sensitivity and low interference. It is especially suitable for the detection of mercury and arsenic in water and its detect limit is around ppb range (Liu et al., 2008). However, this method also has disadvantages. For instance, AFS requires high operational skill and it can only be used for the detection of limited types of heavy metals.

Yuan et al. (2018b) developed an ultraviolet (UV) atomization atomic fluorescence spectrometry (UV-AFS) system to determine the trace cadmium ions without preconcentration and the limit of detection was 0.006 ppb. Carneado et al. (2015) developed a method for the simultaneous determination of methylmercury ($MeHg^{+}$) and mercury(II) (Hg^{2+}) species in wastewater by using liquid chromatography coupled with UV irradiation and cold vapor atomic fluorescence spectrometry (LC-UV-CV-AFS). The limit of detection for the developed method was 15 and 2 ppt for MeHg and Hg(II), respectively. Luo (2012b) used an improved hydride-generation atomic fluorescence spectrometry (HG-AFS) method to determine the total arsenic (As) in wastewater. The samples were digested completely with mixtures of HNO_3 and $HClO_4$. The detection limit for total As in wastewater was 0.09 ppb. Liu et al. (2014d) proposed matrix-assisted photochemical vapor generation for the direct determination of mercury in domestic wastewater by using AFS. Under the optimized condition, the limit of detection could be 0.1 ppb, which demonstrated that this method is a simple, reagent-free, cost-effective, and green method for mercury determination in domestic wastewater.

4.1.3 Anodic stripping voltammetry

ASV is an electrochemical technique where the detected metal ions are first reduced and dissolved under a certain potential. Then, a reverse voltage is applied to the electrode which will produce an oxidation current. The electrons released by the process will form a peak current. The current will be measured and the corresponding potentical will be recorded. The metal species can be identified according to the potential value generated by the oxidation. The metal ion content can be obtained through comparing the peak height or area of the current with the standard solution under the same conditions. ASV is widely used for the detection of heavy metals, such as copper, zinc, plumbum, cadmium, mercury, and arsenic in water. For some metals, its sensitivity is 10–100 times higher than AAS. Due to the low detection limit of ASW, the preconcentration step is not necessary.

Sonthalia et al. (2004) used differential pulse of ASW (DPASV) to detect Cu(II), Pb(II), Cd(II), and Zn(II) in wastewater. dos Santos and Masini (2006) developed a sequential injection ASW (SI-ASV) method to determine Cd(II), Pb(II), and Cu(II) in coatings industry wastewater after proper acid digestion. The detection limits of the method were 0.06, 0.09, and 0.16 μmol/L for Cd, Pb, and Cu, respectively. Javanbakht et al. (2009) introduced a method for the determination of mercury ions at nanomolar level with the employment of the dipyridyl functionalized nanoporous silica gel-chemically modified carbon paste electrode (DPSG-CPE) by ASW. This method can be successfully used for wastewater and the detection limit was 8 nM. In situ DPASV and polytetrafluorethylene membrane-based liquid three-phase microextraction method was used by Allafchian et al. (2017a) to determine the trace level of lead in wastewater samples, and the detection limit was 0.021 ppb with optimum conditions.

4.1.4 Inductively coupled plasma-mass spectrometry

ICP-MS is a multielement technique that combines ICP technology with mass spectrometry. It uses plasma as the mass spectrometer ion source. After atomized, the element in the sample enters the plasma region in the form of aerosol. After evaporation, dissociation, atomization, ionization, and other processes, it is introduced into the high-mass spectrometry. The filter is separated by the mass-to-mass ratio and then is detected by the ion detector. The concentration of the element in the sample is calculated according to the magnitude of the ion intensity. This technology can provide extremely low detection limits (about ppt range), extremely wide dynamic linear range, simple spectral lines and high analytical sensitivity for almost all elements. ICP-MS has been used for the detection of heavy metals in wastewater. It is suitable for the simultaneous detection of copper, plumbum, zinc, cadmium, chromium, antimony, manganese, cobalt, nickel, arsenic, and antimony in sewage. However, this method also has limitations, such as cumbersome sample pretreatment, high interference, expensive equipment, unsatisfactory utilization effect and requirement of high skilled operation.

ICP-MS is very powerful for simultaneous determination of different forms of heavy metals in the wastewater. For instance, Liu et al. (2014c) developed an efficient online system coupling of capillary electrophoresis to inductively coupled plasma-mass spectrometry (CE-ICP-MS) for simultaneous separation and determination of arsenic and selenium compounds. Using this method, six arsenic species, including arsenite (As(III)), arsenate (As(V)), monomethylarsonic acid (MMA), dimethylarsinic acid (DMA), arsenobetaine (AsB), and arsenocholine (AsC) and five selenium species such as sodium selenite (Se(IV)), sodium selenate (Se(VI)), selenocysteine (SeCys), selenomethionine (SeMet), and Se-methylselenocysteine (MeSeCys) were baseline-separated and determined in a single run within 9 min under the optimized conditions. Castillo et al. (2006a) used ICP-MS to determine Hg(II), $MeHg^+$, $EtHg^+$, and $PhHg^+$ species in water samples. This method allows the simultaneous determination of Hg(II), $MeHg^+$, $EtHg^+$, and $PhHg^+$ in water at a very low concentration and the limit of detection was below 0.03 ppb. Chen et al. (2007b) used ion chromatography (IC) coupled with ICP-MS to determine the speciation of chromium, including Cr(III), $[Cr(EDTA)]^-$, and Cr(VI) in water samples and the detection limits for chromium species were below 0.2 ppb.

4.1.5 Inductively coupled plasma-optical emission spectrometry

Inductively coupled plasma-optical (or atomic) emission spectrometry (ICP-OES or ICP-AES) is a high-throughput technique by which the trace metals dissolved in water samples can be determined. The main difference from ICP-MS is that wavelength selectors instead of quadropole mass spectrometers are used to detect the samples. Compared with other metal determination techniques such as ICP-MS or AAS the advantages of ICP-OES include wide linear dynamic range, high matrix tolerance and enhanced analysis speed. However, ICP-OES is not free of interferences. The wavelength overlap of different metals will induce spectral interferences for ICP-OES.

Combined photochemical vapor generation and pneumatic nebulization in the programmable temperature spray chamber and ICP-OES was used by Giersz et al. (2017) to detect the heavy metals in water samples. The method enabled simultaneous determination of nonvolatile forming elements (Fe, Cu, Mn) and volatile Hg. As low as 2 ppb of Hg can be directly determined in waste water by using this method. Dimpe et al. (2014a) used ICP-OES to determine the total content of As, Cd, Cu, Pb, and Zn in wastewater samples and the limits of detection and limits of quantification ranged from 0.12% to 2.18 ppb and 0.61% to 3.43 ppb, respectively. Sereshti et al. (2011) developed a method combined dispersive liquid—liquid microextraction (DLLME) and ICP-OES for simultaneous preconcentration and trace determination of chromium, copper, nickel, and zinc in water samples. The limits of detection for these elements were 0.23—0.55 ppb.

4.1.6 Laser-induced breakdown spectroscopy

LIBS is an atomic emission spectroscopy technique using highly energetic laser pulses to provoke optical sample excitation. LIBS can provide a simple, fast, and in situ detection with a reasonable precision, detection limits, and cost. In addition, different from conventional spectroscopic analytical techniques, this technique does not require any sample preparation that has expanded the application fields of LIBS (Wang et al., 2014). LIBS can be used to determine various metals in wastewater.

Zhao et al. (2019a) have used LIBS to detect heavy metals (Cd, Cr, Cu, Ni, Pb, Zn) in industrial wastewater and the limits of detection for these metals could reach several ppb. Gondal and Hussain (2007) developed a LIBS system to determine the toxic metals in wastewater from local paint manufacturing plant. The detection limits for Pb, Cu, and Zn were 3, 2, and 5 ppm, respectively. Wang et al. (2015b) developed a new pretreatment method by using the chelating reagent 2,4,6-trimercapto-1,3,5-triazine (TMT) for metal precipitation and using mixed cellulose ester microfiltration membrane for separation. Based on this pretreatment, several metals (Cu, Ag, Mn, and Cr) were simultaneously detected in water samples. The detection limits of Cu, Ag, Mn, and Cr obtained in this study were 2.59, 0.95, 0.96, and 1.29 ppb, respectively.

4.2 Detection techniques of organic HRPs in wastewater

The organic HRPs in wastewater mainly include polychlorinated biphenyls (PCBs) (Rodenburg et al., 2010), polycyclic aromatic hydrocarbons (PAHs) (Zhang et al., 2012), organophosphorus pesticides (OPPs) (Zhang and Pagilla, 2010), disinfection by-products (DBPs) (Watson et al., 2012), and pharmaceutical and personal care products (PPCPs) (Ort et al., 2010). These organic pollutants in wastewater usually have endocrine disrupting toxicity (Chen et al., 2016), reproductive developmental toxicity (Kavlock et al., 2006), neurotoxicity (Gagne et al., 2007), genotoxicity (Wang et al., 2007), and carcinogenic toxicity (Monarca et al., 2000). Once these organic HRPs in the wastewater enter the environment, they will cause secondary pollution for soil and atmosphere. If these organic HRPs enter the food chain it will pose a threat to human health. Therefore, it is of great significance to strengthen the accurate detection of these organic HRPs in the wastewater, which will provide valuable information for the removal of these pollutants (Table 4.2).

Table 4.2 Detection techniques for organic HRPs.

Detection methods	Organic pollutants	Detection limits	References
GC	PAHs,	3–10 pg/mL	Amiri et al. (2019) Toušová et al. (2019)
GC–MS	PAHs, OPPs, PCBs, PPCPs	0.007 –0.022 ng/L	Jillani et al. (2019) Adeyinka et al. (2019) Adeyinka et al. (2018) Cuderman and Heath (2007) Erarpat et al. (2018)
HPLC	PAHs, PCBs, PPCPs	0.030–90 ng/L	Mateos et al. (2019) Yang et al. (2018) Yu et al. (2012)

4.2.1 Wastewater sample preparation

The composition of wastewater is very complicated and many factors would affect the determination of organic pollutants in the wastewater. Sample preparation methods are essential for the quantification of organic pollutants in wastewater. According to various estimates, sample preparation typically accounts for 70%–90% of the analysis time. Thus, a great effort is going into the development of reliable sample preparation procedures characterized by the simplicity of both operations and devices involved in the process (Zuloaga et al., 2012).

4.2.1.1 Sample collection and preservation

The material used in sampling apparatus must be anticorrosion, nonstaining, and nonadhesive. The vessel for preserving samples should be inert material such as glass, Teflon, or stainless steel and be treated strictly before use. Samples should be sealed immediately after collection, kept at low temperature (below 4 °C), and shipped back to the laboratory for analysis as soon as possible. If the analysis is not performed immediately, the water sample must be stored at low temperature (0–40°C) in the dark. All sample extractions should be completed within 7 days and the analysis should be completed within 40 days after the extraction.

4.2.1.2 Extraction and enrichment of samples

The composition of the wastewater sample is complex and the concentrations of target compounds are relatively low. So, sometime the water samples cannot be measured directly. It is necessary to adopt different pretreatment methods to extract and enrich the target compounds in wastewater, which can also eliminate interference, improve sensitivity, and reduce detection limit. In organic pollutants in wastewater, several sample preparation methods have been widely conducted, such as liquid–liquid extraction, solid-phase extraction, solid-phase microextraction, Soxhlet extraction, ultrasonic extraction, microwave extraction, supercritical fluid extraction, and accelerated solvent extraction (Afonso-Olivares et al., 2016).

4.2.1.3 Purification and concentration of samples

To prevent other organic substances from interfering the results, the wastewater samples also need to be purified. The general purification method is column chromatography. Frequently used purifiers include silica gel, florisil, and alumina. The most widely used method is silica gel column chromatography (Koha et al., 2007). The extraction volume of the general extraction method is relatively large, and it needs to be concentrated to a constant volume before the measurement. The concentration of samples mainly relies on the rotary evaporation methods, which are in great advantages of high efficiency, large processing capacity, and convenient operation. High purity nitrogen is also used to concentrate lower volume wastewater.

4.2.2 Gas chromatography

GC is a chromatographic separation analysis method using a gas as a mobile phase. The vaporized sample is carried into the column by the carrier gas (mobile phase). The stationary phase in the column has different molecular forces to the components in the sample that flow out from the column at different times to separate from each other. A chromatogram of the time and concentration of each component flowing out of the column is made using an appropriate identification and recording system. According to the height and area of the peak, the compound can be quantitatively analyzed.

GC is suitable for the analysis of volatile, low molecular weight, and thermally stable organic HRPs in the wastewater. Its advantages include small injection volume, high separation efficiency, fast analysis, high detection sensitivity, good selectivity, and wide application range. The disadvantage of GC is that the identification of the chemicals depends on the corresponding chromatographic peak by known data, or combined with other methods (such as mass spectrometry).

Amiri et al. (2019) used GC combined with FID detection to quantify PAHs in wastewater samples (one wastewater sample was collected from a research laboratory and the other wastewater sample was from industrial wastewater). The limits of detection ranged from 3 to 10 pg/mL. Tousova et al. (2019) used headspace solid-phase microextraction (HS-SPME) method in combination with GC-FID to detect PAHs in industrial wastewater and the limits of detection ranged from 0.027 to 0.041 μg/L.

4.2.3 Gas chromatography—mass spectrometry

GC—MS is a combination of a gas chromatograph and a mass spectrometer. Mass spectrometry can perform qualitative analysis, but it is powerless for the analysis of complex organic compounds; chromatography is an effective separation method for organic compounds, especially suitable for quantitative analysis of organic compounds, but it is difficult to use chromatography for the qualitative analysis. Therefore, the combination of these two techniques can efficiently and quantitatively analyze complex organic compounds in wastewater.

GC—MS is the mainstream technology for the analysis of volatile and semivolatile pollutants in wastewater. It has the advantages of strong separation ability, large peak capacity, and high detection sensitivity. GC—MS not only can detect traditional volatile and semivolatile pollutants, but also can play an essential role in the analysis of persistent organic pollutants, such as dioxins, polychlorinated biphenyls (PCBs), brominated flame retardants, polychlorinated naphthalenes, perfluorosulfonates, amides, perfluorotelomers, neutral perfluorinated compounds, short-chain chlorinated paraffins, environmental endocrine disruptors, sunscreens, and synthetic musk. However, compounds such as organic acids would be too reactive during the heating process of GC-MS. So, these compounds need to be derivatized before the analysis. GC—MS cannot be used to determine the compounds that are neither vaporizable nor esterified.

Jillani et al. (2019) used GC—MS for the determination of PAHs from a local wastewater treatment plant in Saudi Arabia and the detection limits ranged from 0.29 to 8.4 ng/mL. Adeyinka et al. (2019) used GC—MS to detect OCPs in effluent from the Darvill Wastewater Treatment Plant (WWTP) of Pietermaritzburg, South Africa. Adeyinka et al. (2018) detected PCBs in the effluent of wastewater treatment plant using GC—MS and the limits of detection ranged from 0.007 to 0.022 ng/L. Cuderman et al. (2007) applied GC-MS coupled with a series of preparation methods, including acidification, filtration, solid-phase extraction, and derivatization, to analyze UV filters and two common antimicrobial agents, clorophene and triclosan in wastewater samples. By using these methods, the obtained limits of detection were 13—266 ng/L for UV filters, and 10—186 ng/L for triclosan and clorophene. Erarpat et al. (2018) developed an accurate and sensitive analytical method, namely switchable solvent-based liquid-phase microextraction combined with GC—MS, for the simultaneous determination of OCPs in a municipal wastewater sample collected from a biological WWTP. The obtained detection limit was 8.6 ng/mL.

4.2.4 High-performance liquid chromatography

HPLC is an important chromatography for the detection of organic pollutants in waters. The liquid phase is used as the mobile phase. A single solvent with different polarities or a mixed solvent of different proportions, buffer, and other mobile phases are pumped into the stationary phase by the high-pressure infusion system. The components are separated in the column and detected by the detector.

HPLC is suitable for the detection of organic HRPs with poor thermal stability in wastewater. It has the advantages of high separation efficiency, good selectivity, and fast analysis. In addition, HPLC can combine with UV, fluorescence, MS, MS—MS, and conductivity detectors. The disadvantage of HPLC is the "extra-column effect". The sensitivity of HPLC detectors is not as good as that of GC. The cost of determination, instrument, and daily maintenance are also expensive.The analysis process generally needs longer time than GC.

Mateos et al. (2019) used reversed-phase HPLC (RP-HPLC) with fluorescence detection for the quantification of PAHs in wastewater. They obtained the detection limits of 0.7—1.5 μg/L. Yang et al. (2018) used SPME-HPLC-UV to detect PAHs in wastewater samples collected from a WWTP of Anning District in Lanzhou, and the limits of detection ranged from 0.025 to 0.051 mg/L. Yu et al. (2012) used ultrahigh performance liquid chromatography—tandem mass spectrometry (UHPLC—MS/MS) for trace analysis of 11 PPCPs in influent and effluent from municipal WWTPs. The quantification limits for the 11 PPCPs ranged from 0.040 to 88 ng/L and from 0.030 to 90 ng/L for influent and effluent, respectively.

4.3 Detection of biological HRPs

Currently, disinfection unit is fundamental in reclaimed wastewater treatment process, because wastewater contains various biological HRPs, such as viruses, rickettsia, mycoplasma, bacteria, fungi, and parasites. Especially, microbial pathogens are

Table 4.3 Detection techniques for pathogenic microorganisms.

Detection methods	Pathogenic microorganisms	Detection limits	References
PCR	Fungal, microthrix, *Staphylococcus aureus*, *Escherichia coli*, Virus *Naegleria fowleri* (ameba)	4.62×10^4copies/L 10–30 copies of per reaction	Maza-Márquez et al. (2019) Amirsoleimani et al. (2019) Senkbeil et al. (2019) Panda et al. (2015) Kitajima et al. (2014) Barril et al. (2015)
Gene chips	Pathogens, helicobacter, Bacteria	10–100 genomes/μL	Tourlousse et al. (2012) Miller et al. (2008) Kim et al. (2004)
High-throughput sequencing	Pathogens, bacterial, Prokaryotic communities		Cai and Zhang (2013) Lu et al. (2015) Li et al. (2015)
Biosensors	Bacteria, *Salmonella E. coli*, *Enterococcus* spp,	$6–1.9\times10^3$ log CFU/mL	(Tang et al., 2016) Rengaraj et al. (2018) Adkins et al. (2017) Zhang et al. (2019)

one of the major health risks associated with wastewater (Toze, 1999). Determination of biological HRPs in wastewater is regarded as a valuable work for disease prevention and assessment of water sanitation. The widely used detection techniques for pathogenic microorganisms are shown in Table 4.3.

4.3.1 Sample preparation

4.3.1.1 Concentration of pathogenic microorganisms in wastewater samples

Generally, the content of pathogenic microorganisms in wastewater is very low. So, large volume of water samples is needed for concentration before detection. Commonly used concentration methods include precipitation (using inorganic flocculants or organic flocculants), solid-phase membrane adsorption, solid-phase particle adsorption, and antigen capture (Muchesa et al., 2014).

4.3.1.2 Extraction of nucleic acids

The concentrated microbial samples need to be extracted and purified to obtain the nucleic acid substances for subsequent detection. Commonly used extraction methods for nucleic acids include lysate lysis, magnetic bead-oligonucleotide hybridization, column chromatography purification, and antigen capture cleavage (Sano et al., 2004).

4.3.2 Polymerase chain reaction

PCR technique uses nucleotide sequence as a "primer" to amplify the target gene (such as a specific DNA sequence of pathogenic microorganisms) by a series of chain reactions. Based on PCR, various new techniques have been derived, such as reverse transcription-PCR (RT-PCR). As many pathogenic viruses in wastewater are RNA viruses, the viral RNA in the sample has to be transcribed into cDNA, and then the cDNA can be used as the target sequence for PCR amplification reaction. The abundance of virus is detected based on the specific amplified DNA product. Furthermore, to improve the sensitivity and specificity of the detection, nested and seminested PCR are developed. For example, Ulloa-Stanojlovic et al. (2016) used nested PCR to detect and genotype *Cryptosporidium* spp. and *Giardia intestinalis* in wastewater samples obtained from five cities in Brazil. Moreover, to improve the detection efficiency, multiplex PCR can be used to detect multiple viruses and virus subtypes simultaneously. Furthermore, quantitative RT-PCR (RT-qPCR) and digital PCR are novel PCR techniques that can be used to quantify pathogenic microorganisms in wastewater with very high precision.

These PCR techniques are fast, sensitive, and highly specific. At the meanwhile, PCR reaction can be easily interfered by other substances existing in the system. For example, humic acid, fulvic acid, certain ions, and carbohydrates can interfere with the action of Taq polymerase. In addition, some compounds used for concentration, storage, and purification of wastewater samples cause inhibitors, such as EDTA, sodium lauryl sulfate and some mercapto compounds. Furthermore, PCR techniques cannot distinguish living and dead cells. It is also impossible to determine whether the virions are infectious in the water samples by using PCR techniques.

Maza-Marquez et al. (2019) evaluated the abundances of total and metabolically active populations of Candidatus Microthrix and Fungi in three different full-scale WWTPs by using qPCR and retrotranscribed qPCR of ribosomal molecular markers and the limit of quantification was 4.62×10^4copies L^{-1}. Senkbeil et al. (2019) used qPCR assays to quantify H8 and H12 marker genes for the detection of *E. coli* in domestic sewage. The limit of quantification of qPCR assay was determined to be 30 gene copies qPCR reaction for both H8 and H12 assays. Barril et al. (2015) used RT-nested PCR for rotavirus detection and VP7/VP4 characterization and real-time PCR for rotavirus quantification in urban raw sewage. A detection limit of 10 copies of target DNA per reaction was determined. Panda et al. (2015) used a PCR-based approach to screen and document the presence of Naegleria spp., in a variety of water bodies. Kitajima et al. (2014) used TaqMan-based qPCR assays to quantify viruses.

4.3.3 Gene chip

Gene chip is essentially a high-density array of oligonucleotides. It uses *in situ* lithography combined with synthetic chemistry and microelectronic chips to sequentially cure a large number of specific DNA fragments onto glass. The immobilized probe may be not only an oligonucleotide molecule but also a microarray composed with gene fragments, polypeptide molecules or antigen (antibody). The position and sequence of each molecule are known. When the fluorescently labeled target molecule is combined with the probe molecule on the chip, the intensity of the fluorescent signal can be detected by laser confocal fluorescence scanning or charge coupled camera. Then, the hybridization results can be quantified.

Gene chips can measure thousands of genes simultaneously. So far, gene chip has been successfully applied to evaluate complex toxicity of wastewater and screen pathogenic microorganisms in wastewater. Although great progress has been achieved for gene chip technology, this technique also has its own disadvantages, such as high operation cost, low detection sensitivity, poor repeatability, and narrow analysis range. These problems are mainly manifested in the preparation of samples, probe synthesis and immobilization, molecular labeling, and data reading and analysis.

Tourlousse et al. (2012) developed a polymer microfluidic chip for quantitative detection of multiple pathogens using isothermal nucleic acid amplification. The chip was successfully evaluated for rapid analysis of multiple virulence and marker genes of *Salmonella*, *Campylobacter jejuni*, *Shigella*, and *Vibrio cholerae*, enabling detection and quantification of 10−100 genomes per μL in less than 20 min. Miller et al. (2008) designed and validated an in situ-synthesized biochip for detection of 12 microbial pathogens, including a suite of pathogens relevant to water safety. The detection limit is between 0.1% and 0.01% relative abundance, depending on the type of pathogens.

4.3.4 High-throughput sequencing

High-throughput sequencing technology, also known as "next-generation" sequencing technology, is characterized by the ability to sequence hundreds of thousands to millions of DNA molecules in parallel and generally short read length. Currently, high-throughput sequencing platforms for the detection of pathogenic microorganisms in wastewater mainly include 454 method by Roche, Solexa by Illumina (typical sequencing platforms such as Miseq and Hiseq), and SOLiD method by Applied Biosystems (ABI). The GSFLX system of 454 method is based on pyrosequencing and relies on bioluminescence to detect DNA sequences. Under the synergistic action of DNA polymerase, ATP sulfase, luciferase, and diphosphatase, the GSFLX system couples the polymerization of each dNTP on the primer to the release of a single fluorescent signal. The content of the target DNA sequence is determined by detecting the intensity of the fluorescent signal. The core idea of Solexa sequencing technology is sequencing while synthesizing. That is, when a new DNA complementary strand is generated, either the dNTP is added to catalyze the substrate to catalyze the fluorescence by enzymatic cascade reaction, or the fluorescently labeled dNTP or semidegenerate primer is directly added to release a fluorescent signal when synthesizing or ligating to generate a complementary strand. SOLiD method is unique and it is based on the continuous ligation synthesis of four-color fluorescent labeled oligonucleotides SOLiD replaces the traditional polymerase ligation reaction and enables large-scale amplification and high-pass sequencing of single-copy DNA fragments.

High-throughput sequencing technology has been widely used to detect the microbial diversity, bacterial structure, functional microorganisms, and pathogenic microorganisms in wastewater. However, it still has some problems of insufficient sequencing depth, the need for repeated sequencing and double-end sequencing, and high cost. In addition, high-throughput sequencing has a high error rate, and as it only generates short sequence fragments, a known genome template must be used.

Cai and Zhang (2013) established 24 metagenomic DNA datasets derived from a high-throughput shotgun sequencing technique to more accurately and efficiently detect human bacterial pathogens in influent, activated sludge, and effluent of two Hong Kong WWTPs. Lu et al. (2015) used 454 pyrosequencing, Illumina high-throughput sequencing and metagenomic analysis to investigate bacterial pathogens and their potential virulence in a sewage treatment plant applying both conventional and advanced treatment processes. Li et al. (2015) investigated the broad-spectrum profile of bacterial pathogens and their fate in sewage treatment plants using high-throughput sequencing-based metagenomic approach. Bai et al. (2014) used high-throughput sequencing to analyze the prokaryotic community composition and function in river water, treated wastewater, and untreated wastewater.

4.3.5 Other methods

In recent years, a series of detection methods based on PCR, molecular hybridization, and enzyme-linked immunity technologies have been used to develop convenient, fast, sensitive, and on-site techniques for the determination of pathogenic microorganisms in wastewater. For example, enzyme-linked immunosorbent assay (ELISA), final product microwell hybridization-ELISA quantitative assay, and biosensor technology have been developed to detect the pathogenic microorganisms in wastewater. Some of these techniques based on color reaction or luminescence reaction may be suitable for on-line detection.

Rengaraj et al. (2018) developed an innovative, simple and low-cost, and paper-based probe for detection of bacteria in water. The probe was fabricated by screen printing carbon electrodes onto hydrophobic paper and the detection limit was 1.9×10^3 CFU mL^{-1}. Adkins et al. (2017) developed a transparency-based electrochemical and paper-based colorimetric analytic detection platform as a complementary method for food and waterborne bacteria detection. Zhang et al. (2019) constructed a label-free, cascade amplification visualization biosensor for the sensitive and rapid detection of *Salmonella enterica* subsp. *enterica* serovar typhimurium based on the RDTG principle (recombinase polymerase amplification, duplex-specific enzyme cleavage, terminal deoxynucleotidyl transferase extension, and G-quadruplexes output) and the low limit is 6 CFU/mL.

4.4 Summary

With the development of chemical industry and the improvement of people's living standard, more and more chemicals enter wastewater through various ways. These pollutants not only induce new environmental health risks, but also bring new challenges to the detection and analysis of HRPs in wastewater. To deal with these challenges, on the one hand, the detection accuracy and efficiency of existing methods need to be further improved. At the meaning while, the detection flux should be elevated as much as possible to achieve the goal of high efficiency, high sensitivity, and high throughput detection for HRPs. On the other hand, the combined technologies of different detection methods need to be investigated to realize the integration, automation, and intelligence of the detection technology of complex pollutants in wastewater.

References

Adeyinka, G.C., Modley, B., Birungi, G., Ndungu, P., 2019. Evaluation of organochlorinated pesticide (OCP) residues in soil, sediment and water from the Msunduzi River in South Africa. Environmental Earth Sciences 78 (6), 223.

Adeyinka, G.C., Moodley, B., Birungi, G., Ndungu, P., 2018. Quantitative analyses of selected polychlorinated biphenyl (PCB) congeners in water, soil, and sediment during winter and spring seasons from Msunduzi River, South Africa. Environmental Monitoring and Assessment 190, 621.

Adkins, J.A., Boehle, K., Friend, C., Chamberlain, B., Bisha, B., Henry, C.S., 2017. Colorimetric and electrochemical bacteria detection using printed paper- and transparency-based analytic devices. Analytical Chemistry 89 (6), 3613–3621.

Adlnasab, L., Ebrahimzadeh, H., Asgharinezhad, A.A., Aghdam, M.N., Dehghani, A., Esmaeilpour, S., 2014. A preconcentration procedure for determination of ultra-trace mercury (II) in environmental samples employing continuous-flow cold vapor atomic absorption spectrometry. Food Analytical Methods 7 (3), 616–628.

Afonso-Olivares, C., Montesdeoca-Esponda, S., Sosa-Ferrera, Z., Santana-Rodriguez, J.J., 2016. Analytical tools employed to determine pharmaceutical compounds in wastewaters after application of advanced oxidation processes. Environmental Science and Pollution Research 23 (24), 24476–24494.

Allafchian, A., Mirahmadi-Zare, S.Z., Gholamian, M., 2017a. Determination of trace lead detection in a sample solution by liquid three-phase microextraction-anodic stripping voltammetry. IEEE Sensors Journal 17 (9), 2856–2862.

Amiri, A., Ghaemi, F., Maleki, B., 2019. Hybrid nanocomposites prepared from a metal-organic framework of type MOF-199(Cu) and graphene or fullerene as sorbents for dispersive solid phase extraction of polycyclic aromatic hydrocarbons. Microchimica Acta 186, 131.

Bai, Y.H., Qi, W.X., Liang, J.S., Qu, J.H., 2014. Using high-throughput sequencing to assess the impacts of treated and untreated wastewater discharge on prokaryotic communities in an urban river. Applied Microbiology and Biotechnology 98 (4), 1841–1851.

Barril, P.A., Fumian, T.M., Prez, V.E., Gil, P.I., Martinez, L.C., Giordano, M.O., Masachessi, G., Isa, M.B., Ferreyra, L.J., Re, V.E., Miagostovich, M., Pavan, J.V., Nates, S.V., 2015. Rotavirus seasonality in urban sewage from Argentina: effect of meteorological variables on the viral load and the genetic diversity. Environmental Research 138, 409–415.

Bhandari, S.A., Amarasiriwardena, D., 2000. Closed-vessel microwave acid digestion of commercial maple syrup for the determination of lead and seven other trace elements by inductively coupled plasma-mass spectrometry. Microchemical Journal 64 (1), 73–84.

Cadorim, H.R., Schneider, M., Hinz, J., Luvizon, F., Dias, A.N., Carasek, E., Welz, B., 2019. Effective and high-throughput analytical methodology for the determination of lead and cadmium in water samples by disposable pipette extraction coupled with high-resolution continuum source graphite furnace atomic absorption spectrometry (HR-CS GF AAS). Analytical Letters 52 (13), 2133–2149.

Cai, L., Zhang, T., 2013. Detecting human bacterial pathogens in wastewater treatment plants by a high-throughput shotgun sequencing technique. Environmental Science and Technology 47 (10), 5433–5441.

Carneado, S., Pero-Gascon, R., Ibanez-Palomino, C., Lopez-Sanchez, J.F., Sahuquillo, A., 2015. Mercury(II) and methylmercury determination in water by liquid chromatography hyphenated to cold vapour atomic fluorescence spectrometry after online short-column preconcentration. Analytical Methods 7 (6), 2699–2706.

Castillo, A., Roig-Navarro, A.F., Pozo, O.J., 2006a. Method optimization for the determination of four mercury species by micro-liquid chromatography-inductively coupled plasma mass spectrometry coupling in environmental water samples. Analytica Chimica Acta 577 (1), 18–25.

Castillo, A., Roig-Navarro, A.F., Pozo, O.J., 2008. Capabilities of microbore columns coupled to inductively coupled plasma mass spectrometry in speciation of arsenic and selenium. Journal of Chromatography A 1202 (2), 132–137.

Chen, T.H., Chou, S.M., Tang, C.H., Chen, C.Y., Meng, P.J., Ko, F.C., Cheng, J.O., 2016. Endocrine disrupting effects of domestic wastewater on reproduction, sexual behavior, and gene expression in the brackish medaka *Oryzias melastigma*. Chemosphere 150, 566–575.

Chen, Z., Megharaj, M., Naidu, R., 2007a. Speciation of chromium in waste water using ion chromatography inductively coupled plasma mass spectrometry. Talanta 72 (2), 394–400.

Cuderman, P., Heath, E., 2007. Determination of UV filters and antimicrobial agents in environmental water samples. Analytical and Bioanalytical Chemistry 387 (4), 1343–1350.

Dimpe, K.M., Ngila, J.C., Mabuba, N., Nomngongo, P.N., 2014a. Evaluation of sample preparation methods for the detection of total metal content using inductively coupled plasma optical emission spectrometry (ICP-OES) in wastewater and sludge. Physics and Chemistry of the Earth 76–78, 42–48.

dos Santos, A.C.V., Masini, J.C., 2006. Development of a sequential injection anodic stripping voltammetry (SI-ASV) method for determination of Cd(II), Pb(II) and Cu(II) in wastewater samples from coatings industry. Analytical and Bioanalytical Chemistry 385 (8), 1538–1544.

Eaton, A.D., Clesceri, L.S., Rice, E.W., Greenberg, A.E., 2005. Standard Methods for the Examination of Water and Waste Water, Twentyfirst ed. New York.

Erarpat, S., Cağlak, A., Bodur, S., Chormey, S.D., Engin, Ö.G., Bakırdere, S., 2018. Simultaneous determination of fluoxetine, estrone, pesticides, and endocrine disruptors in wastewater by gas chromatography–mass spectrometry (GC–MS) following switchable solvent–liquid phase microextraction (SS–LPME). Analytical Letters 52 (5), 869–878.

Fatima, R.A., Ahmad, M., 2005. Certain antioxidant enzymes of *Allium cepa* as biomarkers for the detection of toxic heavy metals in wastewater. Science of the Total Environment 346 (1–3), 256–273.

Gagne, F., Cejka, P., Andre, C., Hausler, R., Blaise, C., 2007. Neurotoxicological effects of a primary and ozonated treated wastewater on freshwater mussels exposed to an experimental flow-through system. Comparative Biochemistry and Physiology C-Toxicology and Pharmacology 146 (4), 460–470.

Giersz, J., Bartosiak, M., Jankowski, K., 2017. Sensitive determination of Hg together with Mn, Fe, Cu by combined photochemical vapor generation and pneumatic nebulization in the programmable temperature spray chamber and inductively coupled plasma optical emission spectrometry. Talanta 167, 279–285.

Gomez-Nieto, B., Gismera, M.J., Sevilla, M.T., Procopio, J.R., 2013. Simultaneous and direct determination of iron and nickel in biological solid samples by high-resolution continuum source graphite furnace atomic absorption spectrometry. Talanta 116, 860–865.

Gondal, M.A., Hussain, T., 2007. Determination of poisonous metals in wastewater collected from paint manufacturing plant using laser-induced breakdown spectroscopy. Talanta 71 (1), 73–80.

Islam, A., Ahmad, H., Zaidi, N., Kumar, S., 2014a. Graphene oxide sheets immobilized polystyrene for column preconcentration and sensitive determination of lead by flame atomic absorption spectrometry. ACS Applied Materials and Interfaces 6 (15), 13257–13265.

Järvinen, S.T., Saari, S., Keskinen, J., Toivonen, J., 2014. Detection of Ni, Pb and Zn in water using electrodynamic single-particle levitation and laser-induced breakdown spectroscopy. Spectrochimica Acta Part B: Atomic Spectroscopy 99, 9–14.

Javanbakht, M., Khoshsafar, H., Ganjali, M.R., Badiei, A., Norouzi, P., Hasheminasab, A., 2009. Determination of nanomolar mercury(II) concentration by anodic-stripping voltammetry at a carbon paste electrode modified with functionalized nanoporous silica gel. Current Analytical Chemistry 5 (1), 35–41.

Jillani, S.M.S., Sajid, M., Alhooshani, K., 2019. Evaluation of carbon foam as an adsorbent in stir-bar supported micro-solid-phase extraction coupled with gas chromatography–mass spectrometry for the determination of polyaromatic hydrocarbons in wastewater samples. Microchemical Journal 144, 361–368.

Karvelas, M., Katsoyiannis, A., Samara, C., 2003. Occurrence and fate of heavy metals in the wastewater treatment process. Chemosphere 53 (10), 1201–1210.

Kavlock, R., Barr, D., Boekelheide, K., Breslin, W., Breysse, P., Chapin, R., Gaido, K., Hodgson, E., Marcus, M., Shea, K., Williams, P., 2006. NTP-CERHR expert panel update on the reproductive and developmental toxicity of di(2-ethylhexyl) phthalate. Reproductive Toxicology 22 (3), 291–399.

Khayatian, G., Moradi, M., Hassanpoor, S., 2018a. MnO_2/3MgO nanocomposite for preconcentration and determination of trace copper and lead in food and water by flame atomic absorption spectrometry. Journal of Analytical Chemistry 73 (5), 470–478.

Kim, B.C., Park, J.H., Gu, M.B., 2004. Development of a DNA Microarray Chip for the Identification of Sludge Bacteria Using an Unsequenced Random Genomic DNA Hybridization Method. Environmental Science and Technology 38 (24), 6767–6774.

Koha, Y.K.K., Chiu, T.Y., Boobis, A., Cartmell, E., Lester, J.N., Scrimshaw, M.D., 2007. Determination of steroid estrogens in wastewater by high performance liquid chromatography - tandem mass spectrometry. Journal of Chromatography A 1173 (1–2), 81–87.

Kyrisoglou, C., Economou, A., Efstathiou, C.E., 2012. Bismuth-coated iridium microwire electrode for the determination of trace metals by anodic stripping voltammetry. Electroanalysis 24 (9), 1825–1832.

Li, B., Ju, F., Cai, L., Zhang, T., 2015. Profile and fate of bacterial pathogens in sewage treatment plants revealed by high-throughput metagenomic approach. Environmental Science and Technology 49 (17), 10492–10502.

Liu, L., Yun, Z., He, B., Jiang, G., 2014a. Efficient interface for online coupling of capillary electrophoresis with inductively coupled plasma-mass spectrometry and its application in simultaneous speciation analysis of arsenic and selenium. Analytical Chemistry 86 (16), 8167–8175.

Liu, L., Zheng, H., Yang, C., Xiao, L., Zhangluo, Y., Ma, J., 2014b. Matrix-assisted photochemical vapor generation for the direct determination of mercury in domestic wastewater by atomic fluorescence spectrometry. Spectroscopy Letters 47 (8), 604–610.

Liu, Y.X., Wu, Y.P., Wang, L., Huang, Z.Y., 2008. Determination of total arsenic and mercury in the fucoidans by hydride generation atomic fluorescence spectroscopy. Spectroscopy and Spectral Analysis 28 (11), 2691–2694.

Lu, X., Zhang, X.X., Wang, Z., Huang, K.L., Wang, Y., Liang, W.G., Tan, Y.F., Liu, B., Tang, J.Y., 2015. Bacterial pathogens and community composition in advanced sewage treatment systems revealed by metagenomics analysis based on high-throughput sequencing. PLoS One 10 (5), 15.

Luo, G., 2012a. Determination of total arsenic in wastewater and sewage sludge samples by using hydride-generation atomic fluorescence spectrometry under the optimized analytical conditions. Analytical Letters 45 (17), 2493–2507.

Maragou, N.C., Pavlidis, G., Karasali, H., Hatjina, F., 2017. Determination of arsenic in honey, propolis, pollen, and honey bees by microwave digestion and hydride generation flame atomic absorption. Analytical Letters 50 (11), 1831–1838.

Mateos, R., Vera-Lopez, S., Saz, M., Diez-Pascual, A.M., San Andres, M.P., 2019. Graphene/sepiolite mixtures as dispersive solid-phase extraction sorbents for the anaysis of polycyclic aromatic hydrocarbons in wastewater using surfactant aqueous solutions for desorption. Journal of Chromatography A 1596, 30–40.

Maza-Marquez, P., Castellano-Hinojosa, A., Gonzalez-Martinez, A., Juarez-Jimenez, B., Gonzalez-Lopez, J., Rodelas, B., 2019. Abundance of total and metabolically active candidatus microthrix and fungal populations in three full-scale wastewater treatment plants. Chemosphere 232, 26–34.

Miller, S.M., Tourlousse, D.M., Stedtfeld, R.D., Baushke, S.W., Herzog, A.B., Wick, L.M., Rouillard, J.M., Gulari, E., Tiedje, J.M., Hashsham, S.A., 2008. In situ-synthesized virulence and marker gene biochip for detection of bacterial pathogens in water. Applied and Environmental Microbiology 74 (7), 2200–2209.

Mirzaei, M., Behzadi, M., Abadi, N.M., Beizaei, A., 2011. Simultaneous separation/preconcentration of ultra trace heavy metals in industrial wastewaters by dispersive liquid-liquid microextraction based on solidification of floating organic drop prior to determination by graphite furnace atomic absorption spectrometry. Journal of Hazardous Materials 186 (2–3), 1739–1743.

Monarca, S., Feretti, D., Collivignarelli, C., Guzzella, L., Zerbini, I., Bertanza, G., Pedrazzani, R., 2000. The influence of different disinfectants on mutagenicity and toxicity of urban wastewater. Water Research 34 (17), 4261–4269.

Muchesa, P., Mwamba, O., Barnard, T.G., Bartie, C., 2014. Detection of free-living amoebae using amoebal enrichment in a wastewater treatment plant of Gauteng province, South Africa. BioMed Research International. Article ID 575297, 10 pages.

Ort, C., Lawrence, M.G., Reungoat, J., Mueller, J.F., 2010. Sampling for PPCPs in wastewater systems: comparison of different sampling modes and optimization strategies. Environmental Science and Technology 44 (16), 6289–6296.

Panda, A., Khalil, S., Mirdha, B.R., Singh, Y., Kaushik, S., 2015. Prevalence of *Naegleria fowleri* in environmental samples from northern part of India. PLoS One 10 (10), 14.

Peng, H., Zhang, N., He, M., Chen, B., Hu, B., 2016. Multi-wall carbon nanotubes chemically modified silica microcolumn preconcentration/separation combined with inductively coupled plasma optical emission spectrometry for the determination of trace elements in environmental waters. International Journal of Environmental Analytical Chemistry 96 (3), 212–224.

Raposo, J.C., Navarro, P., Felipe, J.I.G., Etxeandia, J., Carrero, J.A., Madariaga, J.M., 2014. Trace element determination in water samples by on-line isotope dilution and inductively coupled plasma with mass spectrometry detection. Microchemical Journal 114, 99–105.

Rengaraj, S., Cruz-Izquierdo, A., Scott, J.L., Di Lorenzo, M., 2018. Impedimetric paper-based biosensor for the detection of bacterial contamination in water. Sensors and Actuators B: Chemical 265, 50–58.

Rodenburg, L.A., Du, S.Y., Fennell, D.E., Cavallo, G.J., 2010. Evidence for widespread dechlorination of polychlorinated biphenyls in groundwater, landfills, and wastewater collection systems. Environmental Science and Technology 44 (19), 7534–7540.

Sano, D., Watanabe, T., Matsuo, T., Omura, T., 2004. Detection of infectious pathogenic viruses in water and wastewater samples from urbanised areas. Water Science and Technology 50 (1), 247–251.

Senkbeil, J.K., Ahmed, W., Conrad, J., Harwood, V.J., 2019. Use of *Escherichia coli* genes associated with human sewage to track fecal contamination source in subtropical waters. Science of the Total Environment 686, 1069–1075.

Sereshti, H., Khojeh, V., Samadi, S., 2011. Optimization of dispersive liquid-liquid microextraction coupled with inductively coupled plasma-optical emission spectrometry with the aid of experimental design for simultaneous determination of heavy metals in natural waters. Talanta 83 (3), 885–890.

Sonthalia, P., McGaw, E., Show, Y., Swain, G.M., 2004. Metal ion analysis in contaminated water samples using anodic stripping voltammetry and a nanocrystalline diamond thin-film electrode. Analytica Chimica Acta 522 (1), 35–44.

Tang, J., Bu, Y., Zhang, X.X., Huang, K., He, X., Ye, L., Shan, Z., Ren, H., 2016. Metagenomic analysis of bacterial community composition and antibiotic resistance genes in a wastewater treatment plant and its receiving surface water. Ecotoxicology and Environmental Safety 132, 260–269.

Tourlousse, D.M., Ahmad, F., Stedtfeld, R.D., Seyrig, G., Tiedje, J.M., Hashsham, S.A., 2012. A polymer microfluidic chip for quantitative detection of multiple water- and foodborne pathogens using real-time fluorogenic loop-mediated isothermal amplification. Biomedical Microdevices 14 (4), 769–778.

Toušová, Z., Vrana, B., Smutná, M., Novák, J., Klučárová, V., Grabic, R., Slobodník, J., Giesy, J.P., Hilscherová, K., 2019. Analytical and bioanalytical assessments of organic micropollutants in the Bosna River using a combination of passive sampling, bioassays and multi-residue analysis. Science of the Total Environment 650 (1), 1599–1612.

Toze, S., 1999. PCR and the detection of microbial pathogens in water and wastewater. Water Research 33 (17), 3545–3556.

Ulloa-Stanojlovic, F.M., Aguiar, B., Jara, L.M., Sato, M.I.Z., Guerrero, J.A., Hachich, E., Matte, G.R., Dropa, M., Matte, M.H., de Araujo, R.S., 2016. Occurrence of *Giardia intestinalis* and *Cryptosporidium sp* in wastewater samples from so paulo state, Brazil, and Lima, Peru. Environmental Science and Pollution Research 23 (21), 22197–22205.

Vellaichamy, S., Palanivelu, K., 2011. Preconcentration and separation of copper, nickel and zinc in aqueous samples by flame atomic absorption spectrometry after column solid-phase extraction onto MWCNTs impregnated with D2EHPA-TOPO mixture. Journal of Hazardous Materials 185 (2–3), 1131–1139.

Vieira dos Santos, A.C., Masini, J.C., 2006. Development of a sequential injection anodic stripping voltammetry (SI-ASV) method for determination of Cd(II), Pb(II) and Cu(II) in wastewater samples from coatings industry. Analytical and Bioanalytical Chemistry 385 (8), 1538–1544.

Wang, L.S., Hu, H.Y., Wang, C., 2007. Effect of ammonia nitrogen and dissolved organic matter fractions on the genotoxicity of wastewater effluent during chlorine disinfection. Environmental Science and Technology 41 (1), 160–165.

Wang, X., Wei, Y., Lin, Q.Y., Zhang, J., Duan, Y.X., 2015b. Simple, fast matrix conversion and membrane separation method for ultrasensitive metal detection in aqueous samples by laser-induced breakdown spectroscopy. Analytical Chemistry 87 (11), 5577–5583.

Wang, Z., Fang, D.M., Li, Q., Zhang, L.X., Qian, R., Zhu, Y., Qu, H.Y., Du, Y.P., 2012. Modified mesoporous silica materials for on-line separation and preconcentration of hexavalent chromium using a microcolumn coupled with flame atomic absorption spectrometry. Analytica Chimica Acta 725, 81–86.

Wang, Z., Yuan, T.B., Hou, Z.Y., Zhou, W.D., Lu, J.D., Ding, H.B., Zeng, X.Y., 2014. Laser-induced breakdown spectroscopy in China. Frontiers of Physics 9 (4), 419–438.

Watson, K., Shaw, G., Leusch, F.D.L., Knight, N.L., 2012. Chlorine disinfection by-products in wastewater effluent: bioassay-based assessment of toxicological impact. Water Research 46 (18), 6069–6083.

Yang, X.Y., Hao, Z.Q., Li, C.M., Li, J.M., Yi, R.X., Shen, M., Li, K.H., Guo, L.B., Li, X.Y., Lu, Y.F., Zeng, X.Y., 2016. Sensitive determinations of Cu, Pb, Cd, and Cr elements in aqueous solutions using chemical replacement combined with surface-enhanced laser-induced breakdown spectroscopy. Optics Express 24 (12), 13410–13417.

Yang, Y., Lei, Y., Zhang, R., Wang, X., Du, X., 2018. Electrochemical fabrication of two-dimensional copper oxide nanosheets on stainless steel as a fiber coating for highly sensitive solid-phase microextraction of ultraviolet filters. Analytical Methods 10 (33), 4044–4052.

Yu, K., Li, B., Zhang, T., 2012. Direct rapid analysis of multiple PPCPs in municipal wastewater using ultrahigh performance liquid chromatography–tandem mass spectrometry without SPE pre-concentration. Analytica Chimica Acta 738, 59–68.

Yuan, X., Yang, L., Liu, S.Y., Yang, H.Y., Tang, Y.Y., Huang, K., Zhang, M., 2018b. An effective analytical system based on an ultraviolet atomizer for trace cadmium determination using atomic fluorescence spectrometry. Analytical Methods 10 (39), 4821–4826.

Zhang, W.H., Wei, C.H., Chai, X.S., He, J.Y., Cai, Y., Ren, M., Yan, B., Peng, P.A., Fu, J.M., 2012. The behaviors and fate of polycyclic aromatic hydrocarbons (PAHs) in a coking wastewater treatment plant. Chemosphere 88 (2), 174–182.

Zhang, Y., Tian, J.J., Li, K., Tian, H.T., Xu, W.T., 2019. Label-free visual biosensor based on cascade amplification for the detection of Salmonella. Analytica Chimica Acta 1075, 144–151.

Zhang, Y., Zhong, C., Zhang, Q., Chen, B., He, M., Hu, B., 2015. Graphene oxide–TiO_2 composite as a novel adsorbent for the preconcentration of heavy metals and rare earth elements in environmental samples followed by on-line inductively coupled plasma optical emission spectrometry detection. RSC Advances 5 (8), 5996–6005.

Zhang, Y.M., Pagilla, K., 2010. Treatment of malathion pesticide wastewater with nanofiltration and photo-Fenton oxidation. Desalination 263 (1–3), 36–44.

Zhao, N.J., Meng, D.S., Jia, Y., Ma, M.J., Fang, L., Liu, J.G., Liu, W.Q., 2019a. On-line quantitative analysis of heavy metals in water based on laser-induced breakdown spectroscopy. Optics Express 27 (8), A495–A506.

Zhou, Q., Zhao, N., Xie, G., 2011. Determination of lead in environmental waters with dispersive liquid-liquid microextraction prior to atomic fluorescence spectrometry. Journal of Hazardous Materials 189 (1–2), 48–53.

Zuloaga, O., Navarro, P., Bizkarguenaga, E., Iparraguirre, A., Vallejo, A., Olivares, M., Prieto, A., 2012. Overview of extraction, clean-up and detection techniques for the determination of organic pollutants in sewage sludge: a review. Analytica Chimica Acta 736, 7–29.

CHAPTER

Ecological safety hazards of wastewater

5

Xiaofeng Jiang, Mei Li, PhD

State Key Laboratory of Pollution Control and Resource Reuse, School of the Environment, Nanjing University, Nanjing, China

Chapter outline

There are growing evidences that a variety of high-risk pollutants (HRPs) persist in nearly all kinds of wastewater (Zheng et al., 2017; Ai et al., 2018; Pullagurala et al., 2018). These HRPs, mainly heavy metals, antibiotics, endocrine disrupting chemicals (EDCs), and pharmaceutical and personal care products (PPCPs) in wastewaters, are considered a great concern to the environment. Even though it has been reported that these HRPs in wastewaters are typically present at low environmental concentrations (ng/L to mg/L range), it is still unclear whether the compounds present in water can cause undesired physiological effects in organisms living in these aquatic ecosystems. Recent studies have indicated that several PPCPs are persistent or pseudopersistent in wastewater treatment systems, thereby posing potential risk when discharged in water (Verlicchi et al., 2012; Petrie et al., 2014). The prioritization of these pollutants for risk assessment is difficult, because concentrations in aquatic environments show huge variations (Verlicchi et al., 2012; Petrie et al., 2014). Although this study focused on the daily loads, the fate of HRPs and

High-Risk Pollutants in Wastewater. https://doi.org/10.1016/B978-0-12-816448-8.00005-8

metabolites in the aqueous phase of wastewater treatment plants (WWTPs) influent and effluent, as well as in surface waters in a river system located upstream and downstream of the plant has not been studied. Therefore, this chapter presents an overview on the HRPs in wastewater, focusing on their exposure pathways and damage to the environment, and the potential changes of functions and services of the ecosystem, which can be introduced to readers in a comprehensive and systematic way.

5.1 Exposure pathways

Wastewater is a complex mixtures of a large number of HRP, such as PPCPs, and EDCs, heavy metals and organic pesticides. Although concentrations of these pollutants in water are low, they tend to be highly toxic and bioaccumulative. Some of them also have carcinogenic, teratogenic, and mutagenic effects. Once these substances enter the environment, they will affect health, normal growth, and reproduction of organisms; cause damage to ecosystem structure and function; endanger the integrity and health of ecosystem and have potential ecological risks (Fig. 5.1).

The presence of HRPs in wastewaters has drawn much public concern and scientific attention in recent years. By searching for terms: "HRPs in wastewater" on the Web of Science, we have retrieved 135 articles from 1999 to 2018 (Fig. 5.2), and the citation frequency is increasing annually. Besides, ecological and human health risks are interdependent, and ecological injuries may result in increased human exposures to contaminants or other stressors. In this context of potential exposure to contaminants, examining the relative contribution of various compounds and

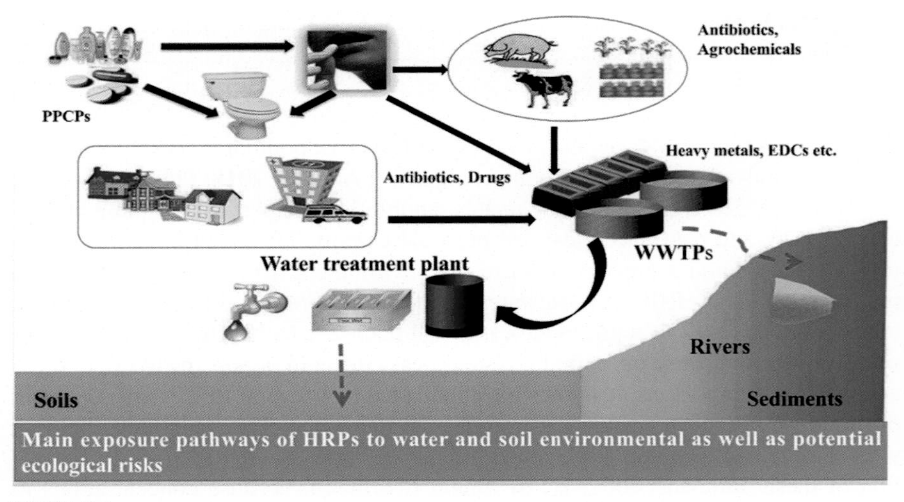

FIGURE 5.1

Conceptual depiction of the HRPs and their routes to the environment.

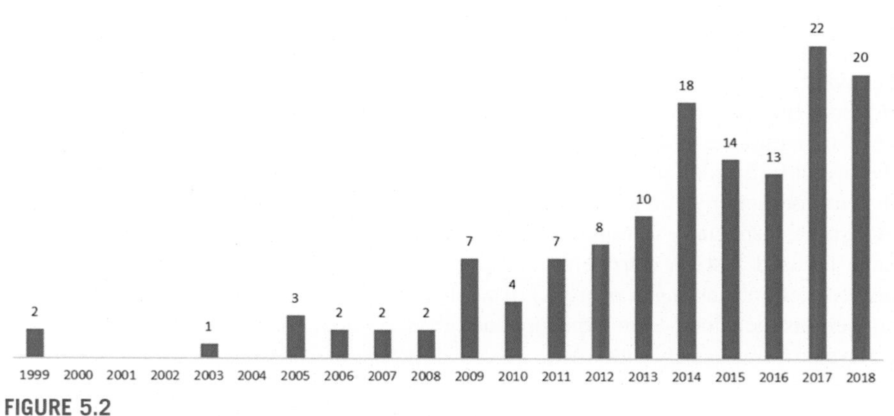

FIGURE 5.2

Annual citations of high-risk pollutants.

pathways should be taken into account when identifying effective risk-management measures.

PPCPs are likely to be found in any water body influenced by raw or treated wastewater, including rivers, lakes, streams, and groundwater, many of which are sources of drinking water (Yang et al., 2011). The major sources of PPCPs in the environment are municipal and industrial WWTPs, and landfill leachates (Daughton and Ternes., 1999). PPCPs are often not completely removed during conventional wastewater treatment processes, and thus are frequently detected in reclaimed surface water at concentrations ranging from ng/L to mg/L (Chen et al., 2013). For example, Nakada et al. (2006) investigated the removal efficiency of drugs and endocrine disruptor chemicals from five municipal WWTPs in Tokyo, and found that after the secondary treatment of activated sludge process, amide-type drugs were detected. In another study, the removal efficiency of ketoprofen and naphthalene in pharmaceutical wastewater after biological treatment process were less than 50%, and the effluent showed marked toxicity to *Anabaena* CPB4337 (Rosal et al., 2010). Gros et al. (2010) also investigated the existence of 73 drugs in the effluent of 7 WWTPs in Ebro River Basin, Spain, and found that the removal effects of different drugs by WWTPs varied distinctly, concentrations of some drugs even increased after the treatment process. The main factors affecting the removal efficiency of drugs in WWTPs include the physicochemical structure of drugs, the temperature during treatment, the redox status, and hydraulic retention time, among which the hydraulic retention time is the most important factor affecting the removal efficiency of drugs in wastewater treatment plants.

EDCs are another group of HRPs that should receive more attention. EDCs are characterized as the artificial chemicals that, when ingested into the body, can either copy or obstruct hormones to affect the body's normal functioning. The Environmental Protection Agency characterizes EDCs as external agents that meddle with the formation, release, transport, attachment, activity, or displacement of body's natural hormones that maintains homeostasis, development, reproduction, and behavior. It is generally acknowledged that the three main classes of EDCs are estrogenic (mimic or alters the functioning of natural estrogens), androgenic (copy or obstruct natural testosterone), and thyroidal (causes immediate or oblique effects to the thyroid) (Snyder et al., 2003). Natural and engineered EDCs are discharged into the environment by human activities, domesticated animals and wild ones, as well as industries, especially through sewage treatment systems. Preponderance studies have focused just on estrogenic compounds. Although EDCs are present in very low concentrations (ng/L or μg/L) in wastewater, these compounds are of profound concern as their long-term exposure and adverse impact on human health as well as ecosystems are largely unknown.

5.2 Damage to organisms

5.2.1 Effects on aquatic organisms

5.2.1.1 PPCPs

In recent years, the pollution of PPCPs in the aquatic environment has become one of the environmental problems that need to be solved urgently (Dai et al., 2015). PPCP is a new type of chemical that is widely used in daily life and has potential ecological effects. These PPCPs include, but are not limited to, a variety of drug compounds and personal care products, such as antibiotics, antiinflammatory painkillers, central stimulants, antiepileptic drugs, contraceptive, daily skin care products and cosmetics, detergents, aromatics, shading agents, and hair dyes admixtures. As Daughton et al. (1999) reported the environmental pollution of PPCPs in 1999, the prevalence, analytical techniques and toxicological effects of PPCPs in the environment have received increasing attention over the last 2 decades.

Fish is one of the most important vertebrates in the surface water environment. Compared with other aquatic living organisms, fish has a higher trophic level and is more closely related to human beings. Therefore, fish plays an extremely important role in PPCPs toxicity evaluation and ecological risk assessment (Du et al., 2014). Some fish species, such as zebrafish, have been selected as important model organisms in the field of environmental toxicology.

5.2.1.1.1 Acute toxicity and embryonic development toxicity

At present, studies on the acute toxicity of PPCPs to fish are increasingly performed. Main PPCPs that have been evaluated include antibiotics, antiinflammatory and analgesic drugs, and disinfectants, which have acute lethal effects on adult fish and embryo (Kim et al., 2009). The embryonic toxicity of PPCPs were mainly caused by the decline of hatching ability, abnormal activity, and teratogenic phenomena, such as bleeding, coagulation, scoliosis, tail bending, and pericardial edema (Fong et al., 2016).

5.2.1.1.2 Endocrine disruption and reproductive toxicity

PPCPs, especially steroid hormones, can interfere with the life activities of fish, such as growth, sex differentiation, sexual maturation, and gametogenesis, resulting in structural abnormalities and functional defects of the reproductive system, and even affecting the quality of life of their offspring (Overturf et al., 2015).

Studies have also shown that exogenous steroids can inhibit gonadotropin releasing hormone and luteinizing hormone through the negative feedback regulation of the hypothalamus—pituitary—gonadal axis. The synthesis and release of follicle-stimulating hormone interfere with the normal development of gonad or results in histological changes of fish gonads, and affect reproductive capacity, spawning quantity, and hatching rate (Li et al., 2011).

5.2.1.1.3 Combined toxicity

Environmental combined exposure assessment is a recent hot topic in toxicology. Many studies on combined pollution have been carried out globally. Combined pollution refers to two or more pollutants of different nature, or different sources of the same pollutants, or two or more types of pollutants in the same environment at the same time to form the phenomenon of environmental pollution, typically with low concentration, long persistence, diversification, and complexity of the characteristics. The interaction between PPCPs and other pollutants or between different PPCPs may occur in water environment, resulting in complex joint toxicity (Zenobio et al., 2014).

In summary, PPCPs are toxic to organisms, complex in mechanism, and may pose a potential threat to human health through biomagnification and bioaccumulation of the food chain; therefore, the PPCPs pollutants in wastewater cannot be ignored.

5.2.1.2 Antibiotics

Over the past decades, antibiotics have been recognized as a new class of water contaminants and have drawn attention due to their adverse effects on aquatic ecosystems (Kummerer, 2009). However, the greatest concern of releasing antibiotics to the aquatic environment is probably connected to the evolution of antibiotic resistance genes (ARGs), which reduce the therapeutic potential against human and animal pathogens. It has been reported that 50%—90% of antibiotics administered by humans or animals are excreted via urine and feces as a mixture of the parent compound and its metabolite forms (e.g., glucuronides) (Kummerer, 2009; Tran et al., 2016). These compounds are then carried to WWTPs where they can be partially eliminated or pass the process unchanged.

5.2.1.2.1 Toxic effects on aquatic organisms

At present, most of the related studies focused on acute toxicity of microbe, algae, cladocera (or "zoea"), and fish. Generally speaking, the effective concentrations of antibiotics on microorganisms and algae are 2–3 orders of magnitude lower than those on high trophic organisms. Chen et al. (2012) showed that the median effect concentration (EC_{50}) of antibiotics on lower aquatic organisms (algae and microorganisms) was generally at the level of 100 μg/L. Robinson et al. (2005) found that the EC_{50} of seven quinolones on five aquatic organisms was 49 μg/L for cyanobacteria (*Microcystis aeruginosa*), 106 μg/L for *Lemna minor* and 7400 μg/L for *Pseudokirchneriella subcapitata*, respectively. However, at a high concentration of 10 mg/L, no significant effect on *Daphnia magna* was observed.

Other studies have also shown that environmentally relevant concentrations of antibiotics have no significant effect on higher aquatic organisms, such as fish (De Liguoro et al., 2009). However, some studies have shown that fish are sensitive to macrolides, and their EC_{50} are less than 100 μg/L. Obviously, the toxic effects of different kinds of antibiotics on the same organism vary greatly.

5.2.1.2.2 Toxic effects on aquatic phytoplankton

As the primary producer in the aquatic environment, phytoplankton play an important role in the whole ecosystem, and when the antibiotic enters the aquatic environment, the phytoplankton is undoubtedly among the first to be affected. Toxicological studies showed that different antibiotics have distinct toxic effect on algae. For example, β-lactam antibiotics mainly inhibit the synthesis of algae cell wall, and aminoglycoside antibiotics mainly inhibit the algae sugar metabolism. Tetracycline antibiotics mainly inhibit the synthesis of algal proteins, and quinolones mainly inhibit the replication of algae DNA and chlorophyll synthesis, while sulfonamides mainly inhibit the metabolism of folic acid (Fu et al., 2017).

Borecka et al. (2016) studied the toxicity of four antibiotics to *Chlorella vulgaris* by 48 and 72 h exposure, and found that sulfamethoxazole and sulfadiazine had the strongest toxicity to *Chlorella vulgaris*. The EC_{50} values for 48 and 72 h exposure were 0.98 mg/L and 1.93 mg/L, respectively. Pan et al. (2017) showed that, when the concentration of norfloxacin was 0.05, 0.5, 5, and 50 mg/L, the algae density of the four-tailed algae decreased by 3.62%, 32.46%, 40.20%, and 50.29%, respectively.

5.2.1.2.3 Persistent organic pollutants

Persistent organic pollutants (POPs) can have carcinogenic, teratogenic, and mutagenic effects, and can have a strong harmful effect on humans and animals' reproduction, genetics, immunity, and nerves endocrine and other systems. Once in the environment, POPs can remain there persistently, resulting in bioaccumulation, and through the biomagnification in the food chain achieve toxic concentrations sufficient to cause serious negative impacts on organisms. At the same time, because in its semivolatility, POPs will lead to global environmental pollution with the flow of air and water and the migration of animals from the pollution source. Presently, there are thousands of POPs substances in the world. POPs are generally divided into three categories: pesticides, industrial chemicals and by-products from industrial processes, and solid waste combustion. Among them, organochlorine pesticides, dioxins, polychlorinated biphenyls (PCBs), polybrominated diphenyl ethers (PBDEs), and brominated flame retardants have been widely studied.

Studies by Waszak et al. (2014) on flounder (*Platichthys flesus*) and sediments near sampling sites in the Vistula (Poland) and Douro (Portugal) rivers found that the contents of 26 PCBs monomers in the Vistula River sediments were higher than those in the Douro River. The content distribution of the fish was also the same, and the composition of the homologues in the fish samples collected from the two rivers was significantly correlated with that in the sediment ($P < .05$), indicating that the halibut had the characteristics of indicating the local pollution level. PBDEs are a type of compounds with excellent flame-retardant properties. Since the 1970s, they have been widely used in the furniture, textile, and electronic industry as a flame retardant (Li et al., 2008). Streets et al. (2006) collected lake salmon (*Opsaridium microlepis*) and water samples from Lake Michigan, and found that the occurrence of BDE monomer in the biological samples and the water phase was significantly correlated ($P < .05$). This indicates that the lake salmon can reflect the pollution of POPs in the surrounding water environment. Gao et al. (2009) studied biological samples (fish, shrimp, and crabs) in the lower reaches of the Yangtze River and found that the composition and structure of the pollutants homologues in the organisms were similar to the sediments. Crucian carp (*Carassius auratus*), a common fish species in the Yangtze River Delta region, has been shown to exhibit a good indicator performance for both PBDEs and PCBs, which can be listed as a candidate for fish monitoring in the Yangtze River Delta region. Wu et al. (2008) also found that there was a good positive correlation between the composition of 18 PBDEs monomers in the northern snakehead fish and the Crucian carp (*Carassius auratus*) in a reservoir in Qingyuan area and the composition of the PBDEs monomers in the aqueous phase ($P < 0.05$).

5.2.2 Effects on terrestrial organisms

5.2.2.1 PPCPs

Most PPCPs are polar and hydrophilic, exhibiting low octanol/water partition coefficient (k_{ow}) and losses as a result of binding to the organic fraction of sludge or suspended sediments are decreased compared to other persistent organic pollutants that are capable of bioaccumulation (Yang et al., 2011). The uptake of PPCPs by plants has been well documented (Yang et al., 2011; Wu et al., 2012). However, PPCPs with strong sorption and recalcitrant to degradation remain in surface soils and have the potential to be taken up by plants (Wu et al., 2010a). Therefore, these PPCPs and several have been found to persist in the environment and exhibit bioaccumulative and endocrine disruptive activity (Clarke and Smith, 2011). Pullagurala et al. (2018) summarized the plant uptake capacity of various contaminants of emerging concern (CEC) in soil (Fig. 5.3), such as pesticides, polycyclic aromatic hydrocarbons, perfluorinated compounds, PPCPs, and engineered nanomaterials,

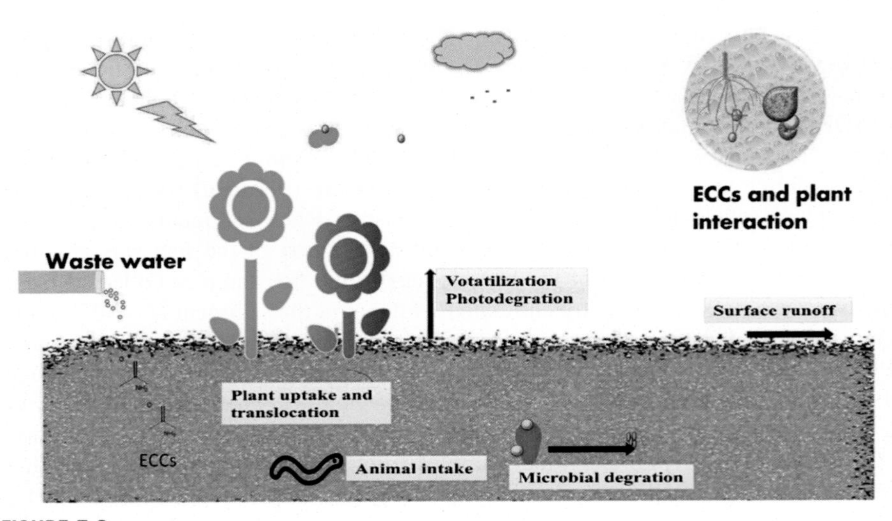

FIGURE 5.3

Environmental fate and transport of CEC in soil.

and among which PPCPs are one of the most studied in recent years. The sorption behavior of pharmaceuticals can be very complex and difficult to assess. These compounds can absorb onto bacterial lipid structures and fat fraction of the sewage sludge through hydrophobic interaction (e.g., aliphatic and aromatic groups).

In recent years, researchers have investigated the distribution of PPCPs in soils and river sediments in some regions. Kinney et al. (2006) detected 19 types of PPCPs in reclaimed water-irrigated soils in Westminster region, USA. The mass ratio of carbamazepine to carbamazepine reached 549 ng/g dry weight, and some PPCPs remained in irrigated area soil for a long time. Gibson et al. (2010) studied that the highest mass ratio of Triclosan used as an antibacterial agent in the reclaimed water-irrigated soil in the Tula Valley region of Mexico is up to 16.7 ng/g dry weight; PPCPs in the soil may also contaminate groundwater by diafiltration.

5.2.2.2 Antibiotics

5.2.2.2.1 Soil microorganisms

Soil microorganism is not only an important part of soil, but also a dynamic source of substances transformation in soil. When antibiotics enter the soil through excrement discharge or sewage irrigation, the residual antibiotics will change the structure community and the activity of microorganism in the soil, and cause various toxic effects. In the microbial community, Kong et al. (2006) found that the community diversity of soil microorganisms decreased with the increase of the concentration of oxytetracycline. When the concentration of the oxytetracycline reaches 11 μmol/L, 20% of the community diversity of the microorganisms is inhibited. Hammesfahr et al. (2008) added sulfanilamide into soil and found that the sulfanilamide can reduce the phospholipid fatty acid in the soil, and that the proportion of the bacteria and fungi was reduced by sulfanilamide even after 2 months. Boleas et al. (2005) found that when the oxytetracycline was 100 mg/kg, 16%– 25% of the microbial respiration was inhibited. When the concentration of oxytetracycline reached 1000 mg/kg, the inhibition rate was 28%–38%. Fang et al. (2014) also found that sulfanilamide and chlortetracycline could inhibit the respiration in the soil.

5.2.2.2.2 Soil plants

The roots of most plants are fixed in the soil, providing nutrients, water, and air to the plants. Some studies have reported enrichment of antibiotics in plants after being discharged into the soil. Migliore et al. (1998) showed that the roots of barley and maize contained sulphadimethoxine in the cultured soil. Kumar et al. (2005) also demonstrated that corn, onions, and cabbage absorb chlortetracycline from the soil. Kong et al. (2007) indicated that when the concentration of the oxytetracycline was higher than 0.002 mmol/L, the growth of the Alfalfa (*Medicago sativa* L.) was inhibited, and the inhibition rate of the oxytetracycline to the stem and root of alfalfa was 61% and 85%, respectively.

5.2.2.2.3 Soil animals

Soil animals play an important role in the soil ecosystems. They not only assimilate the beneficial substances in the soil for their growth, but also return the excreta to the soil. However, there are relatively few studies on the toxicity of antibiotics to soil animals. Žižek and Zidar (2013) studied the ecotoxicity of lasalocid to earthworms (*Eisenia foetida andrei*) and eupolyphaga. The results showed that the mortality rate increased significantly when lasalocid was 163 mg/kg. However, when the concentration of lasalocid reached 202 mg/kg, no obvious effects on the growth and survival of eupolyphaga were observed. Gao et al. (2016) investigated the avoidance behavior of the earthworm in the soil containing oxytetracycline, and no earthworms were found dead in the range of 0–2560 mg/kg of the oxytetracycline after 48 h exposure.

5.2.2.3 Heavy metals

Wastewater irrigation is considered an economic practice for decades in many developed and underdeveloped countries around the world. However, wastewater irrigation may also significantly increase the concentrations of heavy metals in the soil. As outlined in numerous studies, the prospect of heavy metal contamination has been dealt in relation to the generation and consumption of wastewater. Accordingly, sufficiently large mass of heavy metals can be taken up by plants via wastewater irrigation. Such pollution causes possible health risk to consumers of these plants (Mahmood and Malik, 2014; Balkhair and Ashraf, 2016). Investigation of heavy metal contamination on food crop including its risk assessment for human health is still lacking although there are some baseline data from various environmental matrices (Khan et al., 2013a, 2013b; Mahmood and Malik, 2014). The heavy metal in the edible part of the crops grown on the heavy metal polluted soil can enter the human body through the food chain, wherein excess arsenic, mercury, arsenic, lead, may cause great harm to human health.

Heavy metals are harmful to human health because they have the ability to accumulate in all parts of the body. Even at low concentrations, they may have adverse health effects because they are persistent in nature and cannot be degraded (Ikeda et al., 2000; Duruibe et al., 2007). Many edible plants accumulate heavy metals easily, leading to an increase in their concentrations in farm products (Khan et al., 2008). Contaminated soil is one of the main sources of heavy metal food chain translocation and further absorbs heavy metals through the consumption of contaminated vegetables and other crops, which ultimately poses a health risk to human health (Khan et al., 2008). In addition, heavy metals (Cu, Cd, Ni, Pb and Zn) have antagonistic effects on some essential nutrients in the body, which could lead to immune defense failure, intrauterine growth retardation, psychosocial dysfunction, malnutrition, and gastrointestinal neoplasms (Khan et al., 2010).

Cadmium (Cd) has been considered one of the most important heavy metals in food chain pollution (Wang et al., 2011) because it is soluble in water and has very high toxicity in soil. Soluble cadmium in soil is easily absorbed and accumulated in any part of the plant. Accumulation of Cd also leads to growth retardation (Chiraz et al., 2004), and damage to roots and leaves (Souza et al., 2011). Previous studies have shown that vegetation growth is severely depressed when Cd concentrations reach concentrations higher than 10 $mg \cdot kg^{-1}$ soil. In addition, soil properties such as cation exchange capacity, organic matter, and pH value also affect the desorption and diffusion of Cd (Rezaei-Rashti et al., 2014).

5.3 Damage to populations and community

5.3.1 Damage to populations

As described before, wastewater contains many kinds of HRPs, such as heavy metals, antibiotics, PPCPs, and POPs, which have adverse effects on human health and animal reproduction, genetics, immunity, nerves and endocrine as well as other systems. Besides, it is well established that HRPs in wastewater can also contaminate soil. Many researchers have already reported that soil contamination is of great concern in croplands due to potential effects on food safety and soil ecosystems, especially on the microbial communities (Chen et al., 2015; Mekki et al., 2017). The structure, abundance, and activity of soil microbial communities greatly affects the growth and health of plants and soil animals. The disturbance of soil microbes may reduce soil fertility, thus affecting the quality of crop production (Tecon and Or, 2017).

In a recent study, Shen et al. (2019) assessed the characteristics of heavy metal and petroleum contaminants in different land use types after wastewater irrigation. The results indicated that the composition, abundance, and metabolic activity of soil bacteria communities were affected by the wastewater. In addition, wastewater from aquaculture is also a problem worth considering. Aquaculture wastewater can cause severe environmental problems with its low chemical oxygen demand (COD) and high nitrogen and phosphorus contents.

Untreated wastewater is usually rich in nutrients (nitrogen, phosphorus) along with other contaminants (heavy metals, pesticides, etc.). Discharge of untreated sewage wastewater into the water bodies leads to increased nutrient load and may as a result cause formation of algal blooms and imbalance in the ecosystem (Heisler et al., 2008). Microalgae presence in wastewater systems can be used as indicators of water pollution (Torres et al., 2008). Studies on microalgal diversity and their associations in the water bodies as biological indicators are helpful in the assessment of water quality (Shanthala et al., 2009). Renuka et al. (2014) designed an investigation to assess the variation in microalgal diversity vis a vis physicochemical characteristics of sewage wastewater at monthly time intervals. Results indicated that the indices of microalgal diversity showed a positive correlation with nutrients and a negative correlation with COD and heavy metal concentrations, implying the significant role of these factors in influencing the algal population and *Phormidium* sp. that was the dominant genus present throughout the year.

The presence of pharmaceutically active compounds (PhACs) in wastewater treatment plant effluent also poses a potential risk to aquatic ecosystems. Numerous studies have shown that the effluents of conventional WWTPs contain many PhACs, and these waste streams are the most common pathway where PhACs enter freshwater or marine environments (Li et al., 2014; Verlicchi and Zambello, 2015). Although PhACs are usually present at low concentrations (on the order of ng-mg/L), they may cause a wide range of toxic effects in organisms and deleterious effects on ecosystems as a result of their continuous input into aquatic environments, where they act as pseudopersistent pollutants (Corcoran et al., 2010; Deng et al., 2012). PhACs result in several potential adverse biological effects including the feminization of aquatic animals, the development of antibiotic resistant bacteria, and decreasing planktonic diversity (Bouki et al., 2013; Zheng et al., 2017).

5.3.2 Damage to community

Contaminants in cropland soils mainly originate from atmospheric deposition, wastewater irrigation, pesticide spraying, and fertilization (Luo et al., 2009; Xie et al., 2009). Wu et al. (2010b) estimated that nearly one-fifth of Chinese farmland area was contaminated with heavy metals (Cheng et al., 2003). Application of wastewater for irrigation occurs worldwide to enhance crop yields using readily available water that is premixed with carbon, nitrogen, phosphorus, and major and micronutrients (Sridhara et al., 2008). However, wastewater may contain toxic pollutants that contaminate the soil and groundwater (Liu et al., 2005). Hence, long-term wastewater irrigation can introduce toxic substances into the food chain (Sridhara et al., 2008; Lu et al., 2015) and affect soil microbial communities (Zhang et al., 2008). Nowadays, heavy metal pollutants are still an urgent issue, especially in crop-system. Ai et al. (2018) investigated the temporal variations and spatial distributions of heavy metals in a wastewater-irrigated soil-eggplant system and associated influencing factors. Results showed that soil heavy metals had a dominant impact on their accumulations in eggplant fruit, with a variance contribution of 78.0%, while soil properties had a regulatory effect, with a variance contribution of 5.2%.

Because the wastewater contains many nutrients such as nitrogen and phosphorus, a large amount of wastewater discharged into the river may lead to eutrophication of the water body, which may favor cyanobacterial development. Olano et al. (2019) studied the effects of main drinking water source supply in Uruguay (Santa Lucía River, SLR) on cyanobacterial development. The quantitative analysis of the phytoplankton community was carried out in Utermöhl sedimentation chambers by randomly selecting fields to quantify 400 total organisms or 100 organisms of the most abundant taxa. The findings showed that the difference in structure and biomass of the phytoplankton community of SLR and that of the wastewater highlights the vulnerability of the fluvial environments susceptible to invasion by allochthonous phytoplankton.

In recent years, more and more antibiotics are used in humans, livestock, and aquaculture. Because of its low metabolic rate, most of them are discharged into the environment in the form of raw drugs or metabolites through urine and feces, resulting in the residue of antibiotics in environmental media including water and soil. These residual antibiotics can lead to potential environmental risks, the most serious of which is the induction and transmission of various ARGs, which in turn pose a threat to human health. As the antibiotic is designed as an antibacterial drug, the growth of the associated microorganisms in the soil can be directly killed or inhibited, thereby affecting the microbial activity or function. Similar to other contaminants, the microbial toxicity of antibiotics presents a dose-dependent effect (Kong et al., 2006; Kotzerke et al., 2011).

Most studies have shown that low concentrations of antibiotics have no significant toxic effects, which may be closely related to soil adsorption (Thiele-Bruhn et al., 2005). Interestingly, some studies have shown that low-dose antibiotics can even stimulate soil microbes, possibly because they can act as carbon sources for some microbes to promote their growth (Dantas et al., 2008). However, some studies have also shown that microbial activity can be significantly affected under low-concentration antibiotics. For example, Toth et al. (2011) found that sulfamethazine and monensin in environmental concentration ($\leq$200 μg kg^{-1}) have a significant effect on soil iron reduction and nitrification. Molecular fingerprinting techniques (such as DGGE and T-RFLP) were used to study the effects of antibiotics on soil microbial community structure (Demoling et al., 2009). Bands representing different microbial species appear or disappear, indicating that antibiotics have a significant effect on soil microbial community structure. Most antibiotics are bacteriostatic or bactericidal drugs, so the addition of these antibiotics can reduce the number of bacteria in soil and thus increase the ratio of fungi to bacteria in soil (Girardi et al., 2011).

Mohamed et al. (2005) showed that the bacterial population and community structure were also affected at higher concentrations of amphotericin (a fungal inhibitor). Many researchers have observed the effect of antibiotics on the functional diversity of soil microbial communities (Liu et al., 2012). However, until now, there have been a few studies on the functional microbial communities involved in ecosystem processes. Only Schauss et al. (2009) and Kleineidam et al. (2010) studied the effects of sulfadiazine on ammonia oxidation and denitrification of microbial communities in pig dung.

5.4 Damage to ecosystem

As mentioned before, HRPs could lead to many adverse environmental effects on organisms and population as well as communities (Table 5.1). The damage to the ecosystem is also of great concern. Antibiotics are the most popular antimicrobial drug for human disease control and livestock growth promotion for a long time. China is one of the largest producers and consumers of antibiotics in the world. The total usage of antibiotics was 92,700 tons in 2013, and an estimated 54,000 tons of the antibiotics was discharged after human and animals use (Zhang et al., 2015). A large number of these antibiotics cannot be metabolized *in vivo*, 30%−90% of which will be excreted into the environment (Hu et al., 2010; Zhao et al., 2010). Due to the use of organic manure/sludge and irrigation of wastewater or reclaimed water in the agricultural land, a significant amount of antibiotics, such as tetracyclines and quinolones, are detected in soils (Wang et al., 2014; Hou et al., 2015). Many antibiotics are biologically active in the environment (Zhou et al., 2013; Xu et al., 2013), what may pose a pressure on ARGs and bacterial communities (Tello et al., 2012; Xiong et al., 2015). The occurrence of ARGs in sediments,

Table 5.1 Environmental effects of HRPs in wastewater.

Main pollutants	Adverse effect	References
Phthalic acid esters	Fish, daphnia, and green algae ecotoxicity significantly increased.	Liang et al. (2017)
Roxithromycin, clarithromycin, tylosin (antibiotics)	Growth inhibition of algae (*Pseudokirchneriella subcapitata*)	Yang et al. (2008)
Caffeine (stimulant drug)	Endocrine disruption of goldfish (*Carassius auratus*)	Li et al. (2012)
Carbamazepine (antiepileptic drug)	Oxidation stress of rainbow trout (*Oncorhynchus mykiss*)	Li et al. (2010)
HHCB (synthetic musk)	Oxidation stress of goldfish (*Carassius auratus*)	Chen et al. (2012)
Atenolol (antibiotics)	Influence on metabolism and genome changes of crucian carp	Yu et al. (2016)
Diclofenac (personal care products)	Affects the feeding behavior of killifish.	Nassef et al. (2010)

soil, and water has been observed in different countries (Knapp et al., 2010; Chen et al., 2016), which may be enriched by antibiotic residues (Cheng et al., 2016; Yang et al., 2017) and transferred into different environmental media (Chen et al., 2016; Xie et al., 2016). Therefore, antibiotics may cause significant impacts on the environment and human (Tello et al., 2012).

In agricultural soils, ARGs had many sources and could be influenced by antibiotics or heavy metals (Luo et al., 2017). In addition, with more application of manure, sewage sludge and wastewater in agricultural soils, their will be disseminate ARGs from the environment to the agricultural system, such as soils and crops (Zhu et al., 2017). As a result, the antibiotic pollution in soils and their potential damage to ecosystem and human health require more systematic research in the future. Further data on the distribution of ARGs in the studied region are needed to provide more information for the risk management of antibiotics in the environment.

It is worth noting that wastewater treatment facilities are now the most commonly used abatement measures to resolve point-source water pollution, but they also produce a large amount of sewage sludge and treated water. The discharge of large volumes of treated wastewater that contains low levels of chemical constituents may still lead to an excessive input of nutrients in a receiving water body (Muga and Mihelcic, 2008).

Many studies have focused on the analysis of wastewater treatment cost structures (Hernandez-Sancho et al., 2011). However, very few studies link the ecosystem response behavior with the level of wastewater treatment to allow the estimation of economic trade-off associated with setting optimal water quality levels. Some studies analyze the effects of wastewater discharge on lake ecosystems' functioning (Machado and Imberger, 2012). Implications of wastewater on both ecosystem structure and function were, however, until now, mainly assessed in effluent dominated streams of semiarid regions (e.g., Canobbio et al., 2009). Previous studies indicated impairments of secondary treated wastewater in the ecosystem function of leaf litter decomposition in temperate regions (Ladewig et al., 2006; Spänhoff et al., 2007). However, the increasing frequency and intensity of summer droughts, which are potentially due to climate change, have resulted in decreased water levels, particularly in low-before streams, thereby additionally reducing their dilution potential. Aquatic ecosystems (e.g., lake ecosystems) are able to store and absorb waste from human economic activities through dilution, assimilation, and chemical decomposition to a limited extent, acting as "free" water purification plants (De Groot et al., 2002). If the amount of waste exceeds the aquatic ecosystem's purification capacity, the ecosystem will be damaged. On one hand, the over exploitation of the ecosystem capability in attenuating pollution can compromise the long-term functionality of the aquatic ecosystem functionality. On the other hand, if the receiving water system's assimilative capacities are not fully used, higher wastewater treatment costs are wasted.

5.5 Summary

Although the concentration of HRPs in wastewater is low, their potential to harm ecological environment and human health cannot be ignored. Over the past decades, with the continuous detection of HRPs in various wastewaters, pollutants can be transported and transformed in various environmental media, and it is likely that pollutants will be brought into the food chain. Although the existing studies have made it clear that the residues of HRPs in wastewaters can produce certain ecotoxicogical effects, there is a lack of necessary research on the evaluation of their effects, especially the ecotoxicogical effects of long exposure at low doses. In addition, there is still a lack of research on the influence of environmental factors on their ecotoxicities and the regulation mechanisms. At the same time, the detection of new environmental HRPs in the effluent of wastewater treatment plants shows that the wastewater treatment technology is not perfect, and it is urgent to develop and apply more comprehensive wastewater treatment technologies to block the damage to ecosystem induced by the HRPs in wastewater.

References

Ai, S.W., Liu, B.L., Yang, Y., Ding, J., Yang, W.Z., Bai, X.J., Naeem, S., Zhang, Y.M., 2018. Temporal variations and spatial distributions of heavy metals in a wastewater-irrigated soil-eggplant system and associated influencing factors. Ecotoxicology and Environmental Safety 153, 204–214.

Balkhair, K.S., Ashraf, M.A., 2016. Field accumulation risks of heavy metals in soil and vegetable crop irrigated with sewage water in western region of Saudi Arabia. Saudi Journal of Biological Sciences 23, S32–S44.

Boleas, S., Alonso, C., Pro, J., Fernandez, C., Carbonell, G., Tarazona, J.V., 2005. Toxicity of the antimicrobial oxytetracycline to soil organisms in a multi-species-soil system (MS•3) and influence of manure co-addition. Journal of Hazardous Materials 122, 233–241.

Borecka, M., Bialk-Bielinska, A., Halinski, L.P., Pazdro, K., Stepnowski, P., Stolte, S., 2016. The influence of salinity on the toxicity of selected sulfonamides and trimethoprim towards the green algae *Chlorella vulgaris*. Journal of Hazardous Materials 308, 179–186.

Bouki, C., Venieri, D., Diamadopoulos, E., 2013. Detection and fate of antibiotic resistant bacteria in wastewater treatment plants: a review. Ecotoxicology and Environmental Safety 91, 1–9.

Canobbio, S., Mezzanotte, V., Sanfilippo, U., Benvenuto, F., 2009. Effect of multiple stressors on water quality and macro invertebrate assemblages in an effluent dominated stream. Water Air, and Soil Pollution 198, 359–371.

Chen, B.W., Yuan, K., Chen, X., Yang, Y., Zhang, T., Wang, Y.W., Luan, T.G., Zou, S.C., Li, X.D., 2016. Metagenomic analysis revealing antibiotic resistance genes (ARGs) and their genetic compartments in the Tibetan environment. Environmental Science and Technology 50, 6670–6679.

Chen, F., Gao, J., Zhou, Q., 2012. Toxicity assessment of simulated urban runoff containing polycyclic musks and cadmium in *Carassius auratus* using oxidative stress biomarkers. Environmental Pollution 162, 91–97.

Chen, M., Xu, P., Zeng, G., Yang, C., Huang, D., Zhang, J., 2015. Bioremediation of soils contaminated with polycyclic aromatic hydrocarbons, petroleum, pesticides, chlorophenols and heavy metals by composting: applications, microbes and future research needs. Biotechnology Advances 33, 745–755.

Chen, W., Xu, J., Lu, S., Jiao, W., Wu, L., Chang, A.C., 2013. Fates and transport of PPCPs in soil receiving reclaimed water irrigation. Chemosphere 93, 2621–2630.

Cheng, S., 2003. Heavy metal pollution in China: origin, pattern and control. Environmental Science and Pollution Research 10, 192–198.

Cheng, W.X., Li, J.N., Wu, Y., Xu, L.K., Su, C., Qian, Y.Y., Zhu, Y.G., Chen, H., 2016. Behavior of antibiotics and antibiotic resistance genes in eco-agricultural system: a case study. Journal of Hazardous Materials 304, 18–25.

Chiraz, C., Karine, P., Akira, S., Houda, G., Habib, G.M., Céline, M.-D., 2004. Cadmium toxicity induced changes in nitrogen management in *Lycopersicon esculentum* leading to a metabolic safeguard through an amino acid storage strategy. Plant and Cell Physiology 45, 1681–1693.

Clarke, B.O., Smith, S.R., 2011. Review of "emerging" organic contaminants in biosolids and assessment of international research priorities for the agricultural use of biosolids. Environment International 37, 226–247.

Corcoran, J., Winter, M.J., Tyler, C.R., 2010. Pharmaceuticals in the aquatic environment: a critical review of the evidence for health effects in fish. Critical Reviews in Toxicology 40, 287–304.

Dai, G.H., Wang, B., Huang, J., Dong, R., Deng, S., Yu, G., 2015. Occurrence and source apportionment of pharmaceuticals and personal care products in the Beiyun River of Beijing, China. Chemosphere 119, 1033–1039.

Dantas, G., Sommer, M.O.A., Oluwasegun, R.D., Church, G.M., 2008. Bacteria subsisting on antibiotics. Science 320, 100–103.

Daughton, C.G., Ternes, T.A., 1999. Pharmaceuticals and personal care products in the environment: agents of subtle change? Environmental Health Perspectives 107, 907–938.

De Liguoro, M., Fioretto, B., Poltroneri, C., Gallina, G., 2009. The toxicity of sulfamethazine to *Daphnia magna* and its additivity to other veterinary sulfonamides and trimethoprim. Chemosphere 75, 1519–1524.

De Groot, R.S., Wilson, M.A., Boumans, R.M., 2002. A typology for the classification, description and valuation of ecosystem functions, goods and services. Ecological Economics 41, 393–408.

Demoling, L.A., Baath, E., Greve, G., Wouterse, M., Schmitt, H., 2009. Effects of sulfamethoxazole on soil microbial communities after adding substrate. Soil Biology and Biochemistry 41, 840–848.

Deng, Y.Q., Zhang, Y., Gao, Y.X., Li, D., Liu, R.Y., Liu, M.M., Zhang, H.F., Hu, B., Yu, T., Yang, M., 2012. Microbial community compositional analysis for series reactors treating high level antibiotic wastewater. Environmental Science and Technology 46, 795–801.

Du, B., Haddad, S.P., Luek, A., Scott, W.C., Saari, G.N., Kristofco, L.A., Connors, K.A., Rash, C., Rasmussen, J.B., Chambliss, C.K., Brooks, B.W., 2014. Bioaccumulation and trophic dilution of human pharmaceuticals across trophic positions of an effluent-dependent wadeable stream. Philosophical Transactions of the Royal Society of London B Biological Sciences 369 pii: 20140058.

Duruibe, J.O., Ogwuegbu, M.D.C., Egwurugwu, J.N., 2007. Heavy metal pollution and human biotoxic effects. International Journal of the Physical Sciences 2, 112–118.

Fang, H., Han, Y.L., Yin, Y.M., Pan, X., Yu, Y.L., 2014. Variations in dissipation rate, microbial function and antibiotic resistance due to repeated introductions of manure containing sulfadiazine and chlortetracycline to soil. Chemosphere 96, 51–56.

Fong, H.C., Ho, J.C., Cheung, A.H., Lai, K.P., Tse, W.K.F., 2016. Developmental toxicity of the common UV filter, benophenone-2, in zebrafish embryos. Chemosphere 164, 413–420.

Fu, L., Huang, T., Wang, S., Wang, X.H., Su, L.M., Li, C., Zhao, Y.H., 2017. Toxicity of 13 different antibiotics towards freshwater green algae *Pseudokirchneriella subcapitata* and their modes of action. Chemosphere 168, 217–222.

Gao, M., Lyu, M., Han, M., Song, W.H., Wang, D., 2016. Avoidance behavior of *Eisenia fetida* in oxytetracycline-and heavy metal-contaminated soils. Environmental Toxicology and Pharmacology 47, 119–123.

Gao, Z., Xu, J., Xian, Q., Feng, J.F., Chen, X.H., Yu, H.X., 2009. Polybrominated diphenyl ethers (PBDEs) in aquatic biota from the lower reach of the Yangtze River, East China. Chemosphere 75, 1273–1279.

Gibson, R., Duran-alvarez, J.C., Estrada, K.L., Chavez, A., Cisneros, B.J., 2010. Accumulation and leaching potential of some pharmaceuticals and potential endocrine disruptors in soils irrigated with wastewater in the Tula Valley, Mexico. Chemosphere 81, 1437–1445.

Girardi, C., Greve, J., Lamshoft, M., Fetzer, I., Miltner, A., Schaffer, A., Kastner, M., 2011. Biodegradation of ciprofloxacin in water and soil and its effects on the microbial communities. Journal of Hazardous Materials 198, 22–30.

Gros, M., Petrovi, O.M., Ginebreda, A., Barcelo, D., 2010. Removal of pharmaceuticals during wastewater treatment and environmental risk assessment using hazard indexes. Environment International 36, 15–26.

Hammesfahr, U., Heuer, H., Manzke, B., Smalla, K., Thiele-Bruhn, S., 2008. Impact of the antibiotic sulfadiazine and pig manure on the microbial community structure in agricultural soils. Soil Biology and Biochemistry 40, 1583–1591.

Heisler, J., Glibert, P.M., Burkholder, J.M., Anderson, D.M., Cochlan, W., Dennison, W.C., Dortch, Q., Gobler, C.J., Heil, C.A., Humphries, E., Lewitus, A., Magnien, R., Marshall, H.G., Sellner, K., Stockwell, D.A., Stoecker, D.K., Suddleson, M., 2008. Eutrophication and harmful algal blooms: a scientific consensus. Harmful Algae 8, 3–13.

Hernandez-Sancho, F., Molinos-Senante, M., Sala-Garrido, R., 2011. Cost modelling for wastewater treatment processes. Desalination 268, 1–5.

Hou, J., Wan, W.N., Mao, D.Q., Wang, C., Mu, Q.H., Qin, S.Y., Luo, Y., 2015. Occurrence and distribution of sulfonamides, tetracyclines, quinolones, macrolides, and nitrofurans in livestock manure and amended soils of Northern China. Environmental Science and Pollution Research 22, 4545–4554.

Hu, X.G., Zhou, Q.X., Luo, Y., 2010. Occurrence and source analysis of typical veterinary antibiotics in manure, soil, vegetables and groundwater from organic vegetable bases, northern China. Environmental Pollution 158, 2992–2998.

Ikeda, M., Zhang, Z.W., Shimbo, S., Watanabe, T., Nakatsuka, H., Moon, C.S., Matsuda-Inoguchi, N., Higashikawa, K., 2000. Urban population exposure to lead and cadmium in east and south-east Asia. The Science of the Total Environment 249, 373–384.

Khan, S., Cao, Q., Zheng, Y.M., Huang, Y.Z., Zhu, Y.G., 2008. Health risks of heavy metals in contaminated soils and food crops irrigated with wastewater in Beijing, China. Environmental Pollution 152, 686–692.

Khan, S., Rehman, S., Khan, A.Z., Khan, M.A., Shah, M.T., 2010. Soil and vegetables enrichment with heavy metals from geological sources in Gilgit, northern Pakistan. Ecotoxicology and Environmental Safety 73, 1820–1827.

Khan, K., Lu, Y., Khan, H., Ishtiaq, M., Khan, S., Waqas, M., Wei, L., Wang, T., 2013a. Heavy metals in agricultural soils and crops and their health risks in Swat District, northern Pakistan. Food and Chemical Toxicology 58, 449–458.

Khan, M.U., Malik, R.N., Muhammad, S., 2013b. Human health risk from heavy metal via food crops consumption with wastewater irrigation practices in Pakistan. Chemosphere 93, 2230–2238.

Kim, J.W., Ishibashi, H., Yamauchi, R., Ichikawa, N., Takao, Y., Hirano, M., Koga, M., Arizono, K., 2009. Acute toxicity of pharmaceutical and personal care products on freshwater crustacean (*Thamnocephalus platyurus*) and fish (*Oryzias latipes*). Journal of Toxicological Sciences 34, 227–232.

Kinney, C.A., Furlong, E.T., Werner, S.L., Cahill, J.D., 2006. Presence and distribution of wastewater-derived pharmaceuticals in soil irrigated with reclaimed water. Environmental Toxicology and Chemistry 25, 317–326.

Kleineidam, K., Sharma, S., Kotzerke, A., 2010. Effect of sulfadiazine on abundance and diversity of denitrifying bacteria by determining nir K and nir S genes in two arable soils. Microbial Ecology 60, 703–707.

Knapp, C.W., Dolfing, J., Ehlert, P.A.I., Graham, D.W., 2010. Evidence of increasing antibiotic resistance gene abundances in archived soils since 1940. Environmental Science and Technology 44, 580–587.

Kong, W.D., Zhu, Y.G., Fu, B.J., Marschner, P., He, J.Z., 2006. The veterinary antibiotic oxytetracycline and Cu influence functional diversity of the soil microbial community. Environmental Pollution 143, 129–137.

Kong, W.D., Zhu, Y.G., Liang, Y.C., Zhang, J., Smith, F.A., Yang, A., 2007. Uptake of oxytetracycline and its phytotoxicity to alfalfa (*Medicago sativa* L.). Environmental Pollution 147, 187–193.

Kotzerke, A., Hammesfahr, U., Kleineidam, K., Lamshoft, M., Thiele-Bruhn, S., Schloter, M., Wilke, B.M., 2011. Influence of difloxacin-contaminated manure on microbial community structure and function in soils. Biology and Fertility of Soils 47, 177–186.

Kumar, K., Gupta, S.C., Baidoo, S.K., Chander, Y., Rosen, C.J., 2005. Antibiotic uptake by plants from soil fertilized with animal manure. Journal of Environmental Quality 34, 2082–2085.

Kummerer, K., 2009. Antibiotics in the aquatic environment-a review-part I. Chemosphere 75, 417–434.

Ladewig, V., Jungmann, D., Köhler, H., Schirling, M., Triebskorn, R., Nagel, R., 2006. Population structure and dynamics of Gammarus fossarum (*Amphipoda*) upstream and downstream from effluents of sewage treatment plants. Archives of Environmental Contamination and Toxicology 50, 370–383.

Li, H., Helm, P.A., Metcalfe, C.D., 2010. Sampling in the Great Lakes for pharmaceuticals, personal care products and endocrine-disrupting substances using the passive polar organic chemical integrative sampler. Environmental Toxicology & Chemistry 9999, 1–12.

Li, J., Yu, H., Zhao, Y., Zhang, G., Wu, Y.N., 2008. Levels of polybrominated diphenyl ethers (PBDEs) in breast milk from Beijing, China. Chemosphere 73, 182–186.

Li, W., Nanaboina, V., Zhou, Q., Korshin, V.G., 2012. Effects of Fenton treatment on the properties of effluent organic matter and their relationships with the degradation of pharmaceuticals and personal care products. Water Research 46, 403–412.

Li, Y., Zhu, G., Ng, W.J., Tan, S.K., 2014. A review on removing pharmaceutical contaminants from wastewater by constructed wetlands: design, performance and mechanism. The Science of the Total Environment 468–469, 908–932.

Li, Z., Kroll, K.J., Jensen, K.M., 2011. A computational model of the hypothalamic: pituitary: gonadal axis in female fathead minnows (*Pimephales promelas*) exposed to 17α-ethynylestradiol and 17beta-trenbolone. BMC Systems Biology 5, 63–84.

Liang, J.Y., Ning, X.A., Kong, M.Y., Liu, D.H., Wang, G.W., Cai, H.L., Sun, J., Zhang, Y.P., Lu, X.W., Yuan, Y., 2017. Elimination and ecotoxicity evaluation of phthalic acid esters from textile-dyeing wastewater. Environmental Pollution 231, 115–122.

Liu, W.H., Zhao, J.Z., Ouyang, Z.Y., Soderlund, L., Liu, G.H., 2005. Impacts of sewage irrigation on heavy metal distribution and contamination in Beijing, China. Environment International 31, 805–812.

Liu, F., Wu, J., Ying, G.G., Luo, Z.X., Feng, H., 2012. Changes in functional diversity of soil microbial community with addition of antibiotics sulfamethoxazole and chlortetracycline. Applied Microbiology and Biotechnology 95, 1615–1623.

Lu, Y., Song, S., Wang, R., Liu, Z., Meng, J., Sweetman, A.J., Jenkins, A., Ferrier, R.C., Li, H., Luo, W., Wang, T.Y., 2015. Impacts of soil and water pollution on food safety and health risks in China. Environment International 77, 5–15.

Luo, G., Li, B., Li, L.G., Zhang, T., Angelidaki, I., 2017. Antibiotic resistance genes and correlations with microbial community and metal resistance genes in full-scale biogas reactors as revealed by metagenomic analysis. Environmental Science and Technology 51, 4069–4080.

Luo, L., Ma, Y., Zhang, S., Wei, D., Zhu, Y.G., 2009. An inventory of trace element inputs to agricultural soils in China. Journal of Environmental Management 90, 2524–2530.

Machado, D.A., Imberger, J., 2012. Managing wastewater effluent to enhance aquatic receiving ecosystem productivity: a coastal lagoon in Western Australia. Journal of Environmental Management 99, 52–60.

Mahmood, A., Malik, R.N., 2014. Human health risk assessment of heavy metals via consumption of contaminated vegetables collected from different irrigation sources in Lahore, Pakistan. Arabian Journal of Chemistry 7, 91–99.

Mekki, A., Awali, A., Aloui, F., Loukil, S., Sayadi, S., 2017. Characterization and toxicity assessment of wastewater from rock phosphate processing in Tunisia. Mine Water and the Environment 36, 502–507.

Migliore, L., Civitareale, C., Cozzolino, S., Casoria, P., Brambilla, G., Gaudio, L., 1998. Laboratory models to evaluate phytotoxicity of sulphadimethoxine on terrestrial plants. Chemosphere 37, 2957–2961.

Mohamed, M.A.N., Ranjard, L., Catroux, C., Catroux, G., Hartmann, A., 2005. Effect of natamycin on the enumeration, genetic structure and composition of bacterial community isolated from soils and soybean rhizosphere. Journal of Microbiological Methods 60, 31–40.

Muga, H.E., Mihelcic, J.R., 2008. Sustainability of wastewater treatment technologies. Journal of Environmental Management 88, 437–447.

Nakada, N., Tanishima, T., Shinohara, H., Kiri, K., Takada, H., 2006. Pharmaceutical chemicals and endocrine disrupters in municipal wastewater in Tokyo and their removal during activated sludge treatment. Water Research 40, 3297–3303.

Nassef, M., Matsumoto, S., Seki, M., 2010. Acute effects oftriclosan, diclofenac and carbamazepine on feeding performance of Japanese medaka fish (*Oryziaslatipes*). Chemosphere 80, 1095–1100.

Olano, H., Martigani, F., Somma, A., Aubriot, L., 2019. Wastewater discharge with phytoplankton may favor cyanobacterial development in the main drinking water supply river in Uruguay. Environmental Monitoring and Assessment 191, 146.

Overturf, M.D., Anderson, J.C., Pandelides, Z., Beyger, L., Holdway, D.A., 2015. Pharmaceuticals and personal care products: a critical review of the impacts on fish reproduction. Critical Reviews in Toxicology 45, 469–491.

Pan, Y., Liu, C., Li, F., Zhou, C.Q., Yan, S.W., Dong, J.Y., Li, T.R., Duan, C.Q., 2017. Norfloxacin disrupts *Daphnia magna*-induced colony formation in Scenedesmus quadricauda and facilitates grazing. Ecological Engineering 102, 255–261.

Petrie, B., Barden, R., Kasprzyk-Hordern, B., 2014. A review on emerging contaminants in wastewaters and the environment: current knowledge, understudied areas and recommendations for future monitoring. Water Research 72, 3–27.

Pullagurala, V.L.R., Rawat, S., Adisa, I.O., Hernandez-Viezcas, J.A., Peralta-Videa, J.R., Gardea-Torresdey, J.L., 2018. Plant uptake and translocation of contaminants of emerging concern in soil. The Science of the Total Environment 636, 1585–1596.

Renuka, N., Sood, A., Prasanna, R., Ahluwalia, A.S., 2014. Influence of seasonal variation in water quality on the microalgal diversity of sewage wastewater. South African Journal of Botany 90, 137–145.

Rezaei-Rashti, M., Esfandbod, M., Adhami, E., Srivastava, P., 2014. Cadmium desorption behavior in selected sub-tropical soils: effects of soil properties. Journal of Geochemical Exploration 144, 230–236.

Robinson, A.A., Belden, J.B., Lydy, M.J., 2005. Toxicity of fluoroquinolone antibiotics to aquatic organisms. Environmental Toxicology & Chemistry 24, 423–430.

Rosal, R., Rodea-Palomares, I., Bohes, K., Fernandez-Pinas, F., Leganes, F., Gonzalo, S., Petre, A., 2010. Ecotoxicity assessment of lipid regulators in water and biologically treated wastewater using three aquatic organisms. Environmental Science and Pollution Research 17, 135–144.

Schauss, K., Focks, A., Leininger, S., Kotzerke, A., Heuer, H., Thiele-Bruhn, S., Sharma, S., Wilke, B.M., Matthies, M., Smalla, K., Munch, J.C., Amelung, W., Kaupenjohann, M., Schloter, M., Schleper, C., 2009. Dynamics and functional relevance of ammonia-oxidizing archaea in two agricultural soils. Environmental Microbiology 11, 446–456.

Shanthala, M., Hosmani, S.P., Hosetti, B.B., 2009. Diversity of phytoplanktons in a waste stabilization pond at Shimoga town, Karnataka State, India. Environmental Monitoring and Assessment 151, 437–443.

Shen, T.L., Liu, L., Li, Y.C., Wang, Q., Dai, J.L., Wang, R.Q., 2019. Long-term effects of untreated wastewater on soil bacterial communities. The Science of the Total Environment 646, 940–950.

Snyder, S.A., Westerhoff, P., Yoon, Y., Sedlak, D.L., 2003. Pharmaceuticals, personal care products and endocrine disruptors in water: implications for the water industry. Environmental Engineering Science 20, 449–469.

Souza, V.L., Almeida, A.A.F.D., Lima, S.G.C., Cascardo, J.C.D.M., Silva, D.D.C., Mangabeira, P.A.O., Gomes, F.P., 2011. Morphophysiological responses and programmed cell death induced by cadmium in Genipa americana L. (*Rubiaceae*). Biometals 24, 59–71.

Spänhoff, B., Bischof, R., Böhme, A., Lorenz, S., Neumeister, K., Nöthlich, A., 2007. Assessing the impact of effluents from a modern wastewater treatment plant on breakdown of coarse particulate organic matter and benthic macroinvertebrates in a Lowland River. Water Air, and Soil Pollution 180, 119–129.

Sridhara, C.N., Kamala, C.T., Samuel Suman Raj, D., 2008. Assessing risk of heavy metals from consuming food grown on sewage irrigated soils and food chain transfer. Ecotoxicology and Environmental Safety 69, 513–524.

Streets, S.S., Henderson, S.A., Stoner, A.D., 2006. Partitioning and bioaccumulation of PBDEs and PCBs in lake Michigan? Environmental Science and Technology 40, 7263–7269.

Tecon, R., Or, D., 2017. Biophysical processes supporting the diversity of microbial life in soil. FEMS Microbiology Reviews 41, 599–623. https://doi.org/10.1093/femsre/fux039.

Tello, A., Austin, B., Telfer, T.C., 2012. Selective pressure of antibiotic pollution on bacteria of importance to public health. Environmental Health Perspectives 120, 1100–1106.

Thiele-Bruhn, S., Beck, I.C., 2005. Effects of sulfonamide and tetracycline antibiotics on soil microbial activity and microbial biomass. Chemosphere 59, 457–465.

Torres, M.A., Borros, M.P., Campos, S.C.G., Pinto, E., Rajamani, S., Sayre, R.T., Colepicolo, P., 2008. Biochemical biomarkers in algae and marine pollution: a review. Ecotoxicology and Environmental Safety 71, 1–15.

Toth, J.D., Feng, Y., Dou, Z., 2011. Veterinary antibiotics at environmentally relevant concentrations inhibit soil iron reduction and nitrification. Soil Biology and Biochemistry 43, 2470–2472.

Tran, N.H., Chen, H., Reinhard, M., Mao, F., Gin, K.Y.H., 2016. Occurrence and removal of multiple classes of antibiotics and antimicrobial agents in biological wastewater treatment processes. Water Research 104, 461–472.

Verlicchi, P., AlAukidy, M., Zambello, E., 2012. Occurrence of pharmaceutical compounds in urban wastewater: removal, mass load and environmental risk after a secondary treatment-A review. The Science of the Total Environment 429, 123–155.

Verlicchi, P., Zambello, E., 2015. Pharmaceuticals and personal care products in untreated and treated sewage sludge: occurrence and environmental risk in the case of application on soil—a critical review. The Science of the Total Environment 538, 750–767.

Wang, F.H., Qiao, M., Su, J.Q., Chen, Z., Zhou, X., Zhu, Y.G., 2014. High throughput profiling of antibiotic resistance genes in urban park soils with reclaimed water irrigation. Environmental Science and Technology 48, 9079–9085.

Wang, M.Y., Chen, A.K., Wong, M.H., Qiu, R.L., Cheng, H., Ye, Z.H., 2011. Cadmium accumulation in and tolerance of rice (*Oryza sativa* L.) varieties with different rates of radial oxygen loss. Environmental Pollution 159, 1730–1736.

Waszak, I., Dabrowska, H., Komar-Szymczak, K., 2014. Comparison of common persistent organic pollutants (POPs) in flounder (*Platichthys flesus*) from the Vistula (Poland) and Douro (Portugal) River estuaries. Marine Pollution Bulletin 81, 225–233.

Wu, C., Spongberg, A.L., Witter, J.D., Fang, M., Czajkowski, K.P., 2010a. Uptake of pharmaceutical and personal care products by soybean plants from soils applied with biosolids and irrigated with contaminated water. Environmental Science and Technology 44, 6157–6161.

Wu, C., Spongberg, A.L., Witter, J.D., Sridhar, B.B.M., 2012. Transfer of wastewater associated pharmaceuticals and personal care products to crop plants from biosolids treated soil. Ecotoxicology and Environmental Safety 85, 104–109.

Wu, G., Kang, H., Zhang, X., Shao, H., Chu, L., Ruan, C., 2010b. A critical review on the bioremoval of hazardous heavy metals from contaminated soils: issues, progress, ecoenvironmental concerns and opportunities. Journal of Hazardous Materials 174, 1–8.

Wu, J.P., Luo, X.J., Zhang, Y., Luo, Y., Chen, S.J., Mai, B.X., Yang, Z.Y., 2008. Bioaccumulation of polybrominated diphenyl ethers (PBDEs) and polychlorinated biphenyls (PCBs) in wild aquatic species from an electronic waste (e-waste) recycling site in South China. Environment International 34, 1109–1113.

Xie, W.Y., McGrath, S.P., Su, J.Q., Hirsch, P.R., Clark, I.M., Shen, Q., Zhu, Y.G., Zhao, F.J., 2016. Long-term impact of field applications of sewage sludge on soil antibiotic resistome. Environmental Science and Technology 50, 12602–12611.

Xie, W., Zhou, J., Wang, H., Chen, X., Lu, Z., Yu, J., 2009. Short-term effects of copper, cadmium and cypermethrin on dehydrogenase activity and microbial functional diversity in soils after long-term mineral or organic fertilization. Agriculture, Ecosystems and Environment 129, 450–456.

Xiong, W.G., Sun, Y.X., Ding, X.Y., Wang, M.Z., Zeng, Z.L., 2015. Selective pressure of antibiotics on ARGs and bacterial communities in manure-polluted freshwater sediment microcosms. Frontiers in Microbiology 6, 194.

Xu, W.H., Yan, W., Li, X.D., Zou, Y.D., Chen, X.X., Huang, W.X., Miao, L., Zhang, R.J., Zhang, G., Zou, S.C., 2013. Antibiotics in riverine runoff of the Pearl River Delta and Pearl River Estuary, China: concentrations, mass loading and ecological risks. Environmental Pollution 182, 402–407.

Yang, X., Flowers, R.C., Weinberg, H.S., Singer, P.C., 2011. Occurrence and removal of pharmaceuticals and personal care products (PPCPs) in an advanced wastewater reclamation plant. Water Research 45, 5218–5228.

Yang, X., Wu, X., Hao, H., He, Z., 2008. Mechanisms and assessment of water eutrophication. Journal of Zhejiang University – Science B. 9, 197–209.

Yang, Y.Y., Xu, C., Cao, X.H., Lin, H., Wang, J., 2017. Antibiotic resistance genes in surface water of eutrophic urban lakes are related to heavy metals, antibiotics, lake morphology and anthropic impact. Ecotoxicology 26, 831–840.

Yu, H.J., Cao, W.P., 2016. Assessment of pharmaceutical and personal care products (PPCPs) of Dalong Lake in Xuzhou by concentration monitoring and bio-effects monitoring process. Environmental Toxicology and Pharmacology 43, 209–215.

Zenobio, J.E., Sanchez, B.C., Archuleta, L.C., Sepulveda, M.S., 2014. Effects of triclocarban, N,N-diethyl-meta-toluamide, and a mixture of pharmaceuticals and personal care products on fathead minnows (*Pimephales promelas*). Environmental Toxicology and Chemistry 33, 910–919.

Zhang, Q.Q., Ying, G.G., Pan, C.G., Liu, Y.S., Zhao, J.L., 2015. Comprehensive evaluation of antibiotics emission and fate in the river basins of China: source analysis, multimedia modeling, and linkage to bacterial resistance. Environmental Science and Technology 49, 6772–6782.

Zhang, Y.L., Dai, J.L., Wang, R.Q., Zhang, J., 2008. Effects of long-term sewage irrigation on agricultural soil microbial structural and functional characterizations in Shandong, China. European Journal of Soil Biology 44, 84–91.

Zhao, L., Dong, Y.H., Wang, H., 2010. Residues of veterinary antibiotics in manures from feedlot livestock in eight provinces of China. The Science of the Total Environment 408, 1069–1075.

Zheng, J., Su, C., Zhou, J., Xu, L., Qian, Y., Chen, H., 2017. Effects and mechanisms of ultraviolet, chlorination, and ozone disinfection on antibiotic resistance genes in secondary effluents of municipal wastewater treatment plants. Chemical Engineering Journal 317, 309–316.

Zhou, L.J., Ying, G.G., Liu, S., Zhang, R.Q., Lai, H.J., Chen, Z.F., Pan, C.G., 2013. Excretion masses and environmental occurrence of antibiotics in typical swine and dairy cattle farms in China. The Science of the Total Environment 444, 183–195.

Zhu, B.K., Chen, Q.L., Chen, S.C., Zhu, Y.G., 2017. Does organically produced lettuce harbor higher abundance of antibiotic resistance genes than conventionally produced? Environment International 98, 152–159.

Žižek, S., Zidar, P., 2013. Toxicity of the ionophore antibiotic lasalocid to soil-dwelling invertebrates: avoidance tests in comparison to classic sublethal tests. Chemosphere 92, 570–575.

CHAPTER

Human health hazards of wastewater

6

Bing Wu, PhD

State Key Laboratory of Pollution Control and Resource Reuse, School of the Environment, Nanjing University, Nanjing, China

Chapter outline

A growing demand for water to produce food, supply industries, and support other needs of human beings lead to competition for scarce freshwater resource. Under the premise of water shortage, wastewater was considered as replacements and used in a lot of scenarios. Wastewater has become the main source of pollution in the environment, which is commonly discharged into bodies of water with little or no treatment due to the limited availability of treatment facilities in many countries (Qadir et al., 2010). When it comes to the negative influence of wastewater on human health, existing research mainly focused on irrigation usage of wastewater in agriculture, which may lead to accumulation of heavy metals and organic pollutants in food thereby affecting human health. In addition, occurrence of antibiotics and antibiotic resistance genes in wastewater has become the main point worthy to pay attention due to the negative effect on human. Moreover, health risks of oily wastewater

High-Risk Pollutants in Wastewater. https://doi.org/10.1016/B978-0-12-816448-8.00006-X

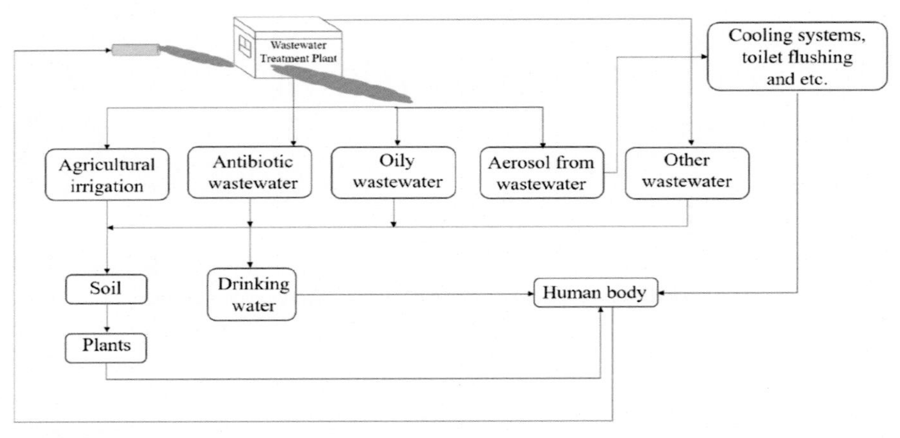

FIGURE 6.1

Exposure pathways of wastewater to human.

and aerosol in reclaimed water have also attracted public attentions. These wastewater exposure pathways to humans are illustrated in (Fig. 6.1). A wide spectrum of health challenges must be carefully evaluated. The main purpose of this chapter is to review the potential human health hazards of different wastewater.

6.1 Agricultural irrigation

6.1.1 Exposure pathways

The use of wastewater for agricultural purposes may be particularly attractive, as this may allow for intensive agriculture while preserving limited resources of high quality water for rapid urban growth. The agricultural sector is currently the largest consumer of water and wastewater globally, accounted for approximately 70% on average (Winpenny et al., 2010). Therefore, the potential risks cannot be ignored. Agricultural soil could be contaminated by toxic elements like heavy metals and organic pollutants, such as cadmium, polychlorinated biphenyls (PCBs) and polycyclic aromatic hydrocarbons (PAHs) during irrigation activities.

A range of exposure pathways associated with wastewater irrigation was identified and illustrated in Fig. 6.2. Farmworker exposure is significant due to multiple pathways, higher contact frequency and longer exposure time. Many studies have evaluated farmworker exposure risks (Forslund et al., 2010; Zhang et al., 2013), and potential impact on family members was also considered (An et al., 2007). Classified by degree of mechanization, direct exposure included planting and weeding (Trang et al., 2007) along with variable exposures were associated with different irrigation methods such as gravity flow irrigation and manual irrigation with buckets (Rutkowski et al., 2007). Enrichment of toxic elements in agricultural soil can be

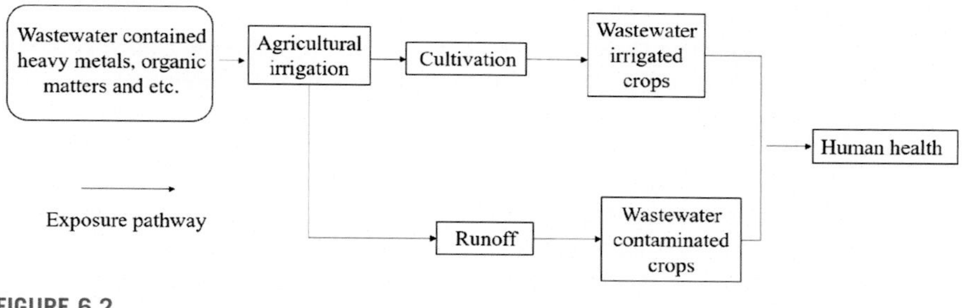

FIGURE 6.2

Exposure pathways associated with wastewater irrigation.

harmful to soil fertility (Maxted et al., 2007). Many studies have shown that soils have been contaminated by heavy metals, which led to the enrichment of heavy metals in food crops such as corn and wheat. Heavy metals transferring from soil into food crops play important roles in increasing exposure of human beings to potential toxic elements. Furthermore, organic contaminants are enriched in the soil that can be transferred to the vegetables through food chain (Haddaoui et al., 2016; Khan et al., 2008). It is worthy to note that consumption of food is the most common exposure for ordinary people.

6.1.2 Impacts of heavy metals

In most of the developing countries, the wastewater discarded from industrial or residential areas is rarely treated or only involves primary processes. Under these circumstances, the effluents are often polluted with heavy metals. Studies of the health impact of wastewater reveal various gastrointestinal problems in the farming communities irrigated with heavy metal wastewater. Heavy metals are highly toxic because of their good solubility in water, long biological half-lives, and the potential to accumulate in different body parts. The toxicity can be acute or chronic dependent upon exposure time and concentration (Dorne et al., 2011; Jarup, 2003).

The most common toxic heavy metals in wastewater include lead (Pb), mercury (Hg), cadmium (Cd), chromium (Cr), copper (Cu), nickel (Ni), silver (Ag), manganese (Mg), and zinc (Zn). Gastrointestinal effects, renal impairment, neurological disorders, cardiovascular troubles, bone problems, convulsions, and paralysis are the typical adverse effects induced by heavy metals in wastewater. Accumulation of heavy metals beyond permissible limits has the potential to affect vital organs such as kidneys, bones, liver, and blood and causes serious health hazards. Ahmad et al. (2018) reported that Pb, Cr, Cu, and Mn in wastewater for agricultural lands presented toxic repercussions by inducing respiratory, digestive, excretory, immune, and glandular malfunctions. Khan et al. (2013) explored the levels of heavy metals in vegetables irrigated with wastewater and found that concentrations of Cu, Cd, Ni, and Mn in wastewater were considerably higher than the permissible limits. Parveen et al. (2012) also examined the effects of heavy metals in wastewater on vegetables in Peshawar, and identified Pb, Ni, and Cd as the core cause of contamination in soil samples of the studied area, which suggested ingestion of these vegetables was unsafe. Accumulation of Cd, particularly in the kidneys, leads to kidney damage and osteoporosis. This is known as itai-itai disease in Japan, where the condition was originally linked to irrigation of rice paddies with highly contaminated water (Jarup, 2003). Some reports have linked skeletal damage (osteoporosis) in humans to heavy metals, such as high levels of selenium. Consumption of vegetables and fish contaminated with the heavy metals such as Cd, Cr, Cu, Hg, Pb, and Zn is the most likely route for human exposure in Tianjin, China (Wang et al., 2005). It will cause a variety of pulmonary adverse health effects, such as lung inflammation, fibrosis, emphysema, and tumors. Furthermore, they investigated that the Cd and Hg are the major contributors counting for 45% and 51% to the target hazard quotients. Balkhair et al. (2016) found that the cultivated okra plant irrigated with wastewater increased the heavy metal contents in both the soil and vegetables. The concentrations of Ni, Pb, Cd, and Cr in the edible portions were above the safe limit in 90%, 28%, 83%, and 63% of the samples, respectively. This could induce chronic human hazards due to consumption of the okra. Furthermore, heavy metals can coexist with other ions and easily form complexes with complexing agents such as ethylene diamine tetraacetic acid, which exacerbated the toxicity (Han et al., 2015) and pose a potential threat to human health (Zuo et al., 2016).

6.1.3 Impacts of organic pollutants

6.1.3.1 Polychlorinated biphenyls

Organic pollutants in wastewater such as PCBs may accumulate in soil during irrigation and magnify through the food chain. PCBs have been classified by the International Agency for Research on Cancer as probably carcinogenic to humans and may have also noncarcinogenic health effects, including dermal lesions, immune system disorders, hepatotoxicity, immunosuppression, reproductive and developmental toxicity, endocrine disruption, neurotoxicity, and carcinogenicity endocrine disruptors (Donato et al., 2006; Borghini et al., 2005; Schuhmacher et al., 2004). PCBs tend to accumulate in fatty tissues in humans because of their nonpolar, lipophilic physical properties and their resistance to biochemical degradation. Li et al. (2008) conducted the risk assessments of PCBs in wastewater, which indicated that intake of fish contaminated with PCBs would not pose a health risk to humans with a consumption of 7.4 ± 8.6 g/person/day according to their acceptable daily intake and minimal risk level. However, the hazardous ratio of the 95th percentile for PCBs in fish from Gaobeidian Lake exceeded 1, which suggested that daily exposure to PCBs had a lifetime cancer risk over 1 in 1,000,000. Other researchers also found that people living in polluted areas have higher values of PCBs in their blood compared to those living in less polluted areas (Franchini et al., 2004; Orloff et al., 2003; Pavuk et al., 2004). High levels of PCBs about 40,000 ng/g of lipids in the contaminated people were higher than, 50 times of the median levels in uncontaminated people (Donato et al., 2006). They found that the main source of human PCBs contamination was the consumption of food contaminated with wastewater contained PCBs.

6.1.3.2 Polycyclic aromatic hydrocarbons

PAHs are mainly derived from incomplete combustion of coal and oil, and wastewater discharge is one of the main channels for PAHs to enter the environment. Long-term irrigation with industrial effluent mixed with municipal wastewater (Cai et al., 2007; Villar et al., 2006) has resulted in the excessive accumulation of PAHs in agricultural soil. In addition, part of PAHs in the air could directly enter the soil by atmospheric deposition. The exposure pathways are illustrated in Fig. 6.3. PAHs in the soils have brought a potential risk to human health along the food chain. Ingestion is considered as the major source of PAHs (McGrath et al., 2007). Vegetables cultivated on the wastewater-contaminated soils may take up these pollutants in sufficient quantities. Numerous studies have demonstrated that vegetables accumulated high concentrations of PAHs that were grown in PAH-contaminated soils (Samsoe-Petersen et al., 2002; Wennrich et al., 2002). Additionally, other pathways, including inhalation and dermal contact, also contribute to the human exposure to environmental carcinogenic PAHs (Chen, Liao, 2006).

PAHs intake could interfere with the cellular membrane functions and enzyme systems associated with the membrane. The major concern of PAHs is about the epoxides and dihydrodiols that can bind with the cellular proteins and DNA leading to biochemical disruptions, cell damage, mutations, developmental malformations, tumors, and cancer. To quantify the carcinogenicity of the selected PAHs, toxic equivalency factor compared to benzo(a)pyrene was applied.

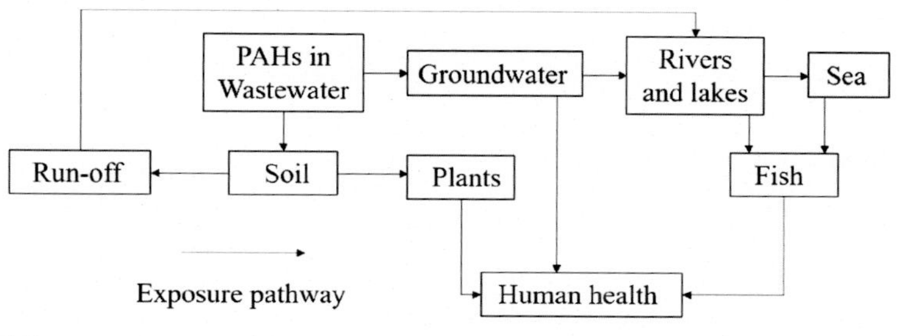

FIGURE 6.3

Exposure pathways of PAHs from wastewater.

6.2 Antibiotic wastewater

6.2.1 Exposure pathways

Antibiotic wastewater is one of the important types of industrial wastewater. With the urgent needs of wastewater reuse, people started to care about the safety of antibiotic wastewater. As we know, frequent use of antibiotics may lead to drug resistance and dysbacteriosis in the body. Some antibiotics, such as penicillin, may cause special toxicity or even death of human. Nowadays, continuous discharge of antibiotics from wastewater treatment plant, animal waste, and other sources raises concerns about their potential adverse ecotoxicological effects and human health risks. Fig. 6.4 depicts the possible routes of antibiotics and antibiotic resistance genes from wastewater treatment plant to consumers. In many cases, discharge points of wastewater treatment plant located upstream of a pumping station, and they were the main source of antibiotics and antibiotic resistance genes in the water supplies. People who live downstream will have great possibility to get in touch with this kind of wastewater. Medical antibiotics consumption and veterinary antibiotics that may enter soil and water with the surface runoff are other sources of antibiotics (Fig. 6.4).

6.2.2 Health hazards of antibiotic wastewater

Current wastewater treatment techniques are unable to completely remove antibiotics. Antibiotics remaining in drinking water will affect body immune system and weaken immunity. For example, β-lactam antibiotics such as penicillin can cause allergic reactions to human bodies. Gentamicin is very toxic to the kidney. Quinolones can increase the sensitivity of human body to light, and tetracyclines

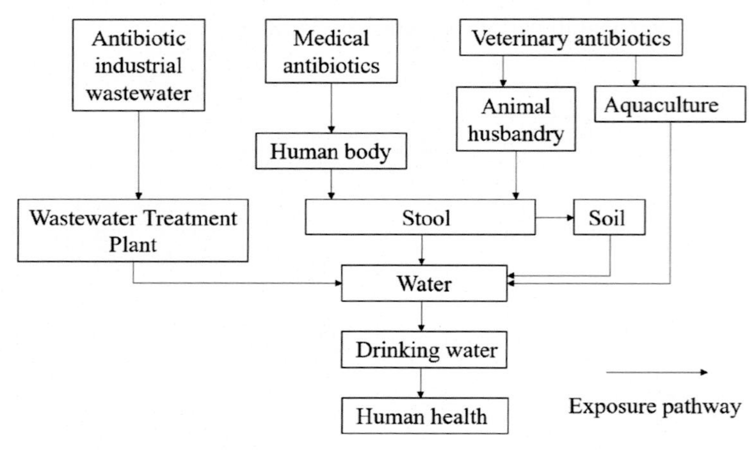

FIGURE 6.4

Exposure pathways of antibiotics and antibiotic resistance genes to human in wastewater.

can seriously affect the development of children's teeth (Kuemmerer, 2009). For children, some important organs are not yet completely mature, and quinolone antibacterials may have adverse effects on bones, liver, and kidney. Although antibiotics in drinking water commonly exist at trace levels, they will still affect immune system and have the potential to weaken immunity (Jones et al., 2005)

Aquatic organisms in the water receiving treated wastewater are exposed to a mixture of antibiotic residues, not just single compound. The presence of antibiotic residues in wastewater, activated sludge, digested sludge, and urban biosolids has become an increasingly recognized environmental risk (Zhang,Li, 2011). Crops absorb antibiotics from polluted soils due to wastewater irrigation or manure amended soils through fertilization. High-level accumulation of antibiotics in food may raise potential human health concerns through the food chain. The concentration of antibiotics can be enriched in high-nutrient organisms and resulted in higher concentrations than in the environment. For human beings, consumption of food contained antibiotics such as fish and vegetables may cause allergic reactions, even food poisoning in severe cases. For example, 634 samples of various animal source foods, including beef, chevon, pork, mutton, and egg, were screened for drug residues and the positive rate was 30.8%, 29.3%, 28.6%, 24%, and 6.8%, respectively (Donkor et al., 2011), indicating that the residue of veterinary drugs would seriously threaten people's lives.

Furthermore, the low level of antibiotics in drinking water contributes to generation of antibiotic resistance genes, especially for children and infants. The danger of drinking water contained antibiotics is not only the antibiotics itself, but also the drug resistance gene produced by antibiotics. Once these resistance genes are transmitted to the pathogenic bacterial, the bacteria may mutate into super bacteria, which can cause infections among human beings, even outbreaks of infectious diseases. Although antibiotic wastewater does not immediately cause illness, it is undoubtedly a potential enemy of human health. For example, antibiotic resistance genes against ampicillin, streptomycin, and tetracycline are known to be transferable to other bacteria (Chopra, Roberts, 2001; Walia et al., 2004). Thus, further investigations on the adverse human health effects of antibiotics and antibiotic resistance genes in wastewater are warranted.

6.3 Oily wastewater

6.3.1 Source of oily wastewater

Oil enters the water environment through different routes and eventually, generating oily wastewater. Oily wastewater is one kind of wastewater with features like large amount, wide surface, complex composition, and serious hazard. It originates from various sources such as the food, vegetable oil, metal, textile and leather industries, oil and gas production, domestic sewage, kitchens, and vehicles (Nishi et al., 2002; Santander et al., 2011).

Oils in wastewater exist in various forms such as fats, lubricants, cutting oils, heavy hydrocarbons, and light hydrocarbons (Srinivasan, Viraraghavan, 2008).

They were transferred to the environment through transportation leaking, refining, spill, and normal discharge (Banerjee et al., 2006; Ibrahim et al., 2010). It is estimated that the total annual oil pollution through transportation of petroleum hydrocarbons alone could reach up to 10 million metric tons (Banerjee et al., 2006). Oily wastewater concentration generated by industrial activities could reach as high as 40,000 mg/L. It has become one of the environmental concerns nowadays. Oil is usually removed from wastewater before discharge to the environment to meet the maximum allowable limit of oil and grease in water as required by local enforcing agency. However, oily wastewater from kitchens and small enterprises is commonly discharged without any prior treatment and hard to administrate. Oil contains many carcinogens, which are enriched by fish, shellfish, and other organisms in the wastewater. They will pose a threat to human health through the food chain ultimately, which is a major cause of increased cancer incidence in human being.

6.3.2 Health hazards of oily wastewater

Oily wastewater, including natural and synthetic sources, could cause health hazards if it is not appropriately treated (Simonovic et al., 2009). When the oil content in seawater reaches 0.01 mg/L, fish, shrimp, and shellfish can be odorous within 24 h. When people consume carcinogens derived from petroleum hydrocarbons, especially those contaminated with polycyclic aromatic hydrocarbons, these carcinogens can endanger human health through the delivery of food chains.

Oil pollutants can affect the normal function organs through inhalation, leading to a variety of diseases. Children who were exposed to oil are under four times higher risk of developing leukemia and seven times higher risks to have acute nonlymphocytic leukemia than normal children. In the vicinity of oil-contaminated pollution, children's skin resistance is significantly weakened, along with fewer white blood cells and increasing blood spray rate. In addition, lung function is affected. The probability of hepatoma in the general population is significantly higher than that in the control area. The mortality rate of malignant tumors, especially the digestive system malignant tumors were significantly higher than that of the control area.

Many studies have been conducted to evaluate the effect of oily wastewater on human health and environment. Most of the studies focused on oil wastewater, which is derived from hydrocarbon oil spill in marine environment. A review by Aguilera et al. (2010) has suggested that oil compounds have the ability for bioaccumulation, as they are transferable along the food chain via oil-contaminated marine food. *In vitro* studies on fuel extract from Erika oil spill showed that consumption of oil-contaminated marine food could cause detrimental effects on human health. For example, Amat-Bronnert et al. (2007) conducted *in vitro* genotoxicity research of Erika fuel extract with human epithelial bronchial cells and human hepatoma cells, and demonstrated that the Erika fuel extract is genotoxic. Another research showed that fuel extracts were the induction of metabolic enzymes involved in carcinogenic processes (Amat-Bronnert et al., 2007) and generated genotoxic damage in consumers (Ha et al., 2012; Lemiere et al., 2005). Aguilera et al.

(2010) also summarized the acute effects and psychological symptoms reported by previous researchers who studied the effect of hydrocarbon oil spill. They found that hydrocarbon oily wastewater did affect human health, especially to those who had direct bare-hand contact with the oil. Studies have also suggested that the urinary volatile organic compounds and PAH metabolite levels in oil spill cleanup volunteers are increased after cleanup (Ha et al., 2012). Long-term bare-handed contact with hydrocarbon oil wastewater leads to an increased risk of developing skin tumors, due to exposure to carcinogenic properties of hydrocarbon oil (Baars, 2002).

6.4 Aerosol from reclaimed water

6.4.1 Generation of aerosol

The reclaimed water has many reuse scenarios such as cooling systems, toilet flushing, spray irrigation, wash vehicles, and so on. It is easily aerosolized and inhaled by surrounding people during these applications. Due to the limitations of existing treatment processes, the reclaimed water still contains various toxic and harmful organic substances, including carcinogenic PAHs, which may threaten public health (Dickin et al., 2016; Hamilton et al., 2018; Sano et al., 2016). Specifically speaking, reclaimed water for cooling system purposes represents the largest industrial wastewater reuse application. Cooling systems may consume 20%–50% of a facility's water usage. Common uses of reclaimed water such as spray irrigation or cooling towers can produce aerosols that are of concern because contaminants can travel beyond the immediate vicinity of application (Li et al., 2011). Additionally, toilet flushing has been identified as a top use of recycled water in a survey of 10 United States recycled water systems (Jjemba et al., 2015). Toilet flushing may generate aerosols associated with health risks (Fewtrell and Kay, 2007; Gerba et al., 1975; Hamilton et al., 2017). Urban recycled wastewater is prone to form aerosol during spraying and generate bioaerosols during wastewater treatment process. The human body is mainly exposed to toxic and harmful substance through inhalation, skin contact absorption, and ingestion through the digestive tract, which may have genotoxic effects and thus its potential health risks have received wide attention.

6.4.2 Effect of aerosols on health

There are many harmful substances in the aerosol generated from the reclaimed water, which may cause inflammatory responses in the lungs after inhalation. Endotoxin, which is a component of the cell wall of gram-negative bacterial, has been identified as the main inducer. Furthermore, *Legionella* spp. grow in biofilms in piping that may slough off and become aerosolized through water fixtures. Inhalation of this bacterial will make people get sick, such as cough, chills, breathing difficulties, fever, muscle pain, and joint ache. Existing researches (Xue et al., 2016; Zhang et al., 2016) have demonstrated that endotoxins in reclaimed water could induce acute lung inflammation, which may result in lung damage under long-term exposure. Zhang et al. (2016) have conducted experiments on mice and found that the acute inhibition exposure of endotoxins in mice could induce tumor necrosis factor-α (TNF-α) and interleukin-6 (IL-6) in the lungs and cause polymorphonuclear cell-dominated inflammation. Persistent inflammation in the lungs has proven to be closely related with other diseases, such as chronic obstructive pulmonary disease, which is a major reason of morbidity and mortality in the United States (Lee et al., 2009). Chronic inflammation indirectly promoted apoptosis resistance and angiogenesis and was ultimately capable of inducing a protumor microenvironment (Montuenga and Pio, 2007; Raghuwanshi et al., 2008). Xue et al. (2016) found that acute inhalation of reclaimed water can induce strong polymorphonuclear cell (PMN)-dominant inflammation in the lungs, and the PMN% in bronchioalveolar lavage fluid is the most sensitive indicator at 3 h post-exposure. Xue et al. (2018) also found the similar inflammation after subchronic exposure of 8 weeks. All of these research proved that induction of aerosol in reclaimed water can produce inflammation in the lungs along with other health hazards, which bring a huge threat to human beings.

Bioaerosols might be a vehicle for the dissemination of human and animal pathogens from wastewater. Their presence in the air might pose a potential epidemiological threat. Researchers found that wastewater treatment plant employees have a markedly higher prevalence than others of the so-called "sewage worker syndrome" including fatigue, headache, dizziness, gastrointestinal symptoms, and respiratory symptoms. Many graywater treatment systems, such as recirculating vertical flow constructed wetlands, can create bioaerosols, which might compromise human health if critical amounts are inhaled, ingested, or come into contact with human skin (Gross et al. 2007, 2008, 2015). Quantitative microbial risk assessment was used to estimate potential risks from possible exposure to graywater aerosols and compare it to the tolerable risk as postulated in the World Health Organization (WHO) guidelines for wastewater reuse. A significant fraction of particles from 0.5 to 10 μm can enter the lung or gastrointestinal tissue (Thomas et al., 2008), consequently, they have been been categorized by the US Environmental Protection Agency (EPA) as particles that may not be filtered by the lungs but rather deposited there or ingested (Olin, 1998). Infectious pathogens, such as viruses and bacteria, are within the size range that can be carried in these aerosols (Bowers et al., 2011). Via exposure to skin, inhalation, or ingestion, they can have a potentially negative impact on human (Jeppesen, 1996; Stellacci et al., 2010).

6.5 Summary

Contaminants in wastewater have the access to the human body through many pathways, such as ingestion through food chain, inhalation by aerosol, and direct contact through drinking and swimming. Therefore, we need to pay special attention to the contaminants in the wastewater. At the same time, we also need to study potential hazards of various types of wastewater on human health. In addition to the wastewater mentioned above, research on the health risks of other wastewater and toxicity throughout the life cycle is also needed.

References

Aguilera, F., Mendez, J., Pasaro, E., Laffon, B., 2010. Review on the effects of exposure to spilled oils on human health. Journal of Applied Toxicology 30 (4), 291−301.

Ahmad, K., Nawaz, K., Iqbal Khan, Z., Nadeem, M., Wajid, K., Ashfaq, A., Munir, B., Memoona, H., Sana, M., Shaheen, F., Kokab, R., Rehman, S., Fahad Ullah, M., Mahmood, N., Muqadas, H., Aslam, Z., Shezadi, M., Noorka, I., Bashir, H., Dogan, Y., 2018. Effect of diverse regimes of irrigation on metals accumulation in wheat crop: an assessment − dire need of the day. Scientific Publications 27, 846−855.

Amat-Bronnert, A., Castegnaro, M., Pfohl-Leszkowicz, A., 2007. Genotoxic activity and induction of biotransformation enzymes in two human cell lines after treatment by Erika fuel extract. Environmental Toxicology and Pharmacology 23 (1), 89−95.

An, Y.-J., Yoon, C.G., Jung, K.-W., Ham, J.-H., 2007. Estimating the microbial risk of E-coli in reclaimed wastewater irrigation on paddy field. Environmental Monitoring and Assessment 129 (1−3), 53−60.

Baars, B.J., 2002. The wreckage of the oil tanker 'Erika' - human health risk assessment of beach cleaning, sunbathing and swimming. Toxicology Letters 128 (1−3), 55−68.

Balkhair, K.S., Ashraf, M.A., 2016. Field accumulation risks of heavy metals in soil and vegetable crop irrigated with sewage water in western region of Saudi Arabia. Saudi Journal of Biological Sciences 23 (1), S32−S44.

Banerjee, S.S., Joshi, M.V., Jayaram, R.V., 2006. Treatment of oil spill by sorption technique using fatty acid grafted sawdust. Chemosphere 64 (6), 1026−1031.

Borghini, F., Grimalt, J.O., Sanchez-Hernandez, J.C., Bargagli, R., 2005. Organochlorine pollutants in soils and mosses from Victoria Land (Antarctica). Chemosphere 58 (3), 271−278.

Bowers, R.M., Sullivan, A.P., Costello, E.K., Collett Jr., J.L., Knight, R., Fierer, N., 2011. Sources of bacteria in outdoor air across cities in the midwestern United States. Applied and Environmental Microbiology 77 (18), 6350−6356.

Cai, Q.-Y., Mo, C.-H., Wu, Q.-T., Zeng, Q.-Y., Katsoyiannis, A., 2007. Occurrence of organic contaminants in sewage sludges from eleven wastewater treatment plants, China. Chemosphere 68 (9), 1751−1762.

Chen, S.-C., Liao, C.-M., 2006. Health risk assessment on human exposed to environmental polycyclic aromatic hydrocarbons pollution sources. Science of the Total Environment 366 (1), 112−123.

Chopra, I., Roberts, M., 2001. Tetracycline antibiotics: mode of action, applications, molecular biology, and epidemiology of bacterial resistance. Microbiology and Molecular Biology Reviews 65 (2), 232−+.

Dickin, S.K., Schuster-Wallace, C.J., Qadir, M., Pizzacalla, K., 2016. A review of health risks and pathways for exposure to wastewater use in agriculture. Environmental Health Perspectives 124 (7), 900−909.

Donato, F., Magoni, M., BergonZi, R., Scarcella, C., Indelicato, A., Carasi, S., Apostoli, P., 2006. Exposure to polychlorinated biphenyls in residents near a chemical factory in Italy: the food chain as main source of contamination. Chemosphere 64 (9), 1562–1572.

Donkor, E.S., Newman, M.J., Tay, S.C.K., Dayie, N.T.K.D., Bannerman, E., Olu-Taiwo, M., 2011. Investigation into the risk of exposure to antibiotic residues contaminating meat and egg in Ghana. Food Control 22 (6), 869–873.

Dorne, J.-L.C.M., Kass, G.E.N., Bordajandi, L.R., Amzal, B., Bertelsen, U., Castoldi, A.F., Heppner, C., Eskola, M., Fabiansson, S., Ferrari, P., Scaravelli, E., Dogliotti, E., Fuerst, P., Boobis, A.R., Verger, P., 2011. Human risk assessment of heavy metals: principles and applications. Metal Ions in Life Sciences 8, 27–60.

Fewtrell, L., Kay, D., 2007. Quantitative microbial risk assessment with respect to *Campylobacter* spp. in toilets flushed with harvested rainwater. Water and Environment Journal 21 (4), 275–280.

Forslund, A., Ensink, J.H.J., Battilani, A., Kljujev, I., Gola, S., Raicevic, V., Jovanovic, Z., Stikic, R., Sandei, L., Fletcher, T., Dalsgaard, A., 2010. Faecal contamination and hygiene aspect associated with the use of treated wastewater and canal water for irrigation of potatoes (*Solanum tuberosum*). Agricultural Water Management 98 (3), 440–450.

Franchini, M., Rial, M., Buiatti, E., Bianchi, F., 2004. Health effects of exposure to waste incinerator emissions:a review of epidemiological studies. Annali dell'Istituto Superiore di Sanita 40 (1), 101–115.

Gerba, C.P., Wallis, C., Melnick, J.L., 1975. Microbiological hazards of household toilets-droplet production and fate of residual organismsI. Applied Microbiology 30 (2), 229–237.

Gross, A., Shmueli, O., Ronen, Z., Raveh, E., 2007. Recycled vertical flow constructed wetland (RVFCW) - a novel method of recycling greywater for irrigation in small communities and households. Chemosphere 66 (5), 916–923.

Gross, A., Sklarz, M.Y., Yakirevich, A., Soares, M.I.M., 2008. Small scale recirculating vertical flow constructed wetland (RVFCW) for the treatment and reuse of wastewater. Water Science and Technology 58 (2), 487–494.

Gross, A., Maimon, A., Alfiya, Y., Friedler, E., 2015. Greywater Reuse. CRC Press, Boca Raton, Washington, D.C.

Ha, M., Kwon, H., Cheong, H.-K., Lim, S., Yoo, S.J., Kim, E.-J., Park, S.G., Lee, J., Chung, B.C., 2012. Urinary metabolites before and after cleanup and subjective symptoms in volunteer participants in cleanup of the Hebei Spirit oil spill. The Science of the Total Environment 429, 167–173.

Haddaoui, I., Mahjoub, O., Mahjoub, B., Boujelben, A., Bella, G.D., 2016. Occurrence and distribution of PAHs, PCBs, and chlorinated pesticides in Tunisian soil irrigated with treated wastewater. Chemosphere 146, 195–205.

Hamilton, K.A., Ahmed, W., Toze, S., Haas, C.N., 2017. Human health risks for Legionella and *Mycobacterium avium* complex (MAC) from potable and non-potable uses of roof-harvested rainwater. Water Research 119, 288–303.

Hamilton, K.A., Hamilton, M.T., Johnson, W., Jjemba, P., Bukhari, Z., LeChevallier, M., Haas, C.N., 2018. Health risks from exposure to *Legionella* in reclaimed water aerosols: toilet flushing, spray irrigation, and cooling towers. Water Research 134, 261–279.

Han, W., Fu, F., Cheng, Z., Tang, B., Wu, S., 2015. Studies on the optimum conditions using acid-washed zero-valent iron/aluminum mixtures in permeable reactive barriers for the removal of different heavy metal ions from wastewater. Journal of Hazardous Materials 302, 437.

Ibrahim, S., Wang, S., Ang, H.M., 2010. Removal of emulsified oil from oily wastewater using agricultural waste barley straw. Biochemical Engineering Journal 49 (1), 78–83.

Jarup, L., 2003. Hazards of heavy metal contamination. British Medical Bulletin 68, 167–182.

Jeppesen, B., 1996. Domestic greywater re-use: Australia's challenge for the future. Desalination 106 (1–3), 311–315.

Jjemba, P.K., Johnson, W., Bukhari, Z., LeChevallier, M.W., 2015. Occurrence and control of *Legionella* in recycled water systems. Pathogens 4 (3), 470–502.

Jones, O.A., Lester, J.N., Voulvoulis, N., 2005. Pharmaceuticals: a threat to drinking water? Trends in Biotechnology 23 (4), 163–167.

Khan, S., Aijun, L., Zhang, S., Hu, Q., Zhu, Y.-G., 2008. Accumulation of polycyclic aromatic hydrocarbons and heavy metals in lettuce grown in the soils contaminated with long-term wastewater irrigation. Journal of Hazardous Materials 152 (2), 506–515.

Khan, A., Javid, S., Muhmood, A., Mjeed, T., Niaz, A., Majeed, A., 2013. Heavy metal status of soil and vegetables grown on peri-urban area of Lahore district. Soil and Environment 32, 49–54.

Kuemmerer, K., 2009. Antibiotics in the aquatic environment – a review – Part I. Chemosphere 75 (4), 417–434.

Lee, G., Walser, T.C., Dubinett, S.M., 2009. Chronic inflammation, chronic obstructive pulmonary disease, and lung cancer. Current Opinion in Pulmonary Medicine 15 (4), 303–307.

Lemiere, S., Cossu-Leguille, C., Bispo, A., Jourdain, M.J., Lanhers, M.C., Burnel, D., Vasseur, P., 2005. DNA damage measured by the single-cell gel electrophoresis (comet) assay in mammals fed with mussels contaminated by the 'Erika' oil-spill. Mutation Research/Genetic Toxicology and Environmental Mutagenesis 581 (1–2), 11–21.

Li, X., Gan, Y., Yang, X., Zhou, J., Dai, J., Xu, M., 2008. Human health risk of organochlorine pesticides (OCPs) and polychlorinated biphenyls (PCBs) in edible fish from Huairou Reservoir and Gaobeidian Lake in Beijing, China. Food Chemistry 109 (2), 348–354.

Li, H., Chien, S.-H., Hsieh, M.-K., Dzombak, D.A., Vidic, R.D., 2011. Escalating water demand for energy production and the potential for use of treated municipal wastewater. Environmental Science and Technology 45 (10), 4195–4200.

Maxted, A.P., Black, C.R., West, H.M., Crout, N.M.J., McGrath, S.P., Young, S.D., 2007. Phytoextraction of cadmium and zinc from arable soils amended with sewage sludge using Thlaspi caerulescens: development of a predictive model. Environmental Pollution 150 (3), 363–372.

McGrath, T.E., Wooten, J.B., Chan, W.G., Hajaligol, M.R., 2007. Formation of polycyclic aromatic hydrocarbons from tobacco: the link between low temperature residual solid (char) and PAH formation. Food and Chemical Toxicology 45 (6), 1039–1050.

Montuenga, L.M., Pio, R., 2007. Tumour-associated macrophages in nonsmall cell lung cancer: the role of interleukin-10. European Respiratory Journal 30 (4), 608–610.

Nishi, Y., Iwashita, N., Sawada, Y., Inagaki, M., 2002. Sorption kinetics of heavy oil into porous carbons. Water Research 36 (20), 5029–5036.

Olin, S.S., 1998. Exposure to Contaminants in Drinking Water: Estimating Uptake through the Skin and by Inhalation. Crc Press. International Life Sciences Institute, Washington, DC, USA.

Orloff, K.G., Dearwent, S., Metcalf, S., Kathman, S., Turner, W., 2003. Human exposure to polychlorinated biphenyls in a residential community. Archives of Environmental Contamination and Toxicology 44 (1), 125–131.

Pavuk, M., Cerhan, J.R., Lynch, C.F., Schecter, A., Petrik, J., Chovancova, J., Kocan, A., 2004. Environmental exposure to PCBs and cancer incidence in eastern Slovakia. Chemosphere 54 (10), 1509–1520.

Perveen, S., Samad, A., Nazif, W., Shah, S., 2012. Impact of sewage water on vegetables quality with respect to heavy metals in peshawar Pakistan. Pakistan Journal of Botany 44 (6), 1923–1931.

Qadir, M., Wichelns, D., Raschid-Sally, L., McCornick, P.G., Drechsel, P., Bahri, A., Minhas, P.S., 2010. The challenges of wastewater irrigation in developing countries. Agricultural Water Management 97 (4), 561–568.

Raghuwanshi, S.K., Nasser, M.W., Chen, X., Strieter, R.M., Richardson, R.M., 2008. Depletion of beta-arrestin-2 promotes tumor growth and angiogenesis in a murine model of lung cancer. The Journal of Immunology 180 (8), 5699–5706.

Rutkowski, T., Raschid-Sally, L., Buechler, S., 2007. Wastewater irrigation in the developing world - two case studies from the Kathmandu Valley in Nepal. Agricultural Water Management 88 (1–3), 83–91.

Samsoe-Petersen, L., Larsen, E.H., Larsen, P.B., Bruun, P., 2002. Uptake of trace elements and PAHs by fruit and vegetables from contaminated soils. Environmental Science and Technology 36 (14), 3057–3063.

Sano, D., Amarasiri, M., Hata, A., Watanabe, T., Katayama, H., 2016. Risk management of viral infectious diseases in wastewater reclamation and reuse: Review. Environment International 91, 220–229.

Santander, M., Rodrigues, R.T., Rubio, J., 2011. Modified jet flotation in oil (petroleum) emulsion/water separations. Colloids and Surfaces A: Physicochemical and Engineering Aspects 375 (1–3), 237–244.

Schuhmacher, M., Nadal, M., Domingo, J.L., 2004. Levels of PCDD/Fs, PCBs, and PCNs in soils and vegetation in an area with chemical and petrochemical industries. Environmental Science and Technology 38 (7), 1960–1969.

Simonović, B.R., Aranđelovic, D., Jovanovic, M., Kovačevic, B., Pezo, L., Jovanovic, A., 2009. Removal of mineral oil and wastewater pollutants using hard coal. Chemical Industry and Chemical Engineering Quarterly 15 (2), 57–62.

Srinivasan, A., Viraraghavan, T., 2008. Removal of oil by walnut shell media. Bioresource Technology 99 (17), 8217–8220.

Stellacci, P., Liberti, L., Notarnicola, M., Haas, C.N., 2010. Hygienic sustainability of site location of wastewater treatment plants A case study. I. Estimating odour emission impact. Desalination 253 (1–3), 51–56.

Thomas, R.J., Webber, D., Sellors, W., Collinge, A., Frost, A., Stagg, A.J., Bailey, S.C., Jayasekera, P.N., Taylor, R.R., Eley, S., Titball, R.W., 2008. Characterization and deposition of respirable large- and small-particle bioaerosols. Applied and Environmental Microbiology 74 (20), 6437–6443.

Trang, D.T., Molbak, K., Cam, P.D., Dalsgaard, A., 2007. Helminth infections among people using wastewater and human excreta in peri-urban agriculture and aquaculture in Hanoi, Vietnam. Tropical Medicine and International Health 12, 82–90.

Villar, P., Callejon, M., Alonso, E., Jimenez, J.C., Guiraum, A., 2006. Temporal evolution of polycyclic aromatic hydrocarbons (PAHs) in sludge from wastewater treatment plants: comparison between PAHs and heavy metals. Chemosphere 64 (4), 535–541.

Walia, S.K., Alan, K., Mohinder, P., G Rasul, C., 2004. Self-transmissible antibiotic resistance to ampicillin, streptomycin, and tetracyclin found in *Escherichia coli* isolates from contaminated drinking water. Environmental Letters 39 (3), 651–662.

Wang, X.L., Sato, T., Xing, B.S., Tao, S., 2005. Health risks of heavy metals to the general public in Tianjin, China via consumption of vegetables and fish. The Science of the Total Environment 350 (1–3), 28–37.

Wennrich, L., Popp, P., Zeibig, M., 2002. Polycyclic aromatic hydrocarbon burden in fruit and vegetable species cultivated in allotments in an industrial area. International Journal of Environmental Analytical Chemistry 82 (10), 677–690.

Winpenny, J., Heinz, I., Koooshima, S., Salgot, M., Collado, J., Hernandez, F., Torricelli, R., 2010. The Wealth of Waste: the Economics of Wastewater Use in Agriculture. Food and Agriculture Organization of the United Nations, Rome.

Xue, J., Zhang, J., Xu, B., Xie, J., Wu, W., Lu, Y., 2016. Endotoxins: the critical risk factor in reclaimed water via inhalation exposure. Environmental Science and Technology 50 (21), 11957–11964.

Xue, J., Zhang, J., Wu, Q.-Y., Lu, Y., 2018. Sub-chronic inhalation of reclaimed water-induced fibrotic lesion in a mouse model. Water Research 139, 240–251.

Zhang, T., Li, B., 2011. Occurrence, transformation, and fate of antibiotics in municipal wastewater treatment plants. Critical Reviews in Environmental Science and Technology 41 (11), 951–998.

Zhang, J., Yang, J.-c., Wang, R.-q., Hou, H., Du, X.-m., Fan, S.-k., Liu, J.-s., Dai, J.-l., 2013. Effects of pollution sources and soil properties on distribution of polycyclic aromatic hydrocarbons and risk assessment. The Science of the Total Environment 463, 1–10.

Zhang, J., Xue, J., Xu, B., Xie, J., Qiao, J., Lu, Y., 2016. Inhibition of lipopolysaccharide induced acute inflammation in lung by chlorination. Journal of Hazardous Materials 303, 131–136.

Zuo, X., Liu, Z., Chen, M., 2016. Effect of H_2O_2 concentrations on copper removal using the modified hydrothermal biochar. Bioresource Technology 207, 262–267.

CHAPTER

Assessment technologies for hazards/risks of wastewater

7

Xiwei He, PhD, Kailong Huang, PhD
State Key Laboratory of Pollution Control and Resource Reuse, School of the Environment, Nanjing University, Nanjing, China

Chapter outline

Wastewater contains a great variety of high-risk pollutants (HRPs) that potentially threat the well-being of ecosystem as well as human health after discharged in the environment. Safeguarding a nontoxic environment and protecting biodiversity and vital ecosystem functions and services by minimizing exposure to mixtures of toxic substances and thus chemical footprints remains a key challenge faced globally (Brack et al., 2017; Reyjol et al., 2014). Current water quality monitoring typically focuses on chemical analysis, which can provide useful information about the concentration and type of HRPs in a sample. However, targeted chemical analysis cannot detect unidentified chemicals and transformation products or account for the mixture effects that can occur between the many HRPs present in water. Organisms in polluted environments are typically exposed to a complex mixture of pollutants and the

High-Risk Pollutants in Wastewater. https://doi.org/10.1016/B978-0-12-816448-8.00007-1

exposure may sometimes cause toxic effects even though the individual stressors are present at concentrations lower than the no-observed effect concentration (NOEC) (Brian et al., 2007; Kortenkamp, 2008). HRPs with similar or different mode of action (MoA) can influence each other's toxicity, resulting in an almost unlimited number of possible additive, synergistic, or antagonistic combinations (Beyer et al., 2014). For a comprehensive view of the wastewater quality assessment, chemical analysis should be combined with bioanalysis through either *in vivo* or *in vitro* tests that can provide information about the joint effect or toxicity of all bioavailable active chemicals present in a sample, with a further alliance of environmental risk assessment toward systematic process of quantitatively describing the probability, extent, time, or characteristic of adverse effects caused by exposure to chemicals as well as pathogenic microorganisms, especially the HRPs, in wastewater through various exposure routes to account for the pollution burden.

7.1 Toxicity evaluation of wastewater

7.1.1 *In vivo* tests

In vivo bioassays refer to experiments in which the severity, time, and dose dependency of toxic effects of various biological entities are tested on multiple standard or nonstandard whole living organisms and communities. The general aim of *in vivo* tests is to assess "integrative" or "apical" effects on, for example, mortality, development, growth, reproduction, or behavior, while the dosing regimens range from acute (exposure time $\leq$ 96 h) to chronic (exposure time >96 h) through life-cycle experiments (Prasse et al., 2015). In the aquatic environment, the route of exposure is usually via food or percutaneous. Testing of single species or biocoenoses is carried out either under laboratory or field conditions (*in situ*). Based on the species of organisms tested, *in vivo* tests can be divided into single-species tests and micro- and mesocosm multispecies tests.

7.1.1.1 Single-species in vivo tests

Single-species tests evaluate the toxic impact of testees on a particular species of interest. They primarily perform their role in the context of regulatory actions and mono-substance testing. In case they are applied for the assessment of complex matrices (i.e., whole effluent testing), the choice of test species has to be done with caution to exclude unintended interference. For instance, when assessing the formation and removal of toxic substance during wastewater treatment, the application of test organisms that are sensitive to nitrogen, salinity, or suspended organic carbon can easily turn into a problem. In addition, test species need to be selected with respect to species' ecology as well as the waterbody section of interest (resident species). The most commonly used test organisms are algae, daphnia, and fish (Table 7.1), which represent the tropical level of producer, primary consumer, and secondary consumer, respectively.

Table 7.1 Commonly used species and standard methods for effluent toxicity testing.

Tropical level	Organism	Species	Standard test methods
Producer	Algae	Freshwater *Desmodesmus subspicatus* *Pseudokirchneriella subcapitata* Saltwater *Skeletonema costatum* *Phaeodactylum tricornutum*	Fresh water algal growth inhibition test with unicellular green algae (ISO 8692) Marine algal growth inhibition test with *Skeletonema* sp. and *Phaeodactylum tricornutum* (ISO 10253) Scientific and technical aspects of batch algae growth inhibition tests (ISO/TR 11044)
Primary consumer	Daphnia	*Daphnia magna straus*	*Daphnia magna* reproduction test (OECD 211) Determination of the inhibition of the mobility of *Daphnia magna* Straus (Cladocera, Crustacea)—acute toxicity test (ISO 6341) Determination of long-term toxicity of substances to *Daphnia magna* Straus (Cladocera, Crustacea) (ISO 10706)
Secondary consumer	Fish	Freshwater Zebrafish *Danio rerio* Rainbow trout *Oncorhynchus mykiss* Fathead minnow *Pimephales promelas* Medaka *Oryzias latipes* Silverside *Menidia* sp. Three-spined Stickleback *Gasterosteus aculeatus* Common carp *Cyprinus carpio* Goldfish *Carassius auratus* Blugill sunfish *Lepomis macrochirus*	Fish, early-life stage toxicity test (OECD 210) Fish short-term reproduction assay (OECD 229) Fish sexual development test (OECD 234) 21-day fish assay: a short-term screening for estrogenic and androgenic activity, and aromatase inhibition (OECD 230) Fish, juvenile growth test (OECD 215) Fish, acute toxicity test (OECD 203) Fish, prolonged toxicity test: 14-day study (OECD 204) Bioaccumulation in fish: aqueous and dietary exposure (OECD 305) Fish, short-term toxicity test on embryo and sac-fry stages (OECD 212) Determination of the acute lethal toxicity of substances to a freshwater fish (ISO 7346)

Continued

Table 7.1 Commonly used species and standard methods for effluent toxicity testing.—*cont'd*

Tropical level	Organism	Species	Standard test methods
		Saltwater Tidewater silverside *Menidia peninsulae* Herring *Clupea harengus* Cod *Gadus morhua* Sheepshead minnow *Cyprinodon variegatus*	Biochemical and physiological measurements on fish (ISO/TS 23893) Determination of toxicity to embryos and larvae of freshwater fish (ISO 12890)

Single-species tests form the basis for general water quality assessment purposes. Similar to chemical environmental hazard assessment, effluent toxicity testing relies on standardized yet somewhat flexible test guidelines, including the OECD and ISO standards (Table 7.1), that accommodate a variety of foreseeable conditions. In general, most approaches measure the effects of an effluent on specific test organism' ability to survive, grow, and reproduce. However, understanding the role of specific and sublethal mode of action (MoA), such as endocrine disruption, is increasingly recognized as equally, if not less, important in regulatory assessments of wastewater effluents (Altenburger et al., 2004).

In many countries, mostly developed countries, toxicity testing of wastewater effluent is mandatory although test requirements vary with respect to species/taxonomic groups used and duration of the test (i.e., acute or chronic). A multitrophic level test battery incorporating the use of algae, invertebrate, or fish is often included in the effluent assessment scheme. For example, in the United States, most point source dischargers must obtain a NPDES (national pollutant discharge elimination system) permit that specify a variety of *in vivo* acute and short-term or chronic aquatic toxicity tests for three trophic levels (i.e., fish, invertebrates, and plants) to control the discharge of toxics (USEPA, 2002); in Canada, three sublethal toxicity tests must be conducted twice a year using tests that measure survival, growth, and/or reproduction endpoints in marine or freshwater plant and invertebrate organisms for the pulp and paper environmental effects monitoring regulations as well as similar sublethal toxicity tests for the metal mining environmental effects monitoring regulations (Environment Canada, 2012a; 2012b). Although various regulatory agencies use different terms, the general goal is consistent: to ensure that effluents being discharged will not harm the environment.

7.1.1.2 Micro-/mesocosm multispecies tests

Micro-/mesocosms are artificial, simplified ecosystems that are used to simulate and predict the behavior of natural ecosystems under controlled conditions. Microcosms typically are open or closed installations that can be set up in laboratories, whereas mesocosms are normally conducted outdoors to incorporate natural variation (e.g., diel cycles), which provide a link between field surveys and highly controlled laboratory experiments. Depending on the test design, aquatic micro-/mesocosms are composed of water enclosures equipped with natural/artificial water, sediment, and biocoenoses of multiple trophic levels, with sizes ranging from 1 to >10,000 L.

Micro-/mesocosm tests using natural communities offer a good compromise between the complexity of the ecosystem and the often highly artificial settings of laboratory experiments. One of the biggest advantages of micro-/mesocosm tests is that they can capture many potential feedbacks in biological systems, especially species interactions that may alter toxic effects of contaminants (Staley et al., 2010). By considering the entire community with many species with different life traits and sensitivities, the micro-/mesocosm approach preserves species interactions and considers the physicochemical environment as well as different exposure routes. In this way, micro-/mesocosm tests obtain data integrative over multiple direct and indirect effects and are thereby more ecologically relevant than single-species tests (SCENIHR, SCCS, SCHER, 2013). However, cost intensity of these tests normally limits the application to assess large number of samples. Standardization of methods and thus repeatability of results across laboratories is another concern (Rohr et al., 2016). Nevertheless, microcosm/mesocosm tests are part of higher tier risk assessment procedures and facilitate greater understanding of toxicological effects on ecological processes (European Commission, 2003).

7.1.2 *In vitro* tests

In vitro tests refer to the scientific analysis of the effects of various biological entities on cultured microorganisms, cells, or molecules outside their normal biological context, usually in wells of microplates. The spectrum of *in vitro* tests range from simple-parameter tests aimed for the assessment of cytotoxicity, sophisticated engineered reporter gene assays to detect receptor interactions and stress response pathways, image-based high-throughput screening tests to simultaneously examine multiple biological endpoints, to cell culture-based biosensing for portable or online monitoring, as well as biomolecule-based in chemico tests for preliminary toxicity screening. Compared to *in vivo* tests, *in vitro* assays are increasingly preferred due to logistical, cost, and time constraints as well as ethical considerations. The simplicity of *in vitro* test systems renders its application in fast and high-throughput screening. Additionally, *in vitro* tests can scan environmental samples for very specific MoAs, which can indicate the presence of specifically acting chemicals in a mixture. However, the extrapolation from *in vitro* test results to toxicological relevance in organisms and ecosystems is fraught with many uncertainties, because *in vitro* assays do not consider counter regulatory processes and systemic effects within whole organisms (Beyer et al., 2014).

7.1.2.1 Conventional single-parameter in vitro assays

Conventional single-parameter *in vitro* assays test an endpoint in procedures consisting of traditional cell culture, dosing, and measuring with or without using an add-mix-measure chemistry. These assays are mostly designed for determining cell viability or cell number, which are typical endpoints for cytotoxicity, a necessary and near compulsory index in *in vitro* risk assessment, after a defined exposure period. Based upon whether a metabolic process of the add-mix-measure chemistry is involved, conventional single-parameter *in vitro* tests can be further divided into metabolism- and nonmetabolic-based assays (Niles et al., 2013).

The metabolism-based assays utilize the nature that reducing capacity of viable cells could be diminished or completely ablated upon contact with cytotoxic substance to measure cell viability (Gonzalez and Tarloff, 2001). Tetrazolium salts (MTT, MTS, and XTT) and resazurin-containing formulations (e.g., alamar blue) are two types of commonly used redox indicator dyes to measure reducing capacity of viable cells. Both of these dyes produce either colorimetric or fluorescent signals that are directly correlated to viability and cellular health after enzymatic reduction. They differ in that the tetrazolium salts are more suited in low-throughput applications due to requisite dye solubilization steps with standardized methods (ISO 10,993-5, 2009), whereas the resazurin-containing formulations have been more widely used for high-through screening (Shum et al., 2008). Besides the measurement of reducing capacity, ATP assays are also considered to be metabolism-based viability assays because ATP is recognized as the basic bioenergetic currency of viable cells and ATP levels are normally proportional to viable cell number (Crouch et al., 1993). Cellular ATP can be conveniently measured using various add-mix-measure formulations of the "firefly" luciferase enzyme reaction.

The nonmetabolic-based assays typically measure the membrane integrity to assess cell viability. The operating premise is that either intact cellular membranes selectively exclude formulations of impermeable dyes or enzymatic activities associated with either viable or nonviable cells can be measured after a cytotoxic insult. Take Gly-Phe-AFC and trypan blue for example, both of which are dyes for determining cell viability, the former specifically measures intracellular protease activities to exclude nonviable cells whereas the latter only binds to genomic DNA from dead cells to distinguish them from the viable ones. Another extensively used nonmetabolic-based method leverages the measurement of leaked lactate dehydrogenase (LDH) from nonviable cells as the biomarker of cytotoxicity. By coupling extracellular LDH activity with a diaphorase enzyme in the presence of a reducible resazurin substrate, this method produces excellent signal windows for detection.

Notably, the aforementioned assays are preferably applied to vertebrate (i.e., mammalian or fish) cell lines. For assessing cytotoxicity to bacteria or algae, the most commonly used single-parameter *in vitro* assays include bioluminescence inhibition tests of naturally bioluminescent bacteria (e.g., *Aliivibrio fischeri*; ISO, 1998), algae growth inhibition tests (e.g., *Selenastrum capricornutum*; ISO, 1998), and photosystem II inhibition tests in algae cells etc.

7.1.2.2 Reporter gene-based in vitro assays

Exposure to toxic chemicals will lead to a change in cellular behavior including a change in gene expression, which is generally decisive in eliciting a toxic response

(Etteieb et al., 2019). By fusing a reporter element (gene encoding easily detectable proteins or enzymes) to the sensing element (in most cases the gene promoter) of a target gene, the activation of the target gene caused by environmental stimulus can thereby be "reported" by the easily detectable proteins or enzymes of the reporter element. This reporter gene technology offers a simple and rapid means for evaluating the potential toxic effects of environmental samples on cellular gene expression, which makes it increasingly popular in environmental toxicology.

The term reporter gene is used to define a gene with a readily measurable phenotype that can be distinguished easily over a background of endogenous proteins (Suto and Ignar, 1997). A number of reporter genes have been developed over the years and some are particularly favored by environmental toxicologists. These include β-Galactosidase (LacZ), firefly luciferase (Luc), bacteria luciferase (Lux), and green fluorescent protein (GFP). They all exhibit good reliability in reporting the expression of the target gene, but differ from each other in regards to sensitivity, dynamic range, or convenience in measurement. A summary of the advantages and disadvantages of these reporter genes is listed in Table 7.2.

Numerous HRPs presented in wastewater have been shown to stimulate cellular gene expression of various kinds. By sophisticated manipulation of the sensing and reporter elements, the reporter gene system can be tailored to target relevant steps in the toxicity pathways including induction of xenobiotic metabolism, specific and reactive MoAs, activation of adaptive stress response pathways, and system response. Associated bioassays commonly applied are listed in Table 7.3. Most of these bioassays were developed from mammalian cells although yeast and bacteria were also encountered. Conductions of these assays are usually in the format of 96- or 384-well plate, which facilitate high-throughput screening when dealing with large amount of testing samples. In a recent article (Escher et al., 2014), Escher and his colleagues proposed a routine test battery of indicator bioassays for water

Table 7.2 Comparison of commonly used reporter genes (Naylor, 1999).

Reporter gene	Advantages	Disadvantages
LacZ	Well characterized; stable; simple colorimetric readouts; sensitive bio- or chemiluminescent assays available.	Endogenous activity (mammalian cells).
Luc	High specific activity; no endogenous activity; broad dynamic range; convenient assays.	Requires substrate (luciferin) and presence of O_2 and ATP.
Lux	Good for measuring/analyzing prokaryotic gene transcription.	Less sensitive than firefly; not suitable for mammalian cells.
GFP	Autofluorescent (no substrate needed); no endogenous activity; mutants with altered spectral qualities available.	Requires posttranslational modification; low sensitivity (no signal amplification).

Table 7.3 Characteristics of bioassays based on reporter-gene technology.

Toxicity pathway	Bioassay	Cell	Reporter gene	Reference
Xenobiotic metabolism: PXR	HG5LN PXR	Hela (mammalian)	Luc	Creusot et al. (2010); Lemaire et al. (2006)
Xenobiotic metabolism: PPARα	HG5LN PPARα	Hela (mammalian)	Luc	le Maire et al. (2009); Seimandi et al. (2005)
	CALUX-PPARα	U2OS (mammalian)	Luc	Piersma et al. (2013)
Xenobiotic metabolism: PPARγ	HG5LN PPARγ	Hela (mammalian)	Luc	le Maire et al. (2009); Seimandi et al. (2005)
	CALUX-PPARγ2	U2OS (mammalian)	Luc	Gijsbers et al. (2013); Piersma et al. (2013)
	PPARγ-GeneBLAzer	HEK 293T (mammalian)	β-lactamase	Huang et al. (2011)
Xenobiotic metabolism: AhR	CAFLUX	H1G1.1c3 (mammalian)	GFP	Nagy et al. (2002)
	H4IIEluc	H4IIE (mammalian)	Luc	Sanderson et al. (1996)
	DR CALUX	H4IIE (mammalian)	Luc	
	AhR-yeast	*Saccharomyces cerevisiae* (yeast)	LacZ	Miller (1999)
Specific modes of action: ER	T47D-KBluc	T47D (mammalian)	Luc	Wilson et al. (2004)
	ERα-CALUX	U2OS (mammalian)	Luc	Sonneveld et al. (2005)
	HELN-ERα	Hela (mammalian)	Luc	Balaguer et al. (1999)
	HELN-ERβ	Hela (mammalian)	Luc	Balaguer et al. (1999)
	ERα-GeneBLAzer	HEK 293T (mammalian)	β-lactamase	Huang et al. (2011)
	YES	*Saccharomyces cerevisiae* (yeast)	LacZ	Routledge and Sumpter (1996)
	BLYES	*Saccharomyces cerevisiae* (yeast)	Lux	Sanseverino et al. (2005)

Table 7.3 Characteristics of bioassays based on reporter-gene technology.—*cont'd*

Toxicity pathway	Bioassay	Cell	Reporter gene	Reference
Specific modes of action: AR	AR-CALUX	U2OS (mammalian)	Luc	Sonneveld et al. (2005)
	PALM	PC-3 (mammalian)	Luc	Terouanne et al. (2000)
	MDA-kb2	MDA-MB-453 (mammalian)	Luc	Wilson et al. (2002)
	YAS	*Saccharomyces cerevisiae* (yeast)	LacZ	Sohoni and Sumpter (1998)
	BLYAS	*Saccharomyces cerevisiae* (yeast)	Lux	Eldridge et al. (2007)
Specific modes of action: GR	GR-CALUX	U2OS (mammalian)	Luc	Sonneveld et al. (2005)
	GR-MDA-kb2	MDA-MB-453 (mammalian)	Luc	Wilson et al. (2002)
	GR-GeneBLAzer	HEK 293T (mammalian)	β-lactamase	Huang et al. (2011)
Specific modes of action: PR	PR-CALUX	U2OS (mammalian)	Luc	Sonneveld et al. (2005)
	PR-GeneBLAzer	HEK 293T (mammalian)	β-lactamase	Huang et al. (2011)
Specific modes of action: TR	TR-CALUX	U2OS (mammalian)	Luc	Sonneveld et al. (2005)
Reactive modes of action	umuC TA1535/pSK1002	*Salmonella. Typhimurium* TA1535 (bacteria)	LacZ	Oda et al. (1985)
Adaptive stress response: Nrf2	Nrf2-CALUX	U2OS (mammalian)	Luc	van der Linden et al. (2014)
	ARE-GeneBLAzer			
Adaptive stress response: p53	P53-CALUX	U2OS (mammalian)	Luc	van der Linden et al. (2014)
	P53-GeneBLAzer	HEK 293T (mammalian)	β-lactamase	Huang et al. (2011)
Adaptive stress response: NF-kB	NF-kB-CALUX	U2OS (mammalian)	Luc	
	NF-kB-GeneBLAzer	HEK 293T (mammalian)	β-lactamase	Huang et al. (2011)
System response: Cytotoxicity	Cytotox-CALUX	U2OS (mammalian)	Luc	van der Linden et al. (2014)
	HEK293-lux	HEK 293T (mammalian)	Lux	Class et al. (2015)

quality assessment after conducting a comprehensive evaluation of 103 unique *in vitro* bioassays. They concluded that a battery of bioassays should include integrative endpoints such as cytotoxicity as well as endpoints specific to relevant steps in the cellular toxicity pathway. As a minimum, indicator bioassays should cover examples found responsive to and representative of "induction of xenobiotic metabolism," "endocrine disruption," and "adaptive stress responses."

7.1.2.3 High-content screening technology

Advanced microscopy and the corresponding image analysis have been developed in recent years into a powerful tool for studying molecular and morphological events in cells and tissues. Cell-based high-content screening (HCS), which combines the efficiency of high-throughput techniques with the ability of cellular imaging to collect quantitative data from complex biological systems, is an upcoming methodology for investigating cellular processes and their alteration by various perturbations. HCS is defined as multiplexed functional screening based on imaging multiple targets in the physiologic context of intact cells by extraction of multicolor fluorescence information (Giuliano et al., 1997). The ability of HCS to define protein functions and dissect signaling pathways is achieved with the aid of chemical genetics, where small organic molecules are used to study biological systems (Stockwell, 2004). Key operations of HCS include experimental design, sample preparation, image acquisition, archiving, processing and analysis, and cellular knowledge mining (Abraham et al., 2004). HCS was preliminarily introduced to meet the need for preclinical drug screening in the pharmaceutical industry, and is now gaining attention in environmental toxicology to facilitate toxicity evaluation and risk assessment of single and mixed chemistries in the environment.

In contrast to traditional target-based high-throughput screening, methods that average the biological response of thousands of cells, HCS utilizes phenotypic screening approaches that acquire functional and morphometric information from collections of individual cells (Zanella et al., 2010). This could effectively avoid false-negative results when contextual phenotypic shift is expected only for a subset of cells: for example, within heterogeneous or cocultured cell cultures where different cell lines may respond discrepantly to a certain stimuli.

One of the most conspicuous features of HCS is its ability to conduct multiple independent measurements simultaneously in intact cells, resulting in the acquisition of multiple dataset indicative of chemical/sample toxicity at the same time. The extraction of toxicity index in fluorescence image depends on image analysis and data mining. The purpose of this process is to segment and classify objects in fluorescence image. Therefore, the toxicity index of HCS not only comes from the location and fluorescence intensity of fluorescent labeling target, but also reacts on the morphology and structure of cells and subcells (Iannetti et al., 2015). For this reason, HCS enables quantitatively measurement of toxicity indices that are difficult to characterize in other toxicity assessment methods, such as formation of apoptotic bodies, nuclear translocation of transcription factors in signaling pathways, cell chemotaxis, and colony formation.

7.1.2.4 Cell culture-based biosensing techniques

A biosensor is a self-contained integrated device that combines a biological recognition element, that is, bioreceptor (e.g., cells or enzymes), immobilized in intimate contact with a physical transducer (e.g., electrochemical, optical, thermal, or piezoelectric) that converts the biological response event into a signal proportional to the initial biorecognition event, and it is thus possible to determine the effective concentration of the ligand of interest (Lagarde and Jaffrezic-Renault, 2011). Biosensors clearly differ from bioassays, in which the transducer is not an integral part of the analytical device. In recent years, biosensors have received particular attention owing to their high sensitivity, low cost, and possible easy adaptation for portable and online measurements.

Early developed biosensors were mainly designed to meet the need of chemical detection, including heavy metals, polycyclic aromatic hydrocarbons (PAHs), or a group of chemical HRPs with similar MoA, in environmental matrix. Microorganisms, specifically bacteria and yeasts, are most commonly used bioreceptor in these biosensors because of their high sensitivity, easy and prolonged preservation, as well as easy amenability to molecular engineering.

Recent advances in cell engineering have allowed the use of vertebrate cells to detect water toxicity by assessing the overall cytotoxic response to external stimuli. Two mainstream cell culture-based biosensing techniques stand out: electric cell-substrate impedance sensing (ECIS) and chromatophore-based approaches. The ECIS utilizes cell culture medium as electrolyte with cells grown on an electrode-embedded culturing surface. By applying a very low, noninvasive alternating current, the cell electric impedance can be measured to indicate the cell status including growth, migration, morphology, cell-to-matrix and cell-to-cell interactions (Jiang, 2012). The chromatophore-based approaches employ neurone-like pigmented cells found in amphibians, fish, and reptiles that are able to respond to stimuli with changing pigment distribution to assess water toxicity. This technique was pioneered by integrating light source, camera and lens, monitor and keyboard, and an image processing software into a portable device (Chaplen et al., 2002). Quantification adopts methods to quantify changes of cell area covered by pigment using photographic recording of cells upon a fixed field of view containing the same number of cells (Mojovic and Jovanovic, 2005).

Notably, although biosensor technology has been studied for decades, most of the developed biosensors are still in the stage of laboratory testing, and there is still a long way to go before they can be applied in routine environmental monitoring. Today, the challenge facing biosensor development is the ability to fulfill the ever-increasing requirements of environmental legislation in data reliability as well as rapidity of response, sensitivity, and cost. To achieve that, key elements, including bioreceptor and transducer, in the biosensor equipment have to be optimized.

7.1.2.5 Toxicity-indicative in-chemico tests

One special approach for toxicity detection is to use the activity of enzymes or reactivity of specific biomolecule *in vitro* as toxicity indexes. The enzymes or biomolecules used are usually key components in the process a specific MoA. The most well-known in-chemico toxicity test is the acetylcholinesterase (AchE) inhibition experiment, which is based on the premise that AchE is a key enzyme in the transmission of nerve signals *in vivo*, the inhibited activity of AchE can be measured *in vitro* as a means of determining the neurotoxicity of a testee (Leusch et al., 2014). Another recently validated in-chemico toxicity test is the N-acetyl-L-cysteine (NAC)-based thiol reactivity analyses, which is used to predict mammalian cell cytotoxicity of water samples (Dong et al. 2017, 2018). This assay is based upon that cysteine thiol in glutathione is the major reductant against reactive toxicants, which can induce adverse biological response if the thiol pool is overwhelmed or depleted (Townsend et al., 2003). Compared with cell culture-based toxicity tests, in-chemico tests are much less laborious and time consuming, making them excellent choices as fast, high-throughput analysis to perform preliminary screening before conducting cell-based *in vitro* or *in vivo* biological assays.

7.2 Health and ecological risk assessment

Risk assessment technology was gradually formed in the early 1980s and widely used in many fields such as environmental protection, management of toxic chemical substances, engineering safety, and prevention and control of disaster accidents. Risk assessment is the process of describing and quantifying the probability of occurrence of undesirable or undesired events. For wastewater risk assessment, it can be defined as the systematic process of quantitative description of the probability and characteristics of adverse effects caused by chemicals and biological HRPs in wastewater through various exposure pathways. The risk assessment of wastewater mainly includes ecological risk assessment and human health risk assessment, which is essential to guarantee human health and ecosystem safety.

7.2.1 Ecological risk assessment

Ecological risk assessment (ERA) aims at assessing the effects of stressors, often chemical HRPs, on the local environment. In ERA, the undesired event often depends on the chemical HRPs of interest and on the risk assessment scenario. This undesired event is usually a detrimental effect on organisms, populations, or ecosystems, especially for the aquatic environment. The ERA methods mainly include the quotient method and the probabilistic method (e.g., joint risk probability distribution curves) (Chen, 2005; Wang et al., 2002), among which the quotient method is the most common and widely used method.

The ecological risk in the investigated area that may pose to the aquatic environment was estimated through their hazard quotients (HQs) at three trophic levels, representative of the aquatic ecosystem (algae, daphnia, and fish). According to EU guidelines (EMEA, 2006), the approach combines the predicted or measured environmental concentrations (PECs or MECs) of the pollutants with toxicity data (predicted no-effect concentration (PNEC)) to evaluate whether a pollutant is likely to pose a risk for aquatic organisms or not (Guillén et al., 2012).

Table 7.4 The recommended assessment factors.

Available experimental toxicity values	Assessment factor
At least one short-term LC50 or EC50 of each of the three trophic levels	1000
A long-term experiment of the no-observed effect concentration (NOEC)	100
A long-term experiment of NOEC with two trophic levels	50
A long-term experiment of NOEC with three trophic levels	10

HQs were determined as the ratio between the MEC and its PNEC, where the MEC represents the highest concentration of the pollutant detected in the samples, while the PNEC is calculated by dividing the lowest acute toxicity value (median EC_{50} or LC_{50}) for the three selected trophic levels by the pertinent assessment factor (AF) (Papageorgiou et al., 2016). When no experimental toxicity values are available in the published literature, EC_{50} or LC_{50} values can be estimated with the ECOSAR predictive model (USEPA, 2012). Due to the uncertainty of the extrapolation process, an appropriate AF needs to be selected when extrapolating PNEC through the EC_{50} or LC_{50} of the three trophic levels. The AF is derived from experience, and the details of recommended AF are shown in Table 7.4.

The potential ecological risk was assessed according to a commonly used risk ranking criterion. A low or negligible risk can be expected if the HQ values are between 0.1 and 1. For HQ values between 1 and 10, a medium risk can be expected, while HQ values greater than 10 indicate a high ecological risk (Mendoza et al., 2015). It should be noted that the ERA method mentioned earlier focus mainly on the toxicity assessment of individual chemicals; however, the mixture or combined exposure of wastewater (the chemical mixtures) may pose potential risks even though each of the pollutants is below its safe concentration. Therefore, new approaches must be developed to evaluate the combined effects in aquatic ecotoxicology in the future (Beyer et al., 2014; Diamond et al., 2018).

Finally, ERA has been widely applied to assess the adverse ecological effects of pollutants (pharmaceuticals and personal care products (Ashfaq et al., 2017; Mater et al., 2014; Molins-Delgado et al., 2016; Rivera-Jaimes et al., 2018), nutrients (Efroymson et al., 2007), endocrine-disrupting parabens (Molins-Delgado et al., 2016), disinfection byproducts (Li et al., 2019), androgens (Zhang et al., 2018), butylparaben and benzylparaben (Yamamoto et al., 2007), as well as chemical mixtures (Beyer et al., 2014; Coors et al., 2018; Diamond et al., 2018) in wastewater treatment plant effluents (Coors et al., 2018; Diamond et al., 2018; Efroymson et al., 2007; Molins-Delgado et al., 2016; Rivera-Jaimes et al., 2018; Yamamoto et al., 2007; Zhang et al., 2018), pharmaceutical wastewater (Ashfaq et al., 2017), hospital wastewater (Mater et al., 2014), chlorinated effluents (Li et al., 2019), and reclaimed wastewater (Zeng et al., 2016)).

7.2.2 Health risk assessment

7.2.2.1 Chemical-based health risk assessment

The chemical-based health risk assessment (HRA) typically includes four steps: data collection and analysis, exposure assessment, toxicity assessment, and risk characterization (Chartres et al., 2019).

(1) Data collection and analysis

The concentrations of target pollutants should be measured with the standard methods (Deng et al., 2018) or collected from published literatures (Wu et al., 2011). Based on the monitoring data, the pathway, frequency, duration, and magnitude of actual human exposure to the contaminant should be estimated. Furthermore, the study should be hypothesized that the health risk is mainly caused by the pollutants and the concentration of contaminant in wastewater are relatively stable.

(2) Exposure assessment

Human exposure to wastewater pollutants mainly through dermal contact of contaminated wastewater and ingestion of food crops irrigated with wastewater (Barker et al., 2013; Jan et al., 2010; Khan et al., 2008).

The exposure dosage through dermal absorption pathway was calculated by Formulas 7.1 and 7.2, which was adapted from USEPA (2004, 2011):

$$DA_{event} = 2FA \times Kp \times C_W \sqrt{(6\tau_{event} \times t_{event})/\pi} \quad (7.1)$$

where DA_{event} is dermal absorbed pollutants per event (μg cm^{-1} $event^{-1}$), FA is the fraction of water absorbed, Kp is the dermal permeability coefficient of pollutants (cm h^{-1}), C_w is the concentration of pollutants in wastewater (μg L^{-1}), τ_{event} is the lag time per event (h $event^{-1}$), and t_{event} is the event duration (h $event^{-1}$).

$$DED_d = (DA_{event} \times SA \times EV \times EF \times ED)/(BW \times AT) \quad (7.2)$$

where DED_d is daily exposure dose through dermal absorption (μg kg^{-1} day^{-1}), SA is the exposed skin area (cm^2), EV is the event frequency (event day^{-1}), EF is exposure frequency (days $year^{-1}$), ED is exposure duration (years), BW is body weight (kg), and AT is the averaging time (day). For noncarcinogenic effects, AT = ED × 365 days $year^{-1}$. For carcinogenic effects, AT = 25,550 days.

The exposure dosage through the pathway of ingestion contaminated crops was calculated by Formulas 7.3 and 7.4, which was adapted from USEPA and IRIS (2006):

$$TF = C_{plant}/C_{soil} \quad (7.3)$$

where TF is soil-to-plant transfer quotient, C_{plant} is the concentration of pollutant in plants (mg kg^{-1}), and C_{soil} is the concentration of pollutant in soil (mg kg^{-1}).

$$DI = C_{pollutant} \times TF \times D_{food\ intake}/BW \quad (7.4)$$

where DI is daily intake of pollutant through ingestion contaminated crops, $C_{pollutant}$ is the pollutant concentrations in plants (mg kg^{-1}), TF is soil-to-plant transfer quotient, $D_{food\ intake}$ is the daily intake of contaminated vegetables (kg $person^{-1}$ day^{-1}), and BW is body weight (kg). The average daily food intake for adults and children was considered to be 0.345 and 0.232 kg $person^{-1}$ day^{-1}, respectively.

(3) Toxicity assessment

The dermal absorption slope factor (SF_d) and the reference dermal absorption dose (RFD_d) of the pollutants were calculated by dividing the oral slope factor (SF_o) and reference oral dose (RFD_o) by the gastrointestinal absorption factor (ABS_{GI}) (USEPA, 2004).

(4) Risk characterization

Risk characterization summarizes and combines outputs of the exposure and toxicity assessment to analyze baseline risk of the pollutant in qualitative statements and quantitative expressions.

The noncarcinogenic risk of single and multiple substances was calculated by the HQ and the hazard index (HI) according to Formulas 7.5 and 7.6 (USEPA, 2004).

$$HQ = DED_d / RFD_d \tag{7.5}$$

$$HI = \sum HQ \tag{7.6}$$

The carcinogenic risk (CR) and risk index (RI) associated with dermal exposure were calculated by Formulas 7.7 and 7.8 (USEPA, 1989; 2004).

$$CR = DED_d \times SF_d \tag{7.7}$$

$$RI = \sum CR \tag{7.8}$$

where CR is the probability of developing cancer over a lifetime as a result of exposure to a pollutant, and RI is the sum of carcinogenic risk of the individual carcinogenic pollutants.

The risk caused by the consumption of contaminated crops was calculated by the health risk index (HRI) and hazard quotient (HQ) according to Formulas 7.9 and 7.10:

$$HRI = DI / RFD_o \tag{7.9}$$

where DI represents the daily intake of pollutants, and RFD_o represents reference oral dose. If the value of HRI is less than 1 then the exposed population will pose no risk.

$$HQ = [W] \times [M] / RFD \times BW \tag{7.10}$$

where [W] is the dry weight of contaminated plant material consumed (mg day^{-1}), [M] is concentration of pollutant in vegetables (mg kg^{-1}), RFD is the food reference dose for the pollutant (mg day^{-1}), and BW is the body weight (kg). The population will experience health risk if the ratio is equal or greater than 1.

To enhance the credibility of risk analysis, uncertainty analysis should be taken into consideration during HRA process (Deng et al., 2018; Landis and Wiegers, 2007). Monte Carlo simulation ($n = 10,000$) was performed to quantify the uncertainties and its impact on the risk assessment. Meanwhile, a sensitivity analysis was carried out to identify the significance of input parameters by calculating rank correlation coefficients between input and output values during Monte Carlo simulations. Sensitivity and uncertainty analysis were conducted to identify critical input variables and quantity uncertainties during the risk assessment process.

Nowadays, probabilistic HRA has been widely applied to assess the adverse health effects of HRPs (PAHs (Cao et al., 2018; Zhang et al., 2012), heavy metals (Duan et al., 2017), pharmaceuticals and personal care products (Christou et al., 2017; Prosser and Sibley, 2015; Semerjian et al., 2018), volatile organic compounds (Widiana et al., 2019; Yang et al., 2012), odors (Niu et al., 2014), semivolatile organic compounds and N-nitrosamines (Deng et al., 2018), and phthalate esters (Olujimi et al., 2017)) in the wastewater (wastewater treatment plant effluents (Cao et al., 2018; Niu et al., 2014; Olujimi et al., 2017; Widiana et al., 2019), sewage sludge (Duan et al., 2017), reclaimed wastewater (Deng et al., 2018; Semerjian et al., 2018; Wu et al., 2014), coking wastewater (Zhang et al., 2012), and irrigation wastewater (Christou et al., 2017; Prosser and Sibley, 2015; Rashid et al., 2018)) for different populations (e.g., on-site workers and residents with different age groups).

7.2.2.2 Microbial-based health risk assessment

Quantitative microbial risk assessment (QMRA) was first proposed for use in the treatment of water in microbiological risk management in the 1990s (Rose et al., 1991). The application of QMRA has largely altered the view that risk assessment can only be used to assess chemical hazards, it was then proposed as a new method to determine the microbial safety of food, water and other environmental matrix (Haas et al., 2014). QMRA aims to provide the best available information to understand the nature of the potential effects from microbial exposure (Haas et al., 1999). Although QMRA works officially as six steps (Abt et al., 2010), most studies follow a four steps procedure (Hamouda et al., 2016): (1) hazard identification, (2) exposure assessment, (3) dose-response analysis, and (4) risk characterization.

The purpose of hazard identification is to identify the potential hazards to human beings caused by certain pathogens, such as bacteria, protozoa, or viruses. Exposure assessment refers to the contact between pathogens and organisms, including the sources, modes, and routes of exposure, as well as the degree of contact with relevant pathogens. In general, exposure assessment determines the exposure dose, time, and frequency. This process involves the quantification and characteristics of the exposed population, but conditional influences must also be considered, such as regional differences, immune status, and population distribution. The third step, dose-response analysis, is to determine the probability of infections through the mathematical models. The most commonly used models are Beta-Poisson model

(Formula 7.11) and exponential model (Formula 7.12) (Soller, 2006). In the Beta-Poisson models, the dose that causes half of the exposed population is N_{50} (Formula 7.13), so Formula 7.11 is equal to Formula 7.14.

$$P_{inf,d}(d) = 1 - (1 + d/\beta)^{-\alpha} \tag{7.11}$$

$$P_{inf,d}(d) = 1 - e^{(-kd)} \tag{7.12}$$

$$N_{50} = \beta\left(2^{1/\alpha} - 1\right) \tag{7.13}$$

$$P_{inf,d}(d) = 1 - \left[1 + \frac{d}{N_{50}}\left(2^{1/\alpha} - 1\right)\right] \tag{7.14}$$

$$P_{inf,y} = 1 - \left(1 - P_{inf,d}\right)^{365} \tag{7.15}$$

Beta-Poisson models are considered to be more suitable to evaluate enteropathogenic bacteria and viruses, while exponential models are suitable for enteropathogenic protozoa (Haas and Rose, 1994; Haas, 1983). The parameters in the models are compiled from a large amount of clinical and pathological data and most of these parameters can refer to epidemical survey reports from the U.S. EPA and McBride (Mcbride et al., 2002). Pathogens are described in detail in the QMRA Wiki database at the University of Michigan (http://qmrawiki.canr.msu.edu/index.php?title=Table_of_Recommended_Best-Fit_Parameters). Furthermore, some typical pathogens found in wastewater and their dose-response parameters are listed in Table 7.5. Risk characterizations are usually made using the established dose-response models (Haas, 1983; Regli et al., 1991). Formula 7.15 can be used to calculate the annual risk of infection. The single data entry can be used for deterministic descriptions, or statistical distributions, while probabilistic description reflects the randomness of variability, so are more suitable to evaluate wastewater conditions. Regardless of the description chosen, risk characterization requires a clear description of the uncertainty and variability of the sources. Output of a detailed figure through the earlier steps can provide accurate information for better solving the risk management challenges compared to qualitative or semiquantitative approaches. In the past two decades, QMRA has already developed into a scientific discipline and was included in WHO's water-related guidelines (WHO, 2016). Recently, QMRA has been used to evaluate the health influence of wastewater effluent discharge on water quality of the receiving water (Xiao et al., 2018). Moreover, combined use of QMRA and life-cycle assessment of wastewater management can reveal the relative importance of pathogen risk in relation to other potential impacts on human health (Heimersson et al., 2014).

7.3 Summary

Toxicological evaluation and risk assessment serve as vital complementation to chemical analysis to constitute a more sophisticated integrity toward comprehensive evaluation of the quality of wastewater. Toxicity testing of wastewater employs either *in vivo*

Table 7.5 Typical pathogens identified in wastewater and their dose-response parameters.

Pathogens		Parameters			References
		α	N_{50}	k	
Bacteria	*Burkholderia pseudomallei*	3.28E-01	5.43E+03		Brett and Woods (1996)
	Campylobacter jejuni	1.44E-01	8.9E+02		Black et al. (1988)
	E.coli enterohemorrhagic		3.18E+03	2.18E-04	Cornick and Helgerson (2004)
	E.coli	1.55E-01	2.11E+06		Dupont et al. (1971)
	E. coli O157:H7	2.09E-01	1.12E+03		Westrell (2004)
	L. pneumophila		1.16E+01	5.99E-02	Muller et al. (1983)
	Listeria monocytogenes	2.53E-01	2.77E+02		Golnazarian et al. (1989)
	Mycobacterium avium			6.93E-04	O'Brien et al. (2006)
	Salmonella typhi	1.75E-01	1.11E+06		Hornick et al. (1966) Hornick et al. (1970)
	Salmonella serotype			3.97E-06	McCullough and Eisele (1951)
	Shigella	2.65E-01	1.48E+03		DuPont et al. (1972)
	Vibrio cholerae	2.50E-01	2.43E+02		Hornick et al. (1971)
	Campylobacter	1.45E-01	8.96E+02		Haas et al. (1999) Westrell (2004)
	P. aeruginosa	1.90E-01	1.85E+04		Lawinbrussel et al. (1993)
Viruses	*Echovirus*	1.06E+00	9.22E+02		Schiff et al. (1984)
	Enteroviruses			3.74E-03	Cliver (1981)
	Rotavirus	2.53E-01	6.17E+00		Ward et al. (1986)
Protozoa	*Cryptosporidium*			5.72E-02	Messner et al. (2001)
	Giardia			1.99E-02	Rendtorff (1954)
	Ascaris			1.00E+00	Westrell (2004)

or *in vitro* tests depending on the objectives, typically with *in vivo* tests for the assessment of "integrative" or "apical" effects on mortality, development, growth, reproduction, or behavior of single- and/or multispecies, while *in vitro* tests can scan wastewater samples for very specific biological endpoints or MoAs with fundamentally simplified procedures. Based on the results of toxicity, the risk assessment of the HRPs in wastewater mainly includes ecological risk assessment and human health risk assessment (chemical-based or microbial-based). Risk analysis revealed that the exposure to the HRPs in wastewater (especially for the combined exposure) may imply an ecological risk and/or a human health risk (carcinogenic or noncarcinogenic risk). Notwithstanding the uncertainties, it should be informed the development of more effective strategies to improve the wastewater treatment processes and reduce toxicity as well as the ecological and health risks induced by the HRPs in wastewater.

References

Abraham, V.C., Taylor, D.L., Haskins, J.R., 2004. High content screening applied to large-scale cell biology. Trends in Biotechnology 22 (1), 15–22.

Abt, E., Rodricks, J.V., Levy, J.I., Zeise, L., Burke, T.A., 2010. Science and decisions: advancing risk assessment. Risk Analysis. An International Journal 30 (7), 1028–1036.

Altenburger, R., Walter, H., Grote, M., 2004. What contributes to the combined effect of a complex mixture? Environmental Science and Technology 38 (23), 6353–6362.

Ashfaq, M., Nawaz Khan, K., Saif Ur Rehman, M., Mustafa, G., Faizan Nazar, M., Sun, Q., Iqbal, J., Mulla, S.I., Yu, C.P., 2017. Ecological risk assessment of pharmaceuticals in the receiving environment of pharmaceutical wastewater in Pakistan. Ecotoxicology and Environmental Safety 136, 31–39.

Balaguer, P., Francois, F., Comunale, F., Fenet, H., Boussioux, A.M., Pons, M., Nicolas, J.C., Casellas, C., 1999. Reporter cell lines to study the estrogenic effects of xenoestrogens. The Science of the Total Environment 233 (1–3), 47–56.

Barker, S.F., O'Toole, J., Sinclair, M.I., Leder, K., Malawaraarachchi, M., Hamilton, A.J., 2013. A probabilistic model of norovirus disease burden associated with greywater irrigation of home-produced lettuce in Melbourne, Australia. Water Research 47 (3), 1421–1432.

Beyer, J., Petersen, K., Song, Y., Ruus, A., Grung, M., Bakke, T., Tollefsen, K.E., 2014. Environmental risk assessment of combined effects in aquatic ecotoxicology: a discussion paper. Marine Environmental Research 96, 81–91.

Black, R.E., Levine, M.M., Clements, M.L., Hughes, T.P., Blaser, M.J., 1988. Experimental *Campylobacter jejuni* infection in humans. The Journal of Infectious Diseases 157 (3), 472–479.

Brack, W., Dulio, V., Agerstrand, M., Allan, I., Altenburger, R., Brinkmann, M., 2017. Towards the review of the European Union water framework directive: recommendations for more efficient assessment and management of chemical contamination in European surface water resources. The Science of the Total Environment 576, 720–737.

Brett, P.J., Woods, D.E., 1996. Structural and immunological characterization of *Burkholderia pseudomallei* O-polysaccharide-flagellin protein conjugates. Infection and Immunity 64 (7), 2824–2828.

Brian, J.V., Harris, C.A., Scholze, M., Kortenkamp, A., Booy, P., Lamoree, M., 2007. Evidence of estrogenic mixture effects on the reproductive performance of fish. Environmental Science and Technology 41, 337–344.

Cao, W., Qiao, M., Liu, B., Zhao, X., 2018. Occurrence of parent and substituted polycyclic aromatic hydrocarbons in typical wastewater treatment plants and effluent receiving rivers of Beijing, and risk assessment. J. Environ. Sci. Health 53 (11), 992–999.

Chaplen, F.W.R., Upson, R.H., McFadden, P.N., Kolodziej, W., 2002. Fish chromatophores as cytosensors in a microscale device: detection of environmental toxins and bacterial pathogens. Pigment Cell Research 15 (1), 19–26.

Chartres, N., Bero, L.A., Norris, S.L., 2019. A review of methods used for hazard identification and risk assessment of environmental hazards. Environment International 123, 231–239.

Chen, C.S., 2005. Ecological risk assessment for aquatic species exposed to contaminants in Keelung River, Taiwan. Chemosphere 61 (8), 1142–1158.

Christou, A., Karaolia, P., Hapeshi, E., Michael, C., Fatta-Kassinos, D., 2017. Long-term wastewater irrigation of vegetables in real agricultural systems: concentration of pharmaceuticals in soil, uptake and bioaccumulation in tomato fruits and human health risk assessment. Water Research 109, 24–34.

Class, B., Thorne, N., Aguisanda, F., Southall, N., McKew, J.C., Zheng, W., 2015. High-throughput viability assay using an autonomously bioluminescent cell line with a bacterial Lux reporter. Journal of Laboratory Automation 20 (2), 164–174.

Cliver, D.O., 1981. Experimental infection by waterborne enteroviruses. Journal of Food Protection 44 (11), 861–865.

Coors, A., Vollmar, P., Sacher, F., Polleichtner, C., Hassold, E., Gildemeister, D., Kuhnen, U., 2018. Prospective environmental risk assessment of mixtures in wastewater treatment plant effluents – theoretical considerations and experimental verification. Water Research 140, 56–66.

Cornick, N.A., Helgerson, A.F., 2004. Transmission and infectious dose of *Escherichia coli* O157: H7 in swine. Applied and Environmental Microbiology 70 (9), 5331–5335.

Creusot, N., Kinani, S., Balaguer, P., Tapie, N., LeMenach, K., Maillot-Marechal, E., Porcher, J.M., Budzinski, H., Ait-Aissa, S., 2010. Evaluation of an hPXR reporter gene assay for the detection of aquatic emerging pollutants: screening of chemicals and application to water samples. Analytical and Bioanalytical Chemistry 396 (2), 569–583.

Crouch, S., Kozlowski, R., Slater, K., Fletcher, J., 1993. The use of ATP bioluminescence as a measure of cell proliferation and cytotoxicity. Journal of Immunological Methods 160 (1), 81–88.

Deng, Y., Bonilla, M., Ren, H., Zhang, Y., 2018. Health risk assessment of reclaimed wastewater: a case study of a conventional water reclamation plant in Nanjing, China. Environment International 112, 235–242.

Diamond, J., Altenburger, R., Coors, A., Dyer, S.D., Focazio, M., Kidd, K., Koelmans, A.A., Leung, K.M.Y., Servos, M.R., Snape, J., Tolls, J., Zhang, X., 2018. Use of prospective and retrospective risk assessment methods that simplify chemical mixtures associated with treated domestic wastewater discharges. Environmental Toxicology & Chemistry 37 (3), 690–702.

Dong, S., Masalha, N., Plewa, M.J., Nguyen, T.H., 2017. Toxicity of wastewater with elevated bromide and iodide after chlorination, chloramination, or ozonation disinfection. Environmental Science and Technology 51 (16), 9297–9304.

Dong, S., Page, M.A., Wagner, E.D., Plewa, M.J., 2018. Thiol reactivity analyses to predict mammalian cell cytotoxicity of water samples. Environmental Science and Technology 52 (15), 8822–8829.

Duan, B., Zhang, W., Zheng, H., Wu, C., Zhang, Q., Bu, Y., 2017. Comparison of health risk assessments of heavy metals and as in sewage sludge from wastewater treatment plants (WWTPs) for adults and children in the urban district of Taiyuan, China. International Journal of Environmental Research and Public Health 14 (10), 1194.

Dupont, H.L., Formal, S.B., Hornick, R.B., Snyder, M.J., Libonati, J.P., Sheahan, D.G., Labrec, E.H., Kalas, J.P., 1971. Pathogenesis of *Escherichia coli* diarrhea. New England Journal of Medicine 285 (1), 1–9.

DuPont, H.L., Hornick, R.B., Snyder, M.J., Libonati, J.P., Formal, S.B., Gangarosa, E.J., 1972. Immunity in *Shigellosis*. II. protection induced by oral live vaccine or primary infection. The Journal of Infectious Diseases 125 (1), 12–16.

Efroymson, R.A., Jones, D.S., Gold, A.J., 2007. An ecological risk assessment framework for effects of onsite wastewater treatment systems and other localized sources of nutrients on aquatic ecosystems. Human and Ecological Risk Assessment 13 (3), 574–614.

Eldridge, M.L., Sanseverino, J., Layton, A.C., Easter, J.P., Schultz, T.W., Sayler, G.S., 2007. *Saccharomyces cerevisiae* BLYAS, a new bioluminescent bioreporter for detection of androgenic compounds. Applied and Environmental Microbiology 73 (19), 6012–6018.

EMEA, 2006. EMEA-European Agency for the Evaluation of Medicinal Products. European Guideline on the Environmental Risk Assessment of Medicinal Products for Human Use-Committee for Medical Products for Human Use; EMEA/CHMP/SWP/4447/00 Corr 2. 2006. London, available online at: https://goo.gl/FapAzh.

Environment Canada, 2012a. Metal Mining Environmental Effects Monitoring (EEM). Gatineau, QC.

Environment Canada, 2012b. Status Report on the Pulp and Paper Effluent Regulations. Gatineau, QC.

Escher, B.I., Allinson, M., Altenburger, R., Bain, P.A., Balaguer, P., Busch, W., Crago, J., Denslow, N.D., Dopp, E., Hilscherova, K., Humpage, A.R., Kumar, A., Grimaldi, M., Jayasinghe, B.S., Jarosova, B., Jia, A., Makarov, S., Maruya, K.A., Medvedev, A., Mehinto, A.C., Mendez, J.E., Poulsen, A., Prochazka, E., Richard, J., Schifferli, A., Schlenk, D., Scholz, S., Shiraish, F., Snyder, S., Su, G.Y., Tang, J.Y.M., van der Burg, B., van der Linden, S.C., Werner, I., Westerheide, S.D., Wong, C.K.C., Yang, M., Yeung, B.H.Y., Zhang, X.W., Leusch, F.D.L., 2014. Benchmarking organic micropollutants in wastewater, recycled water and drinking water with in vitro bioassays. Environmental Science and Technology 48 (3), 1940–1956.

Etteieb, S., Tarhouni, J., Isoda, H., 2019. Cellular stress response biomarkers for toxicity potential assessment of treated wastewater complex mixtures. Water and Environment Journal 33 (1), 4–13.

European Commission (EC), 2003. Technical Guidance Document on Risk Assessment: Part II. European Commission-Joint Research Centre, Institute for Health and Consumer Protection, European Chemicals Bureau (ECB).

Gijsbers, L., van Eekelen, H., de Haan, L.H.J., Swier, J.M., Heijink, N.L., Kloet, S.K., Man, H.Y., Bovy, A.G., Keijer, J., Aarts, J., van der Burg, B., Rietjens, I., 2013. Induction of peroxisome proliferator-activated receptor gamma (PPAR gamma)-mediated gene expression by tomato (*Solanum lycopersicum* L.) extracts. Journal of Agricultural and Food Chemistry 61 (14), 3419–3427.

Giuliano, K.A., DeBiasio, R.L., Dunlay, R.T., Gough, A., Volosky, J.M., Zock, J., Pavlakis, G.N., Taylor, D.L., 1997. High-content screening: a new approach to easing key bottlenecks in the drug discovery process. Journal of Biomolecular Screening 2 (4), 249–259.

Golnazarian, C.A., Donnelly, C.W., Pintauro, S.J., Howard, D.B., 1989. Comparison of infectious dose of Listeria monocytogenes F5817 as determined for normal versus compromised C57B1/6J mice. Journal of Food Protection 52 (10), 696–701.

Gonzalez, R.J., Tarloff, J.B., 2001. Evaluation of hepatic subcellular fractions for Alamar blue and MTT reductase activity. Toxicology in Vitro 15 (3), 257–259.

Guillén, D., Ginebreda, A., Farre, M., Darbra, R.M., Petrovic, M., Gros, M., Barcelo, D., 2012. Prioritization of chemicals in the aquatic environment based on risk assessment: analytical, modeling and regulatory perspective. The Science of the Total Environment 440, 236–252.

Haas, C., Rose, J.B., 1994. Reconciliation of Microbialrisk Models and Outbreak Epidemiology: The Case of the Milwaukee Outbreak.

Haas, C.N., 1983. Estimation of risk due to low doses of microorganisms: a comparison of alternative methodologies. American Journal of Epidemiology 118 (4), 573–582.

Haas, C.N., Rose, J.B., Gerba, C.P., 1999. Quantitative Microbial Risk Assessment. John Wiley & Sons.

Haas, C.N., Rose, J.B., Gerba, C.P., Haas, C.N., Rose, J.B., Gerba, C.P., 2014. Conducting the Dose-Response Assessment.

Hamouda, M.A., Anderson, W.B., Van Dyke, M.I., Douglas, I.P., McFadyen, S.D., Huck, P.M., 2016. Scenario-based quantitative microbial risk assessment to evaluate the robustness of a drinking water treatment plant. Water Quality Research Journal of Canada 51 (2), 81–96.

Heimersson, S., Harder, R., Peters, G.M., Svanstrom, M., 2014. Including pathogen risk in life cycle assessment of wastewater management. 2. Quantitative comparison of pathogen risk to other impacts on human health. Environmental Science and Technology 48 (16), 9446–9453.

Hornick, R.B., Greisman, S.E., Woodward, T.E., Dupont, H.L., Dawkins, A.T., Snyder, M.J., 1970. Typhoid fever: pathogenesis and immunologic control. 2. New England Journal of Medicine 283 (13), 686–691.

Hornick, R.B., Music, S.I., Wenzel, R., Cash, R., Libonati, J.P., Snyder, M.J., Woodward, T.E., 1971. Broad Street pump revisited: response of volunteers to ingested cholera vibrios. Journal of Urban Health 47 (10), 1181.

Hornick, R.B., Woodward, T.E., McCrumb, F.R., Snyder, M.J., Dawkins, A.T., Bulkeley, J.T., De la Macorra, F., Corozza, F.A., 1966. Study of induced typhoid fever in man. I. Evaluation of vaccine effectiveness. Transactions of the Association of American Physicians 79, 361–367.

Huang, R.L., Xia, M.H., Cho, M.H., Sakamuru, S., Shinn, P., Houck, K.A., Dix, D.J., Judson, R.S., Witt, K.L., Kavlock, R.J., Tice, R.R., Austin, C.P., 2011. Chemical genomics profiling of environmental chemical modulation of human nuclear receptors. Environmental Health Perspectives 119 (8), 1142–1148.

Iannetti, E.F., Willems, P., Pellegrini, M., Beyrath, J., Smeitink, J.A.M., Blanchet, L., Koopman, W.J.H., 2015. Toward high-content screening of mitochondrial morphology and membrane potential in living cells. The International Journal of Biochemistry & Cell Biology 63, 66–70.

International Organization for Standardization (ISO), 1998. Water Quality - Determination of the Inhibitory Effect of Water Samples on the Light Emission of Vibrio Fischeri (Luminescent Bacteria Test). ISO no. 11348e3.

International Organization for Standardization (ISO), 1998. Water Quality-Scientific and Technical Aspects of Batch Algae Growth Inhibition Tests. ISO no. 11044.

International Organization for Standardization (ISO), 2009. Biological evaluation of medical devices - Part 5: Tests for in vitro cytotoxicity. ISO no. 10993-5.

Jan, F.A., Ishaq, M., Khan, S., Ihsanullah, I., Ahmad, I., Shakirullah, M., 2010. A comparative study of human health risks via consumption of food crops grown on wastewater irrigated soil (Peshawar) and relatively clean water irrigated soil (lower Dir). Journal of Hazardous Materials 179 (1–3), 612–621.

Jiang, W.G., 2012. Electric Cell-Substrate Impedance Sensing and Cancer Metastasis. Springer Science & Business Media.

Khan, S., Cao, Q., Lin, A.-J., Zhu, Y.-G., 2008. Concentrations and bioaccessibility of polycyclic aromatic hydrocarbons in wastewater-irrigated soil using in vitro gastrointestinal test. Environmental Science & Pollution Research 15 (4), 344–353.

Kortenkamp, A., 2008. Low dose mixture effects of endocrine disrupters: implications for risk assessment and epidemiology. International Journal of Andrology 31, 233–237.

Lagarde, F., Jaffrezic-Renault, N., 2011. Cell-based electrochemical biosensors for water quality assessment. Analytical and Bioanalytical Chemistry 400 (4), 947–964.

Landis, W.G., Wiegers, J.K., 2007. Ten years of the relative risk model and regional scale ecological risk assessment. Hum. Ecol. Risk Assess. 13 (1), 25–38.

Lawinbrussel, C.A., Refojo, M.F., Leong, F.L., Hanninen, L., Kenyon, K.R., 1993. Effect of *Pseudomonas aeruginosa* concentration in experimental contact lens-related microbial keratitis. Cornea 12 (1), 10–18.

le Maire, A., Grimaldi, M., Roecklin, D., Dagnino, S., Vivat-Hannah, V., Balaguer, P., Bourguet, W., 2009. Activation of RXR-PPAR heterodimers by organotin environmental endocrine disruptors. EMBO Reports 10 (4), 367–373.

Lemaire, G., Mnif, W., Pascussi, J.M., Pillon, A., Rabenoelina, F., Fenet, H., Gomez, E., Casellas, C., Nicolas, J.C., Cavailles, V., Duchesne, M.J., Balaguer, P., 2006. Identification of new human pregnane X receptor ligands among pesticides using a stable reporter cell system. Toxicological Sciences 91 (2), 501–509.

Leusch, F.D.L., Khan, S.J., Laingam, S., Prochazka, E., Froscio, S., Trang, T., Chapman, H.F., Humpage, A., 2014. Assessment of the application of bioanalytical tools as surrogate measure of chemical contaminants in recycled water. Water Research 49, 300–315.

Li, Z., Liu, X., Huang, Z., Hu, S., Wang, J., Qian, Z., Feng, J., Xian, Q., Gong, T., 2019. Occurrence and ecological risk assessment of disinfection byproducts from chlorination of wastewater effluents in East China. Water Research 157, 247–257.

Mater, N., Geret, F., Castillo, L., Faucet-Marquis, V., Albasi, C., Pfohl-Leszkowicz, A., 2014. In vitro tests aiding ecological risk assessment of ciprofloxacin, tamoxifen and cyclophosphamide in range of concentrations released in hospital wastewater and surface water. Environment International 63, 191–200.

Mcbride, G., Till, D., Ball, A., 2002. Pathogen Occurrence and Human Health Risk Assessment Analysis.

McCullough, N.B., Eisele, C.W., 1951. Experimental human *Salmonellosis*. I. pathogenicity of strains of Salmonella meleagridis and Salmonella anatum obtained from spray-dried whole egg. The Journal of Infectious Diseases 88 (3), 278–289.

Mendoza, A., Acena, J., Perez, S., Lopez de Alda, M., Barcelo, D., Gil, A., Valcarcel, Y., 2015. Pharmaceuticals and iodinated contrast media in a hospital wastewater: a case study to analyse their presence and characterise their environmental risk and hazard. Environmental Research 140, 225–241.

Messner, M.J., Chappell, C.L., Okhuysen, P.C., 2001. Risk assessment for *Cryptosporidium*: a hierarchical Bayesian analysis of human dose response data. Water Research 35 (16), 3934–3940.

Miller, C.A., 1999. A human aryl hydrocarbon receptor signaling pathway constructed in yeast displays additive responses to ligand mixtures. Toxicology and Applied Pharmacology 160 (3), 297–303.

Mojovic, L.V., Jovanovic, G.N., 2005. Development of a microbiosensor based on fish Chromatophores immobilized on ferromagnetic gelatin beads. Food Technology and Biotechnology 43 (1), 1–7.

Molins-Delgado, D., Diaz-Cruz, M.S., Barcelo, D., 2016. Ecological risk assessment associated to the removal of endocrine-disrupting parabens and benzophenone-4 in wastewater treatment. Journal of Hazardous Materials 310, 143–151.

Muller, D., Edwards, M.L., Smith, D.W., 1983. Changes in iron and transferrin levels and body temperature in experimental airborne legionellosis. The Journal of Infectious Diseases 147 (2), 302–307.

Nagy, S.R., Sanborn, J.R., Hammock, B.D., Denison, M.S., 2002. Development of a green fluorescent protein-based cell bioassay for the rapid and inexpensive detection and characterization of Ah receptor agonists. Toxicological Sciences 65 (2), 200–210.

Naylor, L.H., 1999. Reporter gene technology: the future looks bright. Biochemical Pharmacology 58, 749–757.

Niles, A.L., Moravec, R.A., Worzella, T.J., Evans, N.J., Riss, T.L., 2013. High-throughput screening assays for the assessment of cytotoxicity. In: book: High-Throughput Screening Methods in Toxicity Testing, pp. 107–127.

Niu, Z.G., Xu, S.Y., Gong, Q.C., 2014. Health risk assessment of odors emitted from urban wastewater pump stations in Tianjin, China. Environmental Science and Pollution Research International 21 (17), 10349–10360.

O'Brien, R., Mackintosh, C.G., Bakker, D., Kopecna, M., Pavlik, I., Griffin, J.F.T., 2006. Immunological and molecular characterization of susceptibility in relationship to bacterial strain differences in *Mycobacterium avium* subsp paratuberculosis infection in the red deer (*Cervus elaphus*). Infection and Immunity 74 (6), 3530–3537.

Oda, Y., Nakamura, S., Oki, I., Kato, T., Shinagawa, H., 1985. Evaluation of the new system (UMU-TEST) for the detection of environmental mutagens and carcinogens. Mutation Research 147 (5), 219–229.

Olujimi, O.O., Aroyeun, O.A., Akinhanmi, T.F., Arowolo, T.A., 2017. Occurrence, removal and health risk assessment of phthalate esters in the process streams of two different wastewater treatment plants in Lagos and Ogun States, Nigeria. Environmental Monitoring and Assessment 189 (7), 345.

Papageorgiou, M., Kosma, C., Lambropoulou, D., 2016. Seasonal occurrence, removal, mass loading and environmental risk assessment of 55 pharmaceuticals and personal care products in a municipal wastewater treatment plant in Central Greece. The Science of the Total Environment 543, 547–569.

Piersma, A.H., Bosgra, S., van Duursen, M.B.M., Hermsen, S.A.B., Jonker, L.R.A., Kroese, E.D., van der Linden, S.C., Man, H., Roelofs, M.J.E., Schulpen, S.H.W., Schwarz, M., Uibel, F., van Vugt-Lussenburg, B.M.A., Westerhout, J., Wolterbeek, A.P.M., van der Burg, B., 2013. Evaluation of an alternative in vitro test battery for detecting reproductive toxicants. Reproductive Toxicology 38, 53–64.

Prasse, C., Stalter, D., Schulte-Oehlmann, U., Oehlmann, J., Ternes, T.A., 2015. Spoilt for choice: a critical review on the chemical and biological assessment of current wastewater treatment technologies. Water Research 87, 237–270.

Prosser, R.S., Sibley, P.K., 2015. Human health risk assessment of pharmaceuticals and personal care products in plant tissue due to biosolids and manure amendments, and wastewater irrigation. Environment International 75, 223–233.

Rashid, H., Arslan, C., Khan, S.N., 2018. Wastewater irrigation, its impact on environment and health risk assessment in peri urban areas of Punjab Pakistan-a review. Environ. Contam. Rev. 1 (1), 30–35.

Regli, S., Rose, J.B., Haas, C.N., Gerba, C.P., 1991. Modeling the risk from Giardia and viruses in drinking water. Journal of the American Water Works Association 83 (11), 76–84.

Rendtorff, R.C., 1954. The experimental transmission of human intestinal protozoan parasites. II. *Giardia lamblia* cysts given in capsules. American Journal of Hygiene 59 (2), 209–220.

Reyjol, Y., Argillier, C., Bonne, W., Borja, A., Buijse, A.D., Cardoso, A.C., 2014. Assessing the ecological status in the context of the European water framework directive: where do we go now? The Science of the Total Environment 497, 332–344.

Rivera-Jaimes, J.A., Postigo, C., Melgoza-Aleman, R.M., Acena, J., Barcelo, D., Lopez de Alda, M., 2018. Study of pharmaceuticals in surface and wastewater from Cuernavaca, Morelos, Mexico: occurrence and environmental risk assessment. The Science of the Total Environment 613–614, 1263–1274.

Rohr, J.R., Salice, C.J., Nisbet, R.M., 2016. The pros and cons of ecological risk assessment based on data from different levels of biological organization. Critical Reviews in Toxicology 46 (9), 756–784.

Rose, J.B., Haas, C.N., Regli, S., 1991. Risk assessment and control of waterborne giardiasis. American Journal of Public Health 81 (6), 709–713.

Routledge, E.J., Sumpter, J.P., 1996. Estrogenic activity of surfactants and some of their degradation products assessed using a recombinant yeast screen. Environmental Toxicology & Chemistry 15 (3), 241–248.

Sanderson, J.T., Aarts, J., Brouwer, A., Froese, K.L., Denison, M.S., Giesy, J.P., 1996. Comparison of Ah receptor-mediated luciferase and ethoxyresorufin-O-deethylase induction in H4IIE cells: implications for their use as bioanalytical tools for the detection of polyhalogenated aromatic hydrocarbons. Toxicology and Applied Pharmacology 137 (2), 316–325.

Sanseverino, J., Gupta, R.K., Layton, A.C., Patterson, S.S., Ripp, S.A., Saidak, L., Simpson, M.L., Schultz, T.W., Sayler, G.S., 2005. Use of *Saccharomyces cerevisiae* BLYES expressing bacterial bioluminescence for rapid, sensitive detection of estrogenic compounds. Applied and Environmental Microbiology 71 (8), 4455–4460.

SCENIHR, SCCS, SCHER, 2013. Adressing the New Challenges for Risk Assessment.

Schiff, G.M., Stefanovic, G.M., Young, E.C., Sander, D.S., Pennekamp, J.K., Ward, R.L., 1984. Studies of Echovirus-12 in volunteers: determination of minimal infectious dose and the effect of previous infection on infectious dose. The Journal of Infectious Diseases 150 (6), 858–866.

Seimandi, M., Lemaire, G., Pillon, A., Perrin, A., Carlavan, I., Voegel, J.J., Vignon, F., Nicolas, J.C., Balaguer, P., 2005. Differential responses of PPAR alpha, PPAR delta, and PPAR gamma reporter cell lines to selective PPAR synthetic ligands. Analytical Biochemistry 344 (1), 8–15.

Semerjian, L., Shanableh, A., Semreen, M.H., Samarai, M., 2018. Human health risk assessment of pharmaceuticals in treated wastewater reused for non-potable applications in Sharjah, United Arab Emirates. Environment International 121, 325–331.

Shum, D., Radu, C., Kim, E., Cajuste, M., Shao, Y., Seshan, V.E., Djaballah, H., 2008. A high density assay format for the detection of novel cytotoxic agents in large chemical libraries. Journal of Enzyme Inhibition and Medicinal Chemistry 23 (6), 931–945.

Sohoni, P., Sumpter, J.P., 1998. Several environmental oestrogens are also anti-androgens. Journal of Endocrinology 158 (3), 327–339.

Soller, J.A., 2006. Use of microbial risk assessment to inform the national estimate of acute gastrointestinal illness attributable to microbes in drinking water. Journal of Water and Health 4 (Suppl. 2), 165–186.

Sonneveld, E., Jansen, H.J., Riteco, J.A.C., Brouwer, A., van der Burg, B., 2005. Development of androgen- and estrogen-responsive bioassays, members of a panel of human cell line-based highly selective steroid-responsive bioassays. Toxicological Sciences 83 (1), 136–148.

Staley, Z.R., Rohr, J.R., Harwood, V.J., 2010. The effect of agrochemicals on indicator bacteria densities in outdoor mesocosms. Environmental Microbiology 12 (12), 3150–3158.

Stockwell, B.R., 2004. Exploring biology with small organic molecules. Nature 432 (7019), 846–854.

Suto, C.M., Ignar, D.M., 1997. Selection of an optimal reporter gene for cell-based high throughput screening assays. Journal of Biomolecular Screening 2 (1), 7–9.

Terouanne, B., Tahiri, B., Georget, V., Belon, C., Poujol, N., Avances, C., Orio, F., Balaguer, P., Sultan, C., 2000. A stable prostatic bioluminescent cell line to investigate androgen and antiandrogen effects. Molecular and Cellular Endocrinology 160 (1–2), 39–49.

Townsend, D.M., Tew, K.D., Tapiero, H., 2003. The importance of glutathione in human disease. Biomedicine & Pharmacotherapy 57 (3–4), 145–155.

USEPA, 2012. USEPA-US Environmental Protection Agency. Ecological Structure Activity Relationships (ECOSAR) Predictive Model v1.11. 2012. Available at: https://goo.gl/xBM2VN.

USEPA, 2002. Guidelines Establishing Test Procedures for the Analysis of Pollutants; Whole Effluent Toxicity Test Methods; final rule. U.S. Environmental Protection Agency. Fed Reg 67, 69951–69972.

USEPA and IRIS, United States, Environmental Protection Agency, Integrated Risk Information System. http://www.epa.gov/iris/subst (December, 2006), 2006.

USEPA, 2004. Risk Assessment Guidance for Superfund Volume I: Human Health Evaluation Manual (Part E, Supplemental Guidance for Dermal Risk Assessment). U.S. Environmental Protection Agency, Washington, DC.

USEPA, Risk Assessment Guidance for Superfund Volume I Human Health Evaluation Manual (Part A), 1989, U.S. Environmental Protection Agency, Washington, DC.

US-EPA, 2011. Exposure Factors Handbook. National Center for Environmental Assessment Office of Research and Development. U.S. Environmental Protection Agency, Washington, DC. Available at: https://cfpub.epa.gov/ncea/risk/recordisplay.cfm?deid=236252. accessed 21.10.2016.

van der Linden, S.C., von Bergh, A.R.M., van Vught-Lussenburg, B.M.A., Jonker, L.R.A., Teunis, M., Krul, C.A.M., van der Burg, B., 2014. Development of a panel of high-throughput reporter-gene assays to detect genotoxicity and oxidative stress. Mutation Research/Genetic Toxicology and Environmental Mutagenesis 760, 23–32.

Wang, X.L., Tao, S., Dawson, R.W., Xu, F.L., 2002. Characterizing and comparing risks of polycyclic aromatic hydrocarbons in a Tianjin wastewater-irrigated area. Environmental Research 90, 201–206.

Ward, R.L., Bernstein, D.I., Young, E.C., Sherwood, J.R., Knowlton, D.R., Schiff, G.M., 1986. Human rotavirus studies in volunteers: determination of infectious dose and serological response to infection. The Journal of Infectious Diseases 154 (5), 871–880.

Westrell, T., 2004. Microbial Risk Assessment and its Implications for Risk Management in Urban Water Systems. Linköping University Electronic Press.

WHO, 2016. WHO-world health organization. Quantitative Microbial Risk Assessment: Application for Water Safety Management.

Widiana, D.R., Wang, Y.-F., You, S.-J., Yang, H.-H., Wang, L.-C., Tsai, J.-H., Chen, H.-M., 2019. Air pollution profiles and health risk assessment of ambient volatile organic compounds above a municipal wastewater treatment plant, Taiwan. Aerosol Air Qual. Res. 19 (2), 375–382.

Wilson, V.S., Bobseine, K., Gray, L.E., 2004. Development and characterization of a cell line that stably expresses an estrogen-responsive luciferase reporter for the detection of estrogen receptor agonist and antagonists. Toxicological Sciences 81 (1), 69–77.

Wilson, V.S., Bobseine, K., Lambright, C.R., Gray, L.E., 2002. A novel cell line, MDA-kb2, that stably expresses an androgen- and glucocorticoid-responsive reporter for the detection of hormone receptor agonists and antagonists. Toxicological Sciences 66 (1), 69–81.

Wu, B., Zhang, Y., Zhang, X.X., Cheng, S.P., 2011. Health risk assessment of polycyclic aromatic hydrocarbons in the source water and drinking water of China: quantitative analysis based on published monitoring data. The Science of the Total Environment 410–411, 112–118.

Wu, W.L., Huang, Y.D., Hsu, K.E., Wang, Y.H., Huang, H.H., Hsiung, W.C., Chen, S.M., Chang, H.S., Chu, C.P., Chung, Y.J., Huang, Y.T., 2014. A health risk assessment of reclaimed municipal wastewater for industrial and miscellaneous use. Water Science and Technology 70 (4), 750–756.

Xiao, S., Hu, S., Zhang, Y., Zhao, X., Pan, W., 2018. Influence of sewage treatment plant effluent discharge into multipurpose river on its water quality: a quantitative health risk assessment of *Cryptosporidium* and *Giardia*. Environment and Pollution 233, 797–805.

Yamamoto, H., Watanabe, M., Hirata, Y., Nakamura, Y., Nakamura, Y., Kitani, C., Sekizawa, J., Uchida, M., Nakamura, H., Kagami, Y., Koshio, M., Hirai, N., Tatarazako, N., 2007. Preliminary ecological risk assessment of butylparaben and benzylparaben – 1. Removal efficiency in wastewater treatment, acute/chronic toxicity for aquatic organisms, and effects on medaka gene expression. Environmental Sciences 14 (Suppl. S), 73–87.

Yang, W.B., Chen, W.H., Yuan, C.S., Yang, J.C., Zhao, Q.L., 2012. Comparative assessments of VOC emission rates and associated health risks from wastewater treatment processes. Journal of Environmental Monitoring 14 (9), 2464–2474.

Zanella, F., Lorens, J.B., Link, W., 2010. High content screening: seeing is believing. Trends in Biotechnology 28 (5), 237–245.

Zeng, S., Huang, Y., Sun, F., Li, D., He, M., 2016. Probabilistic ecological risk assessment of effluent toxicity of a wastewater reclamation plant based on process modeling. Water Research 100, 367–376.

Zhang, J.N., Ying, G.G., Yang, Y.Y., Liu, W.R., Liu, S.S., Chen, J., Liu, Y.S., Zhao, J.L., Zhang, Q.Q., 2018. Occurrence, fate and risk assessment of androgens in ten wastewater treatment plants and receiving rivers of South China. Chemosphere 201, 644–654.

Zhang, W., Wei, C., Feng, C., Yan, B., Li, N., Peng, P., Fu, J., 2012. Coking wastewater treatment plant as a source of polycyclic aromatic hydrocarbons (PAHs) to the atmosphere and health-risk assessment for workers. The Science of the Total Environment 432, 396–403.

CHAPTER

Physicochemical technologies for HRPs and risk control

8

Haidong Hu, Ke Xu

State Key Laboratory of Pollution Control and Resource Reuse, School of the Environment, Nanjing University, Nanjing, China

Chapter outline

High-Risk Pollutants in Wastewater. https://doi.org/10.1016/B978-0-12-816448-8.00008-3

In the past few decades, many researchers have studied the efficiency of different biological and physicochemical wastewater treatment processes to remove high-risk pollutants (HRPs) and control their environmental risks (Flox et al., 2006; Hu et al., 2017; Lei et al., 2016, Shi et al., 2017). At present, many of the physicochemical technologies have been successfully applied to practical wastewater treatment and have been shown to be effective in removing HRPs from wastewater (Akbal and Camcı, 2010; Moreno-Piraján et al., 2010). Compared to biological processes, physicochemical processes that are mainly used to transfer pollutants from one phase to another to achieve the purpose of controlling pollution are relatively simple, in which pollutants may change during the process, such as advanced oxidation, or do not change at all, such as membrane separation and adsorption. However, the results of different physicochemical processes acting on different pollutants are not the same, which requires us to fully understand the working principle and efficiency of different physicochemical technologies and choose appropriate methods to deal with different HRPs. In this chapter, we summarized the advantages and disadvantages of various available technologies for the removal of HRPs, including adsorption, advanced oxidation, membrane separation, combination process and emerging technologies, and proposed the development perspectives of these technologies based on the investigation of their current status.

8.1 Adsorption technologies

In recent years, the application of adsorption technology in wastewater treatment has attracted great concerns from the scientific community. Because of its convenience, simple design, and ease of operation, adsorption technology is considered to have great potential for the removal of HRPs. The adsorption process contains many different adsorption forces, and only specific adsorption forces can effectively adsorb specific HRPs. Therefore, the preparation of adsorbent materials with targeted adsorption is the key to the removal of HRPs by adsorption technology. At present, more and more different kinds of adsorption materials (e.g., carbon-based adsorbents, nanoadsorbents, resins, and modified and composite adsorbents) have been developed to remove HRPs in wastewater. They are developed on the basis of different adsorption principles, such as pore-filling, H-bonding, hydrophobic interaction, and ion exchange, and be targeted to effectively remove different HRPs. In this section, we mainly introduce the principle of adsorption technology and the research progress of different adsorption materials applied in HRPs removal. By reviewing different adsorption materials and their design principles, the advantages and disadvantages of adsorption technology to remove HRPs are comprehensively discussed.

8.1.1 Main concepts and bases of adsorption

Adsorption is a mass transfer process that is a phenomenon of sorption of gases or solutes by solid or liquid surfaces. The adsorption on the solid surface is that the molecules or atoms on the solid surface have residual surface energy due to unbalanced forces. When some substances collide with the solid surface, they are attracted by these unbalanced forces and stay on the solid surface. According to the different adsorption forces, the adsorption process can be divided into two categories: physical adsorption and chemical adsorption (Table 8.1). Physical adsorption is produced by the interaction of intermolecular forces (i.e., van der Waals forces), for example, the adsorption of activated carbon for gas. Physical adsorption is generally carried out at a low temperature, and fast adsorption rate, low adsorption heat, and nonselective. As the effect of intermolecular attraction is weak, the structure of the adsorbate molecules hardly changes, the adsorption energy is small, and the adsorbed substance is easily separated again. The adsorption due to the action of chemical bonds is chemical adsorption. Chemical adsorption process includes the formation and destruction of chemical bonds. The absorption or release of adsorption heat is larger, and the activation energy required is also larger. Physical adsorption and chemical adsorption are not isolated and often occur together. In wastewater treatment technology, most of the adsorption is the result of several kinds of adsorption processes. Due to the influence of adsorbents, adsorbates, and other factors, some kind of adsorption may play a leading role.

Activated carbon is the most commonly used adsorbent material in wastewater treatment, and it can be produced by pyrolysis of almost all carbonaceous organic

Table 8.1 Comparison of physical adsorption with chemical adsorption.

	Adsorption categories	
	Physical adsorption	**Chemical adsorption**
Adsorption force	Van der Waals force	Chemical bond force
Selectivity	Nonselective adsorption	Selective adsorption
Adsorption layer	Single or multiple layers	Single layer
Adsorption heat	Low	High
Adsorption rate	Fast	Slow
Stability	Instable	Stable

materials such as coal, wood, husks, coconut shells, and walnut shells. Due to its abundant microporous structure, large specific surface area, and high hydrophobicity, activated carbon exhibits excellent adsorption capacity for most pollutants. To further improve the practical utilization advantages of adsorption technology, various new adsorbent materials have been proposed to replace activated carbon, including new carbon-based adsorbents, nanoadsorbents, metal oxides, and hydroxides-based adsorbents, resins, and modified and composite adsorbents.

8.1.2 Carbon-based adsorbents

8.1.2.1 Activated carbon

As a traditional adsorption material that has been widely used, the adsorption characteristics of activated carbon for HRPs in wastewater have been widely concerned. Because of its well-developed pore structure, large specific surface area, and different surface functional groups, activated carbon materials have a certain ability to remove heavy metals from wastewater. Moreno-Piraján et al. investigated the adsorption characteristics of coconut shell activated carbon on four metal ions of manganese, iron, nickel, and copper in wastewater (Moreno-Piraján et al., 2010). They found that the pH of the solution had a significant effect on the adsorption capacity, and the activated carbon had the strongest adsorption capacity for all metal ions at $pH = 5.8$. All adsorption processes are consistent with the Langmuir isotherm adsorption model and the pseudo-second-order kinetic model. Based on the experimental results, they speculated that ion exchange may be the main mechanism for the adsorption of heavy metals for activated carbon. The recent use of agricultural waste to prepare activated carbon materials to reduce the cost of activated carbon production has also become a research hotspot. Guo et al. prepared activated carbon from poultry litter and used it to remove heavy metal ions such as copper, lead, zinc, and cadmium from water (Guo et al., 2010). The study revealed that poultry litter-based activated carbon possessed significantly higher adsorption affinity and capacity for heavy metals than commercial activated carbon derived from bituminous coal and coconut shells. In addition, the use of the poultry litter-based activated carbon does not release nutrients and heavy metals and thereby causes secondary pollution of water bodies.

Activated carbon also has a high affinity for organic HRPs. Adsorption of ibuprofen, ketoprofen, naproxen, and diclofenac onto a low-cost activated carbon, prepared from olive-waste cakes, has been investigated (Baccar et al., 2012). The results showed that the adsorption capacity of these four drugs for activated carbon varied greatly and was linked essentially to their pKa and their octanol/water coefficient. With the increase of pH, the adsorption capacity of activated carbon for all four drugs decreased, and this trend was more obvious under alkaline conditions. Moreover, the temperature has little effect on adsorption. Salman et al. prepared activated carbon from the banana stalk by potassium hydroxide (KOH) and carbon dioxide (CO_2) activation and investigated its adsorption capacity to pesticides (Salman et al., 2011). The study revealed that activated carbon from banana stalk could effectively remove 2,4-dichlorophenoxyacetic acid (2, 4-D) (196.33 mg/g) and bentazon (115.07 mg/g) from aqueous solutions. The adsorption processes both obey the Freundlich model and the pseudo-second-order kinetic model, and are feasible, spontaneous, and exothermic in nature.

8.1.2.2 Biochar

Biochar is a porous carbonaceous material that is produced from a variety of biomass by different thermal decomposition methods including carbonization, hydrothermal treatment, and pyrolysis. Considering the large porous structure, abundant functional groups, low cost, and solid waste recycling, biochar is regarded as a potential alternative adsorbent for use in wastewater treatment. With the increase of research on biochar adsorption technology, the adsorption mechanisms of biochar on various pollutants are becoming clearer (Fig. 8.1). Rational use of these adsorption mechanisms can make biochar more effective in removing HRPs from wastewater. Abdelhafez and Li prepared two biochars using bagasse and orange peel as raw materials and investigated the adsorption characteristics of lead ions by these two biochars (Abdelhafez and Li, 2016). The results showed that the adsorption capacity of these two biochars for lead ions varied greatly. By analyzing the chemical composition and morphological structure of the two biochars, it was found that the major adsorption mechanisms of both biochars were a specific ion-exchange mechanism and surface precipitation. Because bagasse biochar has a larger specific surface area, it has a stronger adsorption capacity than orange peel biochar.

Similar to most porous carbon materials, biochar also has a high adsorption affinity for organic HRPs and can be adsorbed through various mechanisms. For example, as the main site of adsorption of biochar, the pore structure has an

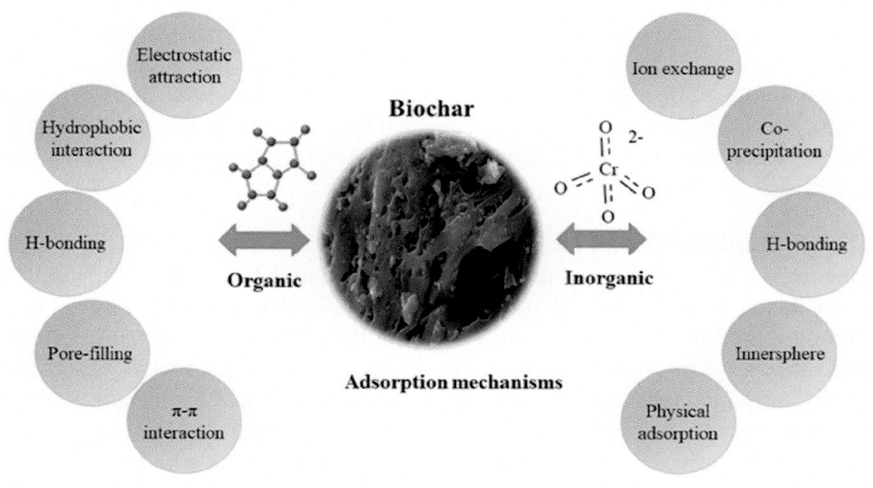

FIGURE 8.1

Adsorption mechanism of organic and inorganic matter by biochar.

important influence on the adsorption of organic HRPs by biochar. Wang et al. evaluated the sorption behavior of acetochlor, dibutyl phthalate, 17α-Ethynyl estradiol, and phenanthrene with biochars produced from three feedstocks (maize straw, pine wood dust, and swine manure) at seven heat treatment temperatures (Wang et al., 2016). The results showed that the organic carbon-normalized CO_2-specific surface area of biochars significantly correlated with the sorption coefficients, suggesting that pore filling could dominate the sorption of tested biochars. In addition, due to abundant surface functional groups, strong hydrophobicity, and graphitized lamellar structure, biochar can also adsorb organic HRPs through hydrophobic action, hydrogen bond, and $\pi-\pi$ interaction. Chen et al. studied the interfacial interaction between carbamazepine and peanut-shell-derived biochar (Chen et al., 2017). They found that the adsorption of carbamazepine (CBZ) on amorphous (loose) carbon was lower than aromatic (condensed) carbon, but the former mainly contributed to the fast adsorption of carbamazepine, indicating the hydrophobic and $\pi-\pi$ interactions were likely the predominant adsorption mechanisms of CBZ on biochar. Through acid treatment experiments, it was proved that the mineral component on the surface of biochar could also generate hydrogen bonding with carbamazepine.

8.1.3 Nanoadsorbents

The application of nanotechnology in wastewater treatment has received extensive attention in the scientific community. Nanomaterials refer to materials with a structural unit size of <100 nm, which are intermediate to microscopic atoms, molecules, and typical macroscopic materials. When the particles of matter enter the nanometer scale, they will show strong small size effects, quantum effects, and huge surface effects. Due to the large specific surface area and more surface atoms of nanomaterials, the coordination of surface atoms is insufficient, and the number of unsaturated bonds is increased. As a result, these surface atoms are highly active, highly unstable and easy to combine with other atoms, thus showing strong adsorption capacity. In addition, on the surface of nanoparticles, there are active groups such as hydroxyl groups, which can bond with heavy metals and organic compounds in wastewater. Therefore, nanomaterials have shown great potential in the preparation of high-performance adsorbents.

8.1.3.1 Graphene-based nanoadsorbent

Graphene is a kind of two-dimensional carbon nanomaterials, which is composed of carbon atoms with sp^2 hybrid orbitals in a hexagonal honeycomb lattice. It is currently the thinnest and most sturdy material with a large specific surface area. Due to its unique physicochemical properties, graphene has been widely used as a potential adsorbent material. The graphene nanosheets prepared by vacuum accelerated low temperature stripping adsorbed lead ions in the water system (Huang et al., 2011). Studies have shown that graphene nanosheets have a good adsorption efficiency for lead ions. Due to its large external surface, the graphene nanosheets exhibit a high initial adsorption rate. However, although the graphene nanosheets exhibit good adsorption properties, it exhibits a strong water repellency and is easily polymerized, which adversely affects its practical use. In the step of preparing graphene by oxidation, the intermediate product graphene oxide has attracted attention. There are a large number of oxygen-containing functional groups on the surface of graphene oxide, including hydroxyl groups, epoxy groups, carbonyl groups, and carboxyl groups. These surface functional groups exhibit good hydrophilicity and can also be adsorption sites for HRPs. Yang et al. prepared graphene oxide following the modified Hummers' method, and used it to remove Cu (II) from water (Yang et al., 2010). They found that graphene oxide could be aggregated by Cu^{2+} in aqueous solution, which was most likely triggered by the coordination between graphene oxide and Cu^{2+}. Graphene oxide has a huge absorption capacity for Cu^{2+}, which is around 10 times that of active carbon. Graphite-based nano-adsorbent can also have high adsorption affinity with organic HRPs through hydrophobic interaction and other interaction. Pei et al. studied the adsorption characteristics of 1,2,4-trichlorobenzene, 2,4,6-trichlorophenol, 2-naphthol, and naphthalene on graphene and graphene oxide (Pei et al., 2013). Through theoretical calculations, the results showed that graphene and graphene oxide adsorbed these organic pollutants via hydrophobic interaction. Fourier infrared spectroscopy results revealed that organic pollutants were adsorbed on graphene oxide mainly via $\pi-\pi$ interaction. In contrast, high adsorption of 2,4,6-trichlorophenol and 2-naphthol on graphene oxide was attributed to the formation of H-bonding.

8.1.3.2 Carbon nanotubes

Carbon nanotubes are one-dimensional quantum materials with a special structure (the radial dimension is on the order of nanometers, the axial dimension is on the order of micrometers, and the ends of the tube are substantially sealed). The carbon nanotubes are mainly composed of hexagonal carbon atoms arranged in a plurality of layers to tens of layers of coaxial tubes, and the layers are kept at a fixed distance from each other. Carbon nanotubes can be regarded as being formed by crimping graphene sheets, so they can be classified into single-walled carbon nanotubes and multiwalled carbon nanotubes according to the number of layers of graphene sheets. Carbon nanotubes have a large specific surface area, high surface energy, and large pore structure, and thus have good adsorption capacity. Abdel Salam investigated the adsorption properties of multiwalled carbon nanotubes on heavy metals in aqueous solutions (Abdel Salam, 2012). Multiwalled carbon nanotubes can effectively remove copper(II), lead(II), chromium(II) and zinc(II) in aqueous solution, and the adsorption process conform to the pseudo-second-order kinetic model. Through the calculation of thermodynamic parameters, the results showed that the adsorption process was a process of endothermic, spontaneous, and entropy increase. The higher the temperature, the more favorable the adsorption reaction occurs, and the stronger the adsorption capacity. The adsorption of ibuprofen and triclosan as representative types of pharmaceutical and personal care products on single-walled carbon nanotubes, multiwalled carbon nanotubes (MWCNTs), and oxidized MWCNTs were studied by Cho et al. (2011). The results showed that these carbon nanotubes could effectively adsorb ibuprofen and triclosan, and the larger the specific surface area, the stronger the adsorption capacity of these two pollutants. In addition, the surface chemistry of carbon nanotubes, the chemical properties of personal care products, and aqueous solution chemistry (pH, ionic strength, fulvic acid) all play an important role in personal care products adsorption onto carbon nanotubes.

8.1.3.3 Nano metal oxide

Iron oxide, copper oxide, and other metal oxides are low-cost adsorbents that can effectively absorb heavy metals. Adsorption is mainly achieved by complexation of metal ions with oxygen atoms of metal oxides. Nanoscale metal oxides have higher adsorption capacity due to their larger specific surface area, shorter diffusion distance and a greater number of surface sites. Jeong et al. evaluated the potential of nano-iron oxide and nano-alumina as arsenic ion adsorbents (Jeong et al., 2007). The studies have shown that by adjusting the pH of the solution and the dosage of nano metal oxide, the removal rate of arsenic can reach 99%. Under the optimum conditions, the maximum adsorption capacity of nano-iron oxide and nano-alumina to arsenic was 0.66 and 0.17 mg/g, respectively. Similar to other porous adsorbent materials, the adsorption properties of nano-metal oxides to arsenic are also mainly controlled by specific surface area. In addition, the combination of the two metal oxides can inherit the properties of the parent metal oxide and also exhibit synergistic effects. Zhang et al. synthesized a novel nanostructured Fe—Cu binary oxide by a facile coprecipitation method (Zhang et al., 2013). The results indicated that the maximal adsorption capacities of the Fe—Cu binary oxide with a Cu: Fe molar ratio of 1:2 for As(V) and As(III) were 82.7 and 122.3 mg/g at pH 7.0, respectively, which has a distinct advantage over other adsorbent materials. Furthermore, this material is easy to regenerate and reuse, and could be a promising adsorbent for both As(V) and As(III) removal from wastewater.

8.1.4 Resins

Resin is a novel organic adsorbent that has a three-dimensional network structure, is a porous sponge, is not melted by heating, is insoluble in common solvents and acids and alkalis, and has a specific surface area of more than 1000 m^2/g. Due to its good adsorption capacity and easy regeneration, some resin materials have been used in actual water treatment processes. At present, some studies have focused on the removal of HRPs by resin materials. Berker et al. synthesized ion-exchange resins using hydroxyethyl cellulose and studied their adsorption characteristics for heavy metal ions such as iron, cobalt, copper, and zinc (Beker et al., 1999). It has been found that pH of the solution has a great influence on the adsorption process. At $pH = 6$, the resin material can effectively remove heavy metal ions in the solution. Robberson et al. investigated the adsorption characteristics of anion-exchange and neutral resins for quinolone antibiotic nalidixic acid (Robberson et al., 2006). They found that the ionic form of nalidixic acid posed a great influence on the adsorption capacity of anion-exchange and neutral resins. In addition, the aromatic structure in the resins is also beneficial to its adsorption of nalidixic acid.

8.1.5 Composite and modified adsorbents

Although the original adsorbent materials have great potential for application, they are always limited by their existence, for example, nonselective adsorption and difficult to recycle. To further improve the application range of these adsorbent materials, the researchers have modified them to strengthen advantages and weaken disadvantages. There are two main types of modification methods: (1) improving the adsorbent structure and increasing the active adsorption sites; (2) composite of two different adsorption materials, facilitating the synergistic effect of the two adsorption materials.

There are many ways to modify adsorbents. In general, grafting active sites and improving the adsorbent structure are the main purpose of modification. For specific HRPs, the adsorption capacity and specificity of the adsorbent can be significantly improved by grafting the corresponding active site. Shi et al. successfully grafted polyethyleneimine onto the surface of biochar produced by the hydrothermal method (Shi et al., 2017). The results showed that the adsorption capacity of modified biochar to chromium(VI) and nickel(II) in aqueous solution was improved obviously. Through X-ray photoelectron spectroscopy analysis, it has been found that polyethyleneimine grafted on biochar plays a key role in improving the adsorption effect. Amino groups in polyethyleneimine molecules can effectively adsorb heavy metal ions through complexation and charge attraction. For some adsorbents, their inherent properties may limit their applications. For example, although graphene materials are used for a large specific surface area, they are prone to agglomeration in water bodies, resulting in low adsorption efficiency. The modification of these adsorbents mainly focuses on the improvement of their structure. Zhao et al. introduced a kind of sulfonated graphene (around 3 nm thick) with high dispersion properties in aqueous solution. In addition to the improvement of dispersion, a large number of parked sp^2-bonded carbon atoms also appeared in the sulfonated graphene structure, suggesting that it could effectively adsorb aromatic and high-risk organic pollutants through $\pi-\pi$ interaction. Experiments confirmed the adsorption capability of the prepared sulfonated graphene for naphthalene and 1-naphthol is about 2.3–2.4 mmol/g, which is one of the highest capabilities of nanomaterials.

Because it can promote the synergistic effect of the two materials, the composite adsorption materials have attracted more and more attention from the scientific community, especially the magnetic composite materials. The difficulty of adsorbent recovery has always been one of the main factors restricting the application of adsorption technology. By combining magnetic nanoparticles with other adsorption materials, other adsorption materials can also be magnetic. In the actual water treatment process, the adsorption materials can be recovered and regenerated from sewage by means of magnetic separation. Jung et al. synthesized cubic spinel-type manganese ferrite ($MnFe_2O_4$)/biochar composites via a one-pot hydrothermal technique (Jung et al., 2018). They found that this composite material not only showed good magnetic properties and was easy to separate, but also exhibited better adsorption capacity for heavy metal ions (Pb(II), Cu(II), and Cd(II)) than the original biochar.

Adsorption technology is an effective means of treating HRPs in wastewater. The application of different adsorbents can specifically remove organic HRPs and reduce their harm. However, as far as the current application of the adsorption technology is concerned, there are certain problems, such as high cost of the adsorbent, difficulty in disposal, and vulnerability to the wastewater environment. Therefore, future studies should focus more on further enhancing the antipollution, selectivity, and economy of adsorbents. In addition, there is a lack of research on the properties of adsorption materials in practical systems. A large number of competitive ions and natural organic matter (NOM) exist in the actual system, which will have a significant impact on the practical application of adsorption materials. However, most of the current studies are in aqueous systems, so it is impossible to know the extent of these effects.

8.2 Advanced oxidation

HRPs can be degraded by advanced oxidation processes (AOPs). Many studies prove that these processes are able to mineralize or convert organic HRPs to less harmful organic matters. Most of AOPs are based on the *in situ* generation of strong oxidants OH-radicals (·OH) for the oxidation of organic compounds (Miklos et al., 2018). The great advantage and guaranty of these AOPs are nonselectivity. Nobody doubts about the potential to degrade HRPs of AOPs. So far, the efficiency of AOPs for HRP removal has been well proven at the laboratory scale. However, more studies at pilot and full scale are needed to determine the optimal operational conditions and elaborate consistent and reliable cost evaluations.

Technologies for AOPs involve widely different methods of activation as well as oxidant generation and can potentially utilize a number of different mechanisms for organic destruction. An overview of different established and emerging AOPs is given in Fig. 8.2. All AOPs comprise two steps: the *in situ* formation of reactive oxidative species and the reaction of oxidants with target contaminants (Miklos et al., 2018). However, in practice, many factors inhibit the production of free radicals, such as pH, temperature, and inorganic anions. Therefore, the application of AOPs in removing HRPs from real wastewater requires more research.

8.2.1 Ozone-based AOPs

Ozone (O_3) is shown to be an effective powerful oxidizer. Ozone-based AOPs can remove many kinds of HRPs in this way. Sometimes ozone alone can be effectively destroyed by molecules of HRPs, but it cannot be effectively mineralized by them. As illustrated, ozonation effectively removed most of the remaining pharmaceutical and personal care products (PPCPs) in the granular activated carbon effluent,

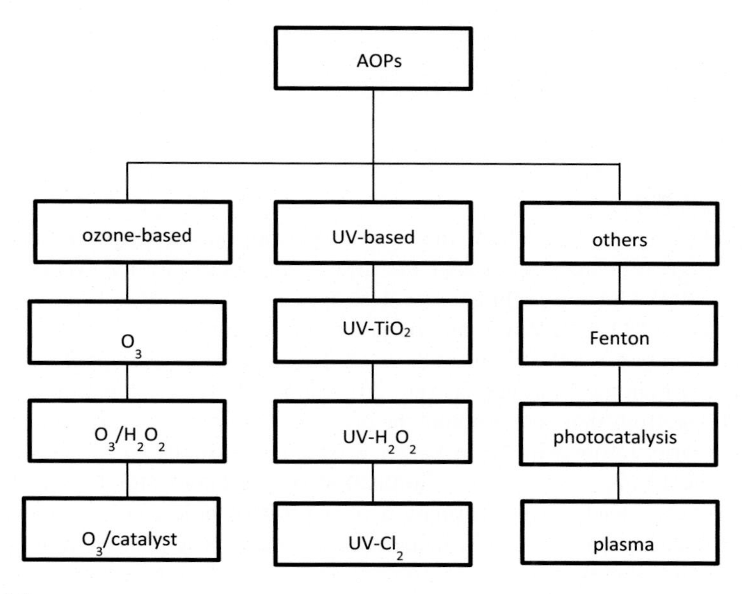

FIGURE 8.2

Broad overview and classification of different AOPs.

although it was expected that most were transformed to other compounds. Sulfamethoxazole, primidone, DEET, and caffeine were the only compounds routinely detected after ozonation; ciprofloxacin was detected on only two occasions. The removal of sulfamethoxazole ranged from 67% to 94% and the removal of trimethoprim and carbamazepine was over 80%. Rapid conversion of carbamazepine, trimethoprim, and sulfamethoxazole during ozonation has been reported by several groups of researchers (Yang et al., 2011). In a full-scale drinking water treatment plants, dilantin, meprobamate, and iopromide were moderately removed (50% or less concentration reduction), while all other analytes were reduced to less than the method reporting limits (Snyder et al., 2006). It is necessary to combine O_3 and catalysts such as ultraviolet (UV), activated charcoal, molecular sieve, and metallic oxide to improve the oxidation performance. In Pan's research, they demonstrated a high-performance of ozone catalyst using spherical Cu–Fe–O nanoparticles (CFO NPs). OH was the main active specie for dimethyl phthalate degradation on the basis of the significant suppression of tert-Butanol. The results indicated that the CFO NPs had a wide pH application range from 3 to 9 (Pan et al., 2018). Ozone-catalyzed oxidation can reduce toxicity. MWCNTs present a higher catalytic performance than common catalyst. For longer ozonation time, the acute toxicity decreases deeply. Simultaneous use of ozone and MWCNTs shows the best results (Goncalves et al., 2012). The addition of H_2O_2 during ozonation can considerably enhance the removal of ozone-resistant pollutants. Generally speaking, the addition H_2O_2 can well enhance the ozone decomposition and ·OH yield in water. Therefore, HRPs with strong ozone resistance can have a more significant enhancement effect during the O_3/H_2O_2 process (Snyder et al., 2006).

8.2.2 UV-based AOPs

8.2.2.1 UV-TiO_2

Titanium dioxide (TiO_2) remains the most promising photocatalyst because of its low cost, high efficiency, chemical inertness, and photostability (Wilcoxon and Thurston, 1999). Organic pollutants are destroyed in the presence of semiconductor photocatalysts (TiO_2 and ZnO), an energetic light source, and an oxidizing agent such as oxygen or air in the photocatalytic oxidation process. Only photons with energies greater than the band-gap energy (ΔE) can result in the excitation of valence band (VB) electrons that then promote the possible reactions. The absorption of photons with energy lower than ΔE or longer wavelengths usually causes energy dissipation in the forms of heat. The illumination of the photocatalytic surface with sufficient energy leads to the formation of a positive hole (hv^+) in the valence band and an electron (e^-) in the conduction band (CB). The positive hole oxidizes either pollutant directly or water to produce ·OH radicals, whereas the electron in the conduction band reduces the oxygen adsorbed on the catalyst (TiO_2) (Ahmed et al., 2010). Several researchers tried the degradation of lindane from drinking water sources using TiO_2 under UV light (Senthilnathan and Philip, 2009; Dionysiou et al., 2000; Zaleska et al., 2000). Zaleska reported that anatase supported on glass hollow microspheres was able to degrade 50% of lindane (40 ppm initial concentration) within 150 min of processing time (Zaleska et al., 2000). Dionysiou reported the degradation of lindane using TiO_2 immobilized on a continuous flow rotating disc and achieved 63% of lindane (initial concentration 0.016 mmol/L) degradation (Dionysiou et al., 2000). Senthilnathan and Philip reported the degradation of lindane using immobilized TiO_2 in single and mixed pesticides systems (Senthilnathan and Philip, 2009). In both cases, complete degradation was achieved; however, the rate of degradation of lindane in single pesticide system (1 mg/L degraded within 230 min) was different from that in mixed (lindane, methyl parathion, and dichlorvos) pesticides system (1 mg/L lindane degraded within 260 min). Although photocatalytic degradation of pesticide using TiO_2 under UV light is possible, it may not be a practical proposition for the treatment of drinking water sources due to the high cost. Efficient utilization of visible and solar light is one of the major goals of modern science and engineering that have a great impact on technological applications. Although AOPs with TiO_2 photocatalysts are shown to be an effective alternative in this regard, the vital snag of TiO_2 semiconductor is that it absorbs a small portion of solar spectrum in the UV region (band-gap energy of TiO_2 is 3.2 eV). To utilize maximum solar energy, it is necessary to shift the absorption threshold toward visible region.

8.2.2.2 UV-H_2O_2

UV/H_2O_2 is originated from the *in situ* generation of $^{\bullet}OH$, which can oxidize a large range of organic pollutants. UV/H_2O_2 treatment—a promising technology—as it can oxidize micropollutants, transform NOM, and inactivate pathogens simultaneously (Audenaert et al., 2013). Different from other advanced treatments (such as Fenton's regent oxidation, membrane filtration, and activated carbon adsorption), UV/H_2O_2 has no need for pH adjustment, sludge disposal, membrane backwash, or adsorbent regeneration. As a result, it is largely used in water and wastewater treatment (Ikehata et al., 2006; Broseus et al., 2009). In addition, water matrix, especially the style and content of natural organic or inorganic matter, can influence the oxidation of organic HRPs by competing for the oxidant or affecting the generation of $^{\bullet}OH$ (Yuan et al., 2013; Buffle et al., 2006; Baeza and Knappe, 2011). Nevertheless, there is little research about the influence of water matrix on the removal of UV/H_2O_2 treatments.

Hydroxyl radical's oxidation potential is 2.8 eV, and it can absolutely remove the pollutants. By the research of Baeza and Knappe (2011), the removal efficiency of Estradiol (EE2) in water by photolysis was increased from 20% to above 90% by addition of 15 mg/L H_2O_2, telling that 98% of Estrone (E1) and Estradiol (E2) in water were removed within 1 h under UV irradiation and the degradation rate increased with the addition of H_2O_2 (Zhang et al., 2007). Chen et al. (2007) also found that UV/H_2O_2 treatment at a H_2O_2 dose of 10 mg/L and a UV fluence of <1000 m^J/cm^2 was able to decrease in vitro estrogenicity of 1-(3-dimethylaminopropyl)-3-ethylcarbodiimide hydrochloride (EDC) mixture.

Although the treatment of EDCs in an aquatic environment with UV/H_2O_2 was documented, and UV or H_2O_2 oxidation was used in pretreat waste activated sludge (WAS) and improve the sludge solubilization (Tokumura et al., 2007, 2009; Erden and Filibeli, 2010; Salihoglu et al., 2012), little information is obtained about the influence of UV/H_2O_2 on the behavior of EDCs in sludge. Different from simple aqueous matrices, WAS is a complex matrix with some species of organic compounds and inorganic substances, which may impact the degradation of micropollutants and cause different incomes from those in water. Niu et al. (2013) also indicated that HA can impact the photodegradation of tetracycline. They found that tetracycline photolysis could be improved at low HA concentration under solar and xenon lamp irradiation, so it could be suppressed at high HA concentration. Sun et al. (1999) found that transition metals can cause positive influence on accelerating the total degree of H_2O_2 oxidation reaction as well as the negative effects of increasing the rate of radical coupling reactions during oxidation. As a result, it is interesting to explore the removal of EDCs during the treatment of WAS with UV/H_2O_2 and the influence of sludge matrix on the degradation of EDCs.

8.2.2.3 UV-Cl_2

Chlorine (Cl_2), a gaseous chlorine or hypochlorite, is widely used in water disinfection (Sharma, 2008; Deborde and von Gunten, 2008), at one or two points of the process: preoxidation (inducing a primary disinfection at the start of the treatment), and posttreatment (maintaining a disinfectant residual in the distribution system). The UV/Cl_2 AOP is used to be efficient in producing reactive species (especially HO· and Cl·; and to a lesser extent) at acidic and neutral pH (Fang et al., 2014). Furthermore, the UV/Cl_2 AOP can be more efficient than the UV/H_2O_2 AOP in the degradation of some emerging contaminants (ECs) such as benzotriazole, tolytriazole, and iopamidole (Sichel et al., 2011). So the UV/Cl_2 treatment can be thought to be an alternative of UV-based AOP for the removal of micropollutants in drinking water and wastewater. Nevertheless, the effectiveness of the UV/Cl_2 process in degrading different groups of recalcitrant ECs (like real surface water or municipal wastewater) is still unclear.

Recently, the combination of chlorine and UV was increasing in wastewater treatment plants (WWTPs). By the results of a survey of UV application in the United States, the orders of the combination treatments in water treatment plants can be UV, followed by chlorine or chlorine followed by UV. Dotson et al. (2012) with the latter application also pointed out an AOP because an exposure of chlorine species to UV can produce radicals (OH• or Cl•) that can improve the disinfection efficiency (Feng et al., 2007). Nevertheless, high fluences (>100 mJ/cm^2) and chlorine doses (>8 mg/L) are widely implied in UV/Cl_2 to get an AOP to remove chemical pollutants (Jin et al., 2011; Wang et al., 2012), raising the issue of whether or not such radicals can exist at the low fluences and chlorine doses applied in water disinfection. Previously, a study investigated the efficiency of a simultaneous application of UV and chlorine and also observed synergistic viral inactivation at a practical UV fluence and chlorine dose of 15 mJ/cm^2 and 1 mg/L, implying the enhancement inactivation by radicals.

Together with the viral control methods, virus inactivation mechanisms are also an important issue to understand. The infection mechanisms of viruses are complicated and usually comprised three main steps: attachment to the host, entry or genome penetration into the host, and genome replication in the host. During the host attachment and entry or genome penetration processes, viral proteins on capsid are a key component for a successive infection, but the viral genome is a very important part during replication. Furthermore, damage to either of these viral components can inactivate the viruses through interrupting the steps in the infectivity processes. Although the individual effects of chlorine and UV on virions have been revealed in a lot of studies (Kim et al., 1980; Eischeid and Linden, 2011), there is no study of virus inactivation mechanisms following the simultaneous application of UV and chlorine. Oxidants such as chlorine or ozone can ruin both the protein components of the viral capsid (Kim et al., 1980) and the viral genome, but UV light induces thymine-dimers in viral genomes, bringing in replication failure and can also alter protein components of the viral capsid (Eischeid and Linden, 2011). Furthermore, it is expected that both capsid and viral genome can be damaged by the UV/Cl_2 because of the oxidative radicals. Taking this background information into consideration, it is vital to prove the existence of radicals in this treatment and investigate the fundamental virus inactivation mechanisms mediated by UV/Cl_2 treatment to understand the fundamental principles of the combined process in a fresh level.

8.2.2.4 Others

Fenton is an efficient advanced oxidation technique. Fenton followed by coagulation is an effective method for the treatment of beet sugar industry wastewater, for removing close to 59% COD as well as 83.9% color (Hajiagha et al., 2018). Photocatalysis is also a promising technology in wastewater treatment. Due to the advances in using photocatalysis in the recalcitrant organic pollutants degradation, this technique was used in photocatalytic membrane reactors in large-scale applications. Semiconductor oxides used in heterogenous photocatalysis are typically either by suspending in the effluent to be treated or immobilizing on a support. Prepared clinoptilolite-supported hybridized PbS−CdS semiconductors for the photocatalytic degradation of a mixture of tetracycline and cephalexin aqueous solution, achieving 92.51% of COD removal efficiency (Azimi and Nezamzadeh-Ejhieh, 2015). Membrane photocatalytic reactor was applied to remove endocrine disrupting chemical, EE2 (17β-oestradiol). About 90% of EE2 was removed in the presence of HA (López et al., 2014).

8.3 Membrane separation technologies

After hundreds of years of evolution, membrane technology has become a new method of separation for handling HRPs. In the face of China's gradual improvement of environmental standards, more and more scholars prefer to use membrane to treat wastewater. Pressure-driven membrane filtration processes such as microfiltration, ultrafiltration, nanofiltration, and reverse osmosis have been used to separate HRPs from wastewater. The separation particles are selected in accordance with the particle diameter, the solution concentration, the charge amount, the applied pressure, and the like. Of course, the membrane can be stimulated by chemical agents to stimulate its filtration mechanism. The following is an introduction to membrane separation technology (Fig. 8.3 and Table 8.2).

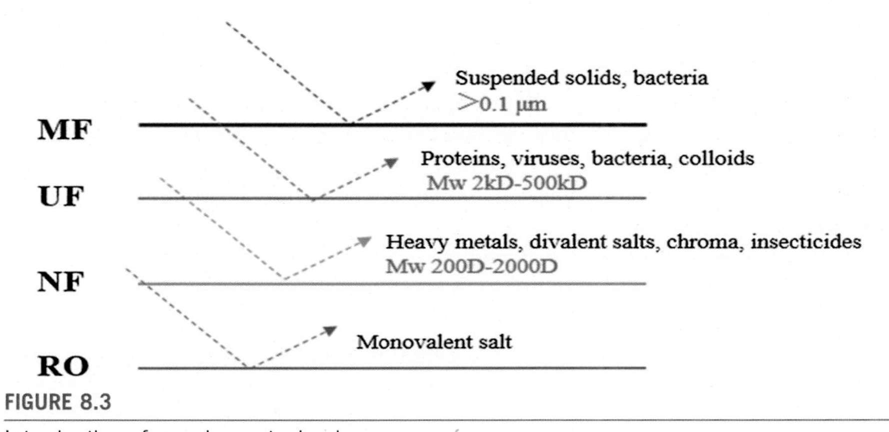

FIGURE 8.3

Introduction of membrane technology.

Table 8.2 Introduction to membrane technology.

Microfiltration (MF)	Microfiltration, also known as microporous filtration, uses porous membrane (microporous membrane) as the filtration medium. Under the pressure of 0.1–0.3 MPa, it can intercept particles such as gravel, silt, and clay in solution, and a large number of solvents, small molecules, and a small amount of macromolecular solutes can pass through the separation process of the membrane.
Ultrafiltration (UF)	Ultrafiltration is one of the membrane separation technologies driven by pressure. Ultrafiltration pore size and molecular weight cut-off (MWCO) range have been defined vaguely. Ultrafiltration membranes are generally considered to have a pore size of 0.001–0.1 μm and a molecular weight cut-off of 1000–1,000,000 Da.
Nanofiltration (NF)	Nanofiltration is a pressure-driven membrane separation process between reverse osmosis and ultrafiltration. The pore size of nanofiltration membrane is about several nanometers.
Reverse osmosis (RO)	Reverse osmosis is a membrane separation operation that separates solvents from solution driven by pressure difference. It is called reverse osmosis because it is opposite to the direction of natural osmosis.
Electrodialysis (ED)	Electrodialysis is a membrane separation operation that uses potential difference as driving force and selective permeability of ion-exchange membrane to remove or enrich electrolytes from solution.

8.3.1 Membrane technology to remove heavy metals

The heavy metal elements in wastewater mainly include Cu, Cd, Hg, Pb, Zn, Ni, Cr, and Cd. Heavy metal ions constitute a hazard to the environment and human health when present in water environments. In this connection, requirements for the quality of wastewater are being continuously tightened, resulting in the increasing costs of drinking and industrial process water. Thus, the need emerges for innovative, science-based approaches to water purification and treatment. Various studies have shown that membrane separation technology can effectively remove heavy metal elements in wastewater. Technologies on the basis of membrane methods hold much promise for resolving most problems.

As early as the 20th century, some scholars studied ultrafiltration (UF) of cobalt(II) and nickel(II) in the presence of micellar solubilized hydrophobic ligands. In aqueous media, metals were intercepted in surfactant micelles containing extractants and effectively intercepted by UF membranes, and their subsequent concentrations were concentrated in suspensions. The results showed that the selective rejection of nickel(II) by cobalt(II) could be achieved by increasing the acidity and alkalinity of the solution and the concentration of extractant, and the separation effect could be improved.

Mohammed et al. (2004) used nanofiltration membrane to treat nickel in industrial electroplating wastewater. The results showed that the rejection rate of nickel was generally high regardless of the presence or absence of other ions.

Reverse osmosis (RO) is a pressure-driven operation, which requires more than 690 kPa of pressure to effectively remove common heavy copper ions in water, especially copper ions in wastewater. Ujang explained the feasibility of using chelating and low pressure reverse osmosis membranes to remove Zn^{2+} and Cu^{2+} (Ujang, et al., 1996). The results showed that the removal efficiency of zinc and copper ions by sulfonated polysulfone reverse osmosis membrane could be improved by using EDTA as chelating agent when the operating pressure was less than 690 kPa. At 450 kPa, the overall rejection rate of Zn^{2+} and Cu^{2+} was 95% at 95°C. In addition, in one study, the authors used the mixed system of electrodialysis-vacuum membrane distillation (ED-VMD) to treat wastewater and recover cerium resources and waste acid. The rejection rate of Ce^{4+} by VMD process is over 99.9%, and it is concentrated to the greatest extent (Ren et al., 2018).

With the advancement of membrane technology and the development of nanomaterials, research on modified membranes has gradually increased. Many scholars have also used modified membranes for the removal of heavy metals and achieved good results.

In one study, two different kinds of filter membranes, cellulose triacetate and polyamide-based film composite membrane, were used to study the performance of positive osmosis (FO) in treating feed water containing different salinity of heavy metal Ni^{2+} with high salinity, in which the removal rate of Ni^{2+} was more than 93% (Zhao et al., 2016).

More and more membrane technology is being used to treat heavy metals in wastewater. Ultrafiltration membranes, nanofiltration membranes, and reverse osmosis membranes each have their own advantages and disadvantages. On the one hand, ultrafiltration membranes require less operating pressure, lower energy consumption, and higher throughput than nanofiltration and reverse osmosis. In addition, ceramic ultrafiltration membranes have higher mechanical strength and chemical stability. On the other hand, nanofiltration and reverse osmosis have higher processing efficiencies. However, these membrane technologies own some drawbacks, such as fouling problem and high energy consumption.

8.3.2 Membrane technology to remove pesticides and drugs

Due to the widespread use of pesticides and biocides, water entering WWTPs will contain pesticide residues and pharmaceuticals. Some common pharmaceuticals and pesticides and their compounds are presented in Tables 8.3 and 8.4. Some scholars (Rodriguez-Mozaz et al., 2015) used MF and RO to treat the existing substances in WWTPs, and achieved the effective removal of most compounds. Almost all compounds were reduced to below 16 ng/L.

Nanofiltration technology can reduce organic and inorganic pollutants, eliminate water hardness, disinfection, and nitrate reduction, and remove trace pollutants, so it is considered as one of the best membrane processes for pesticide removal in recent years. Mahmoudi et al. used nanofiltration to remove pesticides from polluted water and studied the effect of effective parameters on the removal

Table 8.3 Some common pharmaceuticals and their compounds.

Pharmaceuticals	Pharmaceutical compounds
Antibiotics	Azithromycin
	Erythromycin
	Sulfamethoxazole
Analgesics and antiinflammatories	Diclofenac
	Ibuprofen
	Ketoprofen
	Naproxen
β-Blockers	Metoprolol
	Propranolol
	Sotalol
Lipid regulators	Clofibric acid
	Gemfibrozil
	Mevastatin

Table 8.4 Some common pesticides and their compounds.

Pesticides	Pesticide compound
Triazines	Atrazine
	Cyanazine
	Deethylatrazine
	Terbuthylazine
Phenylureas	Diuron
	Isoproturon
	Linuron
	Chlortoluron
Chloroacetanilides	Alachlor
	Metolachlor
Acidic herbicides	Mecoprop
	Bentazon

of nitrate and diazine pesticides from polluted water (Mahmoodi et al., 2015). It was found that increasing the concentration of diazinon and the pH value could increase the removal efficiency by 94%, while increasing the concentration of nitrate could increase the efficiency of commercial nanofilters from 80% to 85%. Under the optimum conditions, when the diazine concentration is 90 ug/L, the nitrate concentration is 80 mg/L and the pH value is 9, the removal rate of diazine is about 93%. Nanofiltration and reverse osmosis technology can effectively remove drugs from wastewater, but there are also problems of high energy consumption and serious membrane fouling. It is hoped that new multifunctional membrane materials can be developed to solve membrane fouling and energy consumption problems.

8.3.3 Membrane technology to remove phenolic compounds

Phenolic wastewater poses a great threat to human health, and its discharge has become an important environmental problem. In one study, an electrodialysis (EED) process was studied to remove phenol molecule in the form of benzene oxide ions and more than 90% of phenol could be removed. In the treatment of low salt and high phenol wastewater, EED process was more energy saving. These results could significantly improve the application prospects of this new separation process (Wu et al., 2019).

Membrane separation technology to remove HRPs has the advantages of high removal efficiency and small footprint, but there are also problems such as serious membrane fouling and expensive membranes. We can choose new membrane materials, further improve the membrane structure, develop antipollution membranes, and improve membrane pretreatment and membrane cleaning methods, so that these problems can be better solved.

8.4 Combination process technologies

The single treatment may not be effective for all HRPs that are known to occur in municipal and industrial wastewater. To overcome this, the low removal efficiency of highly diverse chemical and biological substances with high toxicities necessitates the integration of physicochemical processes to ensure adequate removal of HRPs. Combination process technologies provide an important solution in the reuse and recovery of wastewater. Because combination process technologies offer various advantages such as their rapid process, ease of operation and control, and flexibility to change of temperature. In addition, the treatment system requires a lower space and installation cost. Hence, physicochemical treatments have been found as one of the most suitable treatments for HRPs removal.

In this chapter, many combination process technologies for eliminating the HRPs present in the effluents of WWTPs have been reviewed: NF/AOP, UV/H_2O_2, NF/UV, electro-Fenton, combination of Fenton processes and biotreatment, and so on.

8.4.1 Nanofiltration combined with ozone-based advanced oxidation processes

Nanofiltration (NF) is a kind of membrane filtration process that has received growing concerns due to its advantages such as easily being scaled-up, low capital input, and high quality of products (Acero et al., 2010). Besides, the effluent from

Table 8.5 Antibiotic concentrations detected in secondary effluent, NF concentrate, and the treated samples.

Antibiotic	Concentration in different samples (ng L^{-1})			
	Secondary effluent	NF concentrate	Ozonation treated samples	UV/O_3 treated samples
NOR	221	715	87	41
OFL	253	863	32	–
AZI	296	1060	–	–
ROX	372	1402	–	–

NF is also considered as an important water resource, which may be recycled in agriculture, urban areas, industry, aquifer recharge, etc. (Bixio et al., 2006). NF combined with chemical process selected four antibiotics, ofloxacin (OFL), norfloxacin (NOR), azithromycin (AZI), and roxithromycin (ROX) as processed targets, NF process was implemented to remove targets, and NF concentrate was treated using UV_{254} photolysis, ozonation, and UV/O_3 process, respectively (Liu et al., 2014). Table 8.5 indicated concentrations of antibiotic detected in secondary effluent, filtration for 8 h, NOR and OFL were detected with concentrations below the limit of quantification in the permeate, while AZI and ROX were not found in NF concentrate and the treated samples.

It has been proved that the treated NF concentrate by the UV/O_3 process had a satisfying effect on removing the antibiotics (>87%), reducing dissolved organic carbon (DOC) partially (40%), increasing biodegradability (4.6 times), and reducing ecotoxicity (58%) (Liu et al., 2014). Therefore, nanofiltration is a reliable method and the residual antibiotics in NF concentrate can be effectively eliminated by ozone-based AOPs.

8.4.2 Photocatalytic ozonation

Photoactivated semiconductor in combination with ozone is an advanced oxidation method called photocatalytic ozonation (Agustina et al., 2005; Mehrjouei et al., 2012). This combination is thought to be a promising technique for the decomposition of refractory microorganisms and organic compounds in water (Mehrjouei et al., 2015). Photocatalytic ozonation was tested as an efficient decomposition method for several kinds of pollution (Hsing et al., 2007; Khan et al., 2013; Sanchez et al., 1998; Yildirim et al., 2011). The decomposition effects are primarily attributed to the formation of more reactive but nonselective hydroxyl radicals in the oxidation medium, which react with almost all organic molecules at a rate of $10^6–10^9$ M/s (Andreozzi et al., 1999). Mehrjouei et al. (2015) gave a systematic review on photocatalytic ozonation for the treatment of water and wastewater, in which the mechanisms, kinetics, and economic aspects are discussed, then the synergistic effects produced by applying this oxidation method to the different organic pollutants media. There are plenty of experiments carried out and expressed the same conclusion. The synergistic effects of photocatalytic ozonation are 6.6 times of the sum of photocatalysis and ozonation when the process object is 2-chlorophenol (Oyama et al., 2011); 1.5 times when dealing with 4-chloronitrobenzene (Ardizzone et al., 2011; Klare et al., 1999); 1.4 times when it comes to cyanide ion (Hernandez-Alonso et al., 2007; Oyama et al., 2009); 1.3 times on formic acid (Wang et al., 2002), 1.2 times when dealing with gallic acid (Beltran et al., 2006); 1.5 times and 2.3 times when the objects are methanol (Mena et al., 2012) and methylene blue, respectively (Hur et al., 2005; Pichat et al., 2000). Photocatalytic ozonation showed a great potential in dealing with monochloroacetic acid, and the synergistic effects are 7.5 times more than the sum of photocatalysis and ozonation (Mehrjouei et al., 2014). It is clear that photocatalytic ozonation appears more cost-effective than ozonation and photocatalysis (Beltran et al., 2005; Gimeno et al., 2007; Ochiai et al., 2012; Rodriguez et al., 2013; Zou and Zhu, 2008).

8.4.3 Ozone/H_2O_2 process

AOPs in wastewater and drinking water are widely concerned because of the refractory nature of some micropollutants, while conventional physicochemical and biological treatments are not able to provide adequate elimination of these compounds. Due to the advantages of efficacy and saving energy compared with UV/H_2O_2, ozone is popular. Buffle et al. (2006) optimized AOPs with the addition of H_2O_2 (Buffle et al., 2006), the combination of molecular ozone and the more powerful, nonselective $\cdot OH$ allows for the degradation of more recalcitrant compounds and acceleration of the overall treatment process (Gerrity et al., 2011). To validate the effectiveness of ozone/H_2O_2 process for contaminant oxidation and disinfection in wastewater and water reclamation, extensive testing was brought out at the Reno-Stead Water Reclamation Facility in Reno, Nevada (Gerrity et al., 2011). The author collected over a 5-month period of continuous operation, and three sets of samples were obtained. These samples were analyzed for a suite of several microbial surrogates and trace organic contaminants that include the bacteriophage MS2, total and fecal coliforms, and Bacillus spores. The experimental results indicated that the use of sand filtration and ozone/H_2O_2 may be a viable alternative to the standard configuration. Another study achieved by Gerrity et al. (2011) focused on removing a suite of micropollutants in the process of water reclaiming. The results showed a satisfying efficiency (>90%) for most target pollutants (Gerrity et al., 2011).

8.4.4 Combination of Fenton processes and biotreatment

Fenton process is achieved by Fenton's reagent, which is a mixture of hydrogen peroxide (H_2O_2) and ferrous sulfate ($FeSO_4$). Fenton and related reactions consist of reactions of peroxides (generally hydrogen peroxide (H_2O_2)) with iron ions to generate active oxygen species that can oxidize organic and inorganic compounds (Babuponnusami and Muthukumar, 2014). Irons are used as catalyst to accelerate the decomposition of H_2O_2 to hydroxyl. Recent researchers have attempted to overcome the disadvantages of the Fenton process (such as high concentration of anions and ferrous iron sludge) (Hansson et al., 2012).

Table 8.6 The application of sono-Fenton process in HRP removal in wastewater.

HRP	Usages	Removal efficiency	References
2,4-Dichlorophenol	Precursor of herbicide 2,4-dichlorophenoxyacetic	Lower initial concentrations of Fenton's reagent were used in sono-Fenton process with a promising result. This process was feasible at pH 5.with a high degradation efficiency of 2,4-dichlorophenoxyacetic (77.6).	Ranjit et al. (2008)
2,4-Dinitrophenol	Antiseptic and nonselective bioaccumulation pesticide	Sono-Fenton process can achieve removal efficiency of 98% 2,4-dinitrophenol after reaction time of 1 h.	Guo et al. (2005)
p-Nitrophenol	The intermediate in the synthesis of paracetamol	The enhancement in degradation of 0.5% *p*-nitrophenol (w/v) was achieved by 66.4% degradation.	Pradhan and Gogate (2010)
p-Chlorobenzoic acid	The precursor of drugs and dyes	The degradation rate of *p*-chlorobenzoic rate was enhanced in the presence of ultrasound and Fenton's reagent.	Neppolian et al. (2004)
Carbofuran	Carbamate pesticide	The oxidization efficiency of carbofuran was increased from 40% (solely ultrasonic process in 2 h) to 99% (combined process in 30 min).	Ma and Sung (2010)
Rhodamine B	Dyes	The degradation of rhodamine B was achieved by 46%, 52%, and 51% in sono-Fe^{2+}, sono-Fe^{3+}, and sono-Fe0 process, respectively.	Ai et al. (2007)

8.4.4.1 Sono-Fenton process

The combined treatment by coupling ultrasound with Fenton reagent is named as sono-Fenton process (Pradhan and Gogate, 2010). The SF process can stimulate the production of hydroxyl radicals and has effective removal efficiency on HRP, such as ibuprofen. Water pyrolysis can product hydroxyl radicals in the effect of ultrasound wave, which increase the concentration of hydroxyl radical. This combined process can improve removal of refractory organic pollutants. The research reported the applications of sono-Fenton in HRP removal are shown in Table 8.6.

8.4.4.2 Electro-Fenton process

For a refractory part of HRPs, the combination process of electro-oxidation technology and Fenton's reagent is an effective approach. Refractory HRPs can be mineralized by election transfer reactions in anodic oxidation (Pignatello et al., 2006). The most promising electro-Fenton process is that ferric ion is reduced to ferrous ion at cathode. The impacted factors on electro-Fenton process are pH, electrode nature, electrolytes, dissolved oxygen level, and temperature. The research reported the applications of electro-Fenton in HRP removal, and they are shown in Table 8.7.

8.4.4.3 Combination process of multiple technologies

Several researchers used the combination processes by coupling ultrasound and ultraviolet with Fenton reagent, which is named as sono-photo-Fenton (SPF) process (Méndez-Arriaga et al., 2009; Segura et al., 2009; Wu et al., 2001). The SPF process can significantly enhance the formation of hydroxyl radicals in wastewater.

Table 8.7 The application of electro-Fenton process in HRP removal in wastewater.

HRP	Usages	Removal efficiency	References
Azo benzene	Dyes	In electro-Fenton process, COD as a surrogate for azo benzene was observed that more than 80% COD removal was achieved.	Guivarch et al. (2003)
Clofibric acid	Naturally occurring pharmaceutical compounds	The efficient formation of hydroxyl radicals improved the mineralization efficiency of clofibric acid to 80%.	Sirés et al. (2007)
Indigo carmine	Dyes	The mineralization of indigo carmine was enhanced in the presence of anode and 1 mmol/L Fe^{2+}.	Flox et al. (2006)
Nitro phenols	The precursors of drug and pesticide	The significant enhancement of nitrophenol degradation was observed by 98% removal. The toxicity was eliminated and the nitro phenols biodegradability was improved in the electro-Fenton process.	Yuan et al. (2006)

Compared to Fenton process, the advantage of SPF process is the smaller requirement of the amount of ferrous salt (Ting et al., 2008; Vaishnave et al., 2014). Thus, SPF process is another useful technology for HRP removal. The removal efficiency of SPF process has been investigated. Segura et al. (2009) reported the degradation of phenol by SPF process with 45% TOC removal (the degradation of TOC was 30% and 40% in sono-Fenton and photo-Fenton processes, respectively). Méndez-Arriaga et al. (2009) investigated the utilization of SPF process in the degradation of pharmaceutical micropollutant ibuprofen. The degradation and mineralization of ibuprofen in this process were observed by 95% and 60%, respectively. Vaishnave et al. (2014) compared the efficiency of SPF with photo-Fenton processes on azure-B removal, and the results indicated that azure-B could be completely decomposed into CO_2 and H_2O under the conditions of pH 2.1, H_2O_2 0.5 mL, and Fe^{3+} 5.0×10^{-4} mol/L.

The photoelectro-Fenton (PEF) process is the combination of photochemical and electrochemical process with the Fenton reagent (Boye et al., 2002). Hydroxyl radicals are generated by photolysis through the hemolytic breakdown of the peroxide

Table 8.8 The application of electro-Fenton process in HRP removal in wastewater.

HRP	Usages	Removal efficiency	Reference
Aniline	Dyes	The PEF process can achieve 92% TOC removal.	Brillas et al. (1998)
2,4,5-Trichlorophenoxy acetic acid	Herbicide	A fast and complete removal of herbicides under the conditions of pH 2–4 and 1 mmol/L Fe^{2+}.	Boye et al. (2003)
3,6-Dichloro-2-methoxybenzoic acid	Herbicide	The PEF process showed a faster and more complete degradation of herbicide versus electro-Fenton process (60%–70% mineralization).	Brillas et al. (2003)
Indigo carmine dye	Dyes	Complete mineralization was achieved in the PEF process. The optimum pH was 3.	Flox et al. (2006)
Benzene sulfonic acid	The ingredient of detergent and pharmaceutical	The removal efficiency can be enhanced by 75%, and the maximum TOC removal efficiency was 72%. The PEF process showed a higher removal efficiency of COD (increasing by 14%).	Ting et al. (2008)
Sulfamethoxazole	Antibiotic	A faster and more complete removal degradation of sulfamethoxazole of 80% TOC removal and 63% mineralization.	Wang et al. (2011)

molecule (Siedlecka and Stepnowski, 2005). The oxidative potential was increased because of the additional formation of hydroxyl radicals. However, the applications of PEF process are limited and previous research is associated with the treatment of herbicide and dyes. The research reported the applications of photoelectro-Fenton in HRP removal, and they are shown in Table 8.8.

To achieve removing various HPRs in municipal and industrial wastewater, several combination process technologies are mentioned in this chapter. Nanofiltration combined with ozone-based advanced oxidation processes showed great potential in removing antibiotics, reducing DOC partially, reducing ecotoxicity, and increasing biodegradability. Photocatalytic ozonation was analyzed by multiple researchers in recent years and was proved to be efficient to most of the target pollutants. Ozone/H_2O_2 process, which gave a satisfying efficiency for most target micropollutants in wastewater and drinking water, was developed and optimized. Combination of Fenton processes and biotreatments is a main part in combination process technologies. It could be divided into several kinds. The sono-Fenton process improved removal of refractory organic pollutants by producing hydroxyl radicals; the electro-Fenton process removed refractory HRPs by utilizing election transfer reactions; the combination process of multiple technologies also showed effectiveness.

8.5 Emerging technologies

In the earlier sections, we reviewed a variety of different traditional treatment techniques and their mechanisms of processing. Although these techniques can effectively remove HRPs from wastewater and have a considerable application, they also have many problems such as low efficiency, high cost, and easy to cause secondary pollution. Therefore, to improve the traditional technology or research and development of new processing technology has become the focus of this work. Recently, the scientific community has proposed various new processing technologies to solve the HRPs problem such as photocatalysis, electrosorption, and electroflocculation. These emerging technologies have extensively applied new theoretical systems and clean energy to efficiently and purposefully remove HRPs from wastewater. Therefore, they have advantages over traditional technologies in processing effectiveness and economic applicability. In this section, we mainly introduce the principles of each new processing technology and the current application and put forward the prospect for the future development of emerging technologies.

8.5.1 Photocatalysis technologies

Photocatalytic technology as a green renewable chemical technology has caused a wide range of scientific community. Due to its advantages of low price, excellent performance, energy saving, and environmental protection, it has an important application prospect in the field of environmental governance and renewable and clean energy. Photocatalysis is a new field, the essence of which is the photochemical reaction under the catalyst, thus combining photochemistry and catalytic chemistry. The photocatalytic material is a semiconductor material, and the reaction process mainly depends on the energy band structure of the semiconductor. The energy band structure of the semiconductor photocatalyst is composed of a valence band in which an electron is filled and a conduction band that is not filled with electrons. The bandwidth between the top of the valence band and the bottom of the conduction band is called as forbidden bandwidth. When the light with energy greater than or equal to the forbidden bandwidth irradiates the semiconductor photocatalyst, the electrons in the valence band will be excited to transition to the conduction band, and a hole will be created in the valence band, and an electron–hole pair will be generated in the semiconductor. At this time, the semiconductor material is in an unstable excited state, and to maintain its own stability, the semiconductor material in an excited state will release the absorbed external energy, that is, energy relaxation. The energy relaxation can occur in two main processes: (1) electrons and holes recombine inside the semiconductor, generate heat or release energy in the form of fluorescence; (2) electrons and holes are transferred from the inside of the semiconductor to the surface. The generated electrons have a strong reducing ability, and the holes have a strong oxidizing ability, and holes and electrons on the surface of the semiconductor material undergo a redox reaction with the acceptor substance on the surface of the catalyst. Fig. 8.4 shows the photocatalytic process of titanium dioxide (TiO_2). The holes possess high positive oxidation potentials and therefore can oxidize nearly all chemicals, including oxidizing the water molecules to form

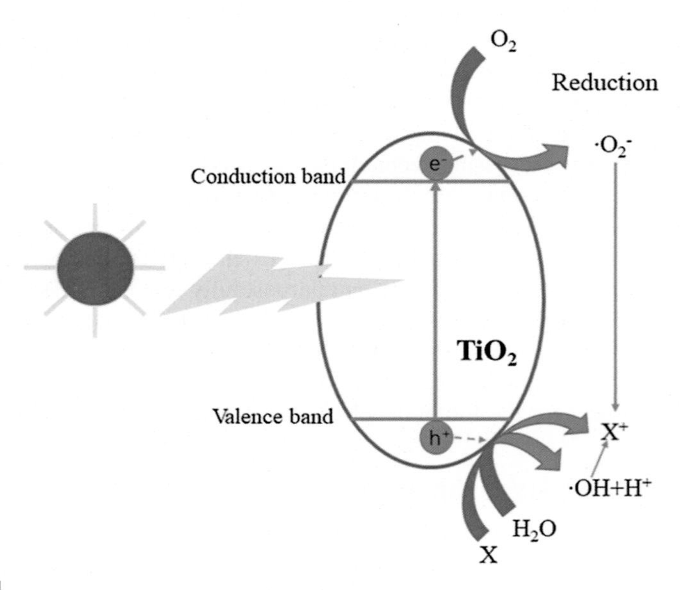

FIGURE 8.4

Photocatalytic mechanism of titanium dioxide.

their practical use. Only when these technologies can effectively remove HRPs and produce significant environmental benefits can they be widely used.

References

Abdel Salam, M., 2012. Removal of heavy metal ions from aqueous solutions with multi-walled carbon nanotubes: kinetic and thermodynamic studies. International Journal of Environmental Science and Technology 10 (4), 677–688.

Abdelhafez, A.A., Li, J., 2016. Removal of Pb(II) from aqueous solution by using biochars derived from sugar cane bagasse and orange peel. Journal of the Taiwan Institute of Chemical Engineers 61, 367–375.

Acero, J.L., Javier Benitez, F., Leal, A.I., Real, F.J., Teva, F., 2010. Membrane filtration technologies applied to municipal secondary effluents for potential reuse. Journal of Hazardous Materials 177, 390–398.

Agustina, T.E., Ang, H.M., Vareek, V.K., 2005. A review of synergistic effect of photocatalysis and ozonation on wastewater treatment. Journal of Photochemistry and Photobiology C: Photochemistry Reviews 6, 264–273.

Ahmed, S., Rasul, M.G., Martens, W.N., Brown, R., Hashib, M.A., 2010. Heterogeneous photocatalytic degradation of phenols in wastewater: a review on current status and developments. Desalination 261, 3–18.

Ai, Z., Lu, L., Li, J., Zhang, L., Qiu, J., Wu, M., 2007. Fe@Fe_2O_3 Core–Shell nanowires as iron reagent. 1. Efficient degradation of rhodamine B by a novel sono-fenton process. The Journal of Physical Chemistry C 111, 4087–4093.

Akbal, F., Camcı, S., 2010. Comparison of electrocoagulation and chemical coagulation for heavy metal removal. Chemical Engineering & Technology 33 (10), 1655–1664.

Andreozzi, R., Caprio, V., Insola, A., Marotta, R., 1999. Advanced oxidation processes (AOP) for water purification and recovery. Catalysis Today 53, 51–59.

Ardizzone, S., Cappelletti, G., Meroni, D., Spadavecchia, F., 2011. Tailored TiO_2 layers for the photocatalytic ozonation of cumylphenol, a refractory pollutant exerting hormonal activity. Chemical Communications 47, 2640–2642.

Audenaert, W.T.M., Vandierendonck, D., Van Hulle, S.W.H., Nopens, I., 2013. Comparison of ozone and HO center dot induced conversion of effluent organic matter (EFOM) using ozonation and UV/H_2O_2 treatment. Water Research 47, 2387–2398.

Azimi, S., Nezamzadeh-Ejhieh, A., 2015. Enhanced activity of clinoptilolite-supported hybridized pbs-cds semiconductors for the photocatalytic degradation of a mixture of tetracycline and cephalexin aqueous solution. Journal of Molecular Catalysis A: Chemical 408, 152–160.

Babuponnusami, A., Muthukumar, K., 2014. A review on Fenton and improvements to the Fenton process for wastewater treatment. Journal of Environmental Chemical Engineering 2, 557–572.

Baccar, R., Sarrà, M., Bouzid, J., Feki, M., Blánquez, P., 2012. Removal of pharmaceutical compounds by activated carbon prepared from agricultural by-product. Chemical Engineering Journal 211–212, 310–317.

Baeza, C., Knappe, D.R.U., 2011. Transformation kinetics of biochemically active compounds in low-pressure UV Photolysis and UV/H_2O_2 advanced oxidation processes. Water Research 45, 4531–4543.

Beker, U.G., Guner, F.S., Dizman, M., Erciyes, A.T., 1999. Heavy metal removal by ion exchanger based on hydroxyethyl cellulose. Journal of Applied Polymer Science 74 (14), 3501–3506.

Beltran, F.J., Gimeno, O., Rivas, F.J., Carbajo, M., 2006. Photocatalytic ozonation of gallic acid in water. Journal of Chemical Technology and Biotechnology 81, 1787–1796.

Beltran, F.J., Rivas, F.J., Gimeno, O., 2005. Comparison between photocatalytic ozonation and other oxidation processes for the removal of phenols from water. Journal of Chemical Technology and Biotechnology 80, 973–984.

Bilgin Simsek, E., 2017. Solvothermal synthesized boron doped TiO_2 catalysts: photocatalytic degradation of endocrine disrupting compounds and pharmaceuticals under visible light irradiation. Applied Catalysis B: Environmental 200, 309–322.

Bixio, D., Thoeye, C., De Koning, J., Joksimovic, D., Savic, D., Wintgens, T., Melin, T., 2006. Wastewater reuse in Europe. Desalination 187, 89–101.

Boye, B., Dieng, M.M., Brillas, E., 2002. Degradation of herbicide 4-chlorophenoxyacetic acid by advanced electrochemical oxidation methods. Environmental Science & Technology 36, 3030–3035.

Boye, B., Morième Dieng, M., Brillas, E., 2003. Anodic oxidation, electro-Fenton and photoelectro-Fenton treatments of 2,4,5-trichlorophenoxyacetic acid. Journal of Electroanalytical Chemistry 557, 135–146.

Brillas, E., Baños, M.Á., Garrido, J.A., 2003. Mineralization of herbicide 3,6-dichloro-2-methoxybenzoic acid in aqueous medium by anodic oxidation, electro-Fenton and photoelectro-Fenton. Electrochimica Acta 48, 1697–1705.

Brillas, E., Mur, E., Sauleda, R., Sànchez, L., Peral, J., Domènech, X., Casado, J., 1998. Aniline mineralization by AOP's: anodic oxidation, photocatalysis, electro-Fenton and photoelectro-Fenton processes. Applied Catalysis B: Environmental 16, 31–42.

Broseus, R., Vincent, S., Aboulfadl, K., Daneshvar, A., Sauve, S., Barbeau, B., Prevost, M., 2009. Ozone oxidation of pharmaceuticals, endocrine disruptors and pesticides during drinking water treatment. Water Research 43, 4707–4717.

Buffle, M.-O., Schumacher, J., Meylan, S., Jekel, M., von Gunten, U., 2006. Ozonation and advanced oxidation of wastewater: effect of O-3 dose, pH, DOM and HO center dot-scavengers on ozone decomposition and HO center dot generation. Ozone Science & Engineering 28, 247–259.

Chen, J., Zhang, D., Zhang, H., Ghosh, S., Pan, B., 2017. Fast and slow adsorption of carbamazepine on biochar as affected by carbon structure and mineral composition. The Science of the Total Environment 579, 598–605.

Chen, P.-J., Rosenfeldt, E.J., Kullman, S.W., Hinton, D.E., Linden, K.G., 2007. Biological assessments of a mixture of endocrine disruptors at environmentally relevant concentrations in water following UV/H_2O_2 oxidation. The Science of the Total Environment 376, 18–26.

Cho, H.H., Huang, H., Schwab, K., 2011. Effects of solution chemistry on the adsorption of ibuprofen and triclosan onto carbon nanotubes. Langmuir 27 (21), 12960–12967.

Deborde, M., von Gunten, U., 2008. Reactions of chlorine with inorganic and organic compounds during water treatment – kinetics and mechanisms: a critical review. Water Research 42, 13–51.

Dionysiou, D., Khodadoust, A.P., Kern, A.M., Suidan, M.T., Baudin, I., Laine, J.M., 2000. Continuous-mode photocatalytic degradation of chlorinated phenols and pesticides in water using a bench-scale TiO_2 rotating disk reactor. Applied Catalysis B: Environmental 24, 139–155.

Dotson, A.O., Rodriguez, C.E., Linden, K.G., 2012. UV disinfection implementation status in US water treatment plants. Journal American Water Works Association 104, 77–78.

Durán-Álvarez, J.C., Avella, E., Ramírez-Zamora, R.M., Zanella, R., 2016. Photocatalytic degradation of ciprofloxacin using mono- (Au, Ag and Cu) and bi- (Au–Ag and Au–Cu) metallic nanoparticles supported on TiO_2 under UV-C and simulated sunlight. Catalysis Today 266, 175–187.

Eischeid, A.C., Linden, K.G., 2011. Molecular indications of protein damage in adenoviruses after UV disinfection. Applied and Environmental Microbiology 77, 1145–1147.

Erden, G., Filibeli, A., 2010. Improving anaerobic biodegradability of biological sludges by Fenton pre-treatment: effects on single stage and two-stage anaerobic digestion. Desalination 251, 58–63.

Eskandarian, M.R., Choi, H., Fazli, M., Rasoulifard, M.H., 2016. Effect of UV-LED wavelengths on direct photolytic and TiO_2 photocatalytic degradation of emerging contaminants in water. Chemical Engineering Journal 300, 414–422.

Fang, J., Fu, Y., Shang, C., 2014. The roles of reactive species in micropollutant degradation in the UV/free chlorine system. Environmental Science & Technology 48, 1859–1868.

Feng, Y., Smith, D.W., Bolton, J.R., 2007. Photolysis of aqueous free chlorine species (NOCl and OCl^-) with 254 nm ultraviolet light. Journal of Environmental Engineering and Science 6, 277–284.

Flox, C., Ammar, S., Arias, C., Brillas, E., Vargas-Zavala, A.V., Abdelhedi, R., 2006. Electro-Fenton and photoelectro-Fenton degradation of indigo carmine in acidic aqueous medium. Applied Catalysis B: Environmental 67, 93–104.

Gerrity, D., Gamage, S., Holady, J.C., Mawhinney, D.B., Quinones, O., Trenholm, R.A., Snyder, S.A., 2011. Pilot-scale evaluation of ozone and biological activated carbon for trace organic contaminant mitigation and disinfection. Water Research 45, 2155–2165.

Gimeno, O., Rivas, F.J., Beltran, F.J., Carbajo, M., 2007. Photocatalytic ozonation of winery wastewaters. Journal of Agricultural and Food Chemistry 55, 9944–9950.

Goncalves, A.G., Orfao, J.J., Pereira, M.F., 2012. Catalytic ozonation of sulphamethoxazole in the presence of carbon materials: catalytic performance and reaction pathways. Journal of Hazardous Materials 239–240, 167–174.

Guivarch, E., Trevin, S., Lahitte, C., Oturan, M.A., 2003. Degradation of azo dyes in water by Electro-Fenton process. Environmental Chemistry Letters 1, 38–44.

Guo, M., Qiu, G., Song, W., 2010. Poultry litter-based activated carbon for removing heavy metal ions in water. Waste Management 30 (2), 308–315.

Guo, Z., Zheng, Z., Zheng, S., Hu, W., Feng, R., 2005. Effect of various sono-oxidation parameters on the removal of aqueous 2,4-dinitrophenol. Ultrasonics Sonochemistry 12, 461–465.

Hajiagha, A.A., Zaeimdar, M., Jozi, S.A., Sadjadi, N., Ghadi, A., 2018. Combination of chemical coagulation and photo-fenton oxidation process for the treatment of beet sugar industry wastewater: optimization of process conditions by response surface methodology. Ozone: Science & Engineering 41, 265–273.

Hansson, H., Kaczala, F., Marques, M., Hogland, W., 2012. Photo-fenton and fenton oxidation of recalcitrant industrial wastewater using nanoscale zero-valent iron. International Journal of Photoenergy 2012, 1–11.

Hernandez-Alonso, M.D., Coronado, J.M., Soria, J., Conesa, J.C., Loddo, V., Addamo, M., Augugliaro, V., 2007. EPR and kinetic investigation of free cyanide oxidation by photocatalysis and ozonation. Research on Chemical Intermediates 33, 205–224.

Hsing, H., Chiang, P., Chang, E., Chen, M., 2007. The decolorization and mineralization of acid orange 6 azo dye in aqueous solution by advanced oxidation processes: a comparative study. Journal of Hazardous Materials 141, 8–16.

Hu, H., Jiang, C., Ma, H., Ding, L., Geng, J., Xu, K., Huang, H., Ren, H., 2017. Removal characteristics of DON in pharmaceutical wastewater and its influence on the *N*-nitrosodimethylamine formation potential and acute toxicity of DOM. Water Research 109, 114–121.

Huang, Z., Lu, L., Cai, Z., Ren, Z.J., 2016. Individual and competitive removal of heavy metals using capacitive deionization. Journal of Hazardous Materials 302, 323–331.

Huang, Z.H., Zheng, X., Lv, W., Wang, M., Yang, Q.H., Kang, F., 2011. Adsorption of lead(II) ions from aqueous solution on low-temperature exfoliated graphene nanosheets. Langmuir 27 (12), 7558–7562.

Hur, J.S., Oh, S.O., Lim, K.M., Jung, J.S., Kim, J.W., Koh, Y.J., 2005. Novel effects of TiO_2 photocatalytic ozonation on control of postharvest fungal spoilage of kiwifruit. Postharvest Biology and Technology 35, 109–113.

Ikehata, K., Naghashkar, N.J., Ei-Din, M.G., 2006. Degradation of aqueous pharmaceuticals by ozonation and advanced oxidation processes: a review. Ozone Science & Engineering 28, 353–414.

Jeong, Y., Fan, M., Singh, S., Chuang, C.-L., Saha, B., Hans van Leeuwen, J., 2007. Evaluation of iron oxide and aluminum oxide as potential arsenic(V) adsorbents. Chemical Engineering and Processing – Process Intensification 46 (10), 1030–1039.

Ji, Q., Hu, C., Liu, H., Qu, J., 2018. Development of nitrogen-doped carbon for selective metal ion capture. Chemical Engineering Journal 350, 608–615.

Jin, J., El-Din, M.G., Bolton, J.R., 2011. Assessment of the UV/Chlorine process as an advanced oxidation process. Water Research 45, 1890–1896.

Jung, K.W., Lee, S.Y., Lee, Y.J., 2018. Facile one-pot hydrothermal synthesis of cubic spinel-type manganese ferrite/biochar composites for environmental remediation of heavy metals from aqueous solutions. Bioresource Technology 261, 1–9.

Khan, M., Jung, H., Lee, W., Jung, J., 2013. Chlortetracycline degradation by photocatalytic ozonation in the aqueous phase: mineralization and the effects on biodegradability. Environmental Technology 34, 495–502.

Kim, C.K., Gentile, D.M., Sproul, O.J., 1980. Mechansim of ozone inactivation of bacteriophage-f^2. Applied and Environmental Microbiology 39, 210–218.

Klare, M., Waldner, G., Bauer, R., Jacobs, H., Broekaert, J.A.C., 1999. Degradation of nitrogen containing organic compounds by combined photocatalysis and ozonation. Chemosphere 38, 2013–2027.

Lei, Z.D., Wang, J.J., Wang, L., Yang, X.Y., Xu, G., Tang, L., 2016. Efficient photocatalytic degradation of ibuprofen in aqueous solution using novel visible-light responsive graphene quantum dot/AgVO3 nanoribbons. Journal of Hazardous Materials 312, 298–306.

Li, P., Gui, Y., Blackwood, D.J., 2018. Development of a nanostructured alpha-MnO2/carbon Paper composite for removal of Ni(2+)/Mn(2+) ions by electrosorption. ACS Applied Materials & Interfaces 10 (23), 19615–19625.

Liu, P., Zhang, H., Feng, Y., Yang, F., Zhang, J., 2014. Removal of trace antibiotics from wastewater: a systematic study of nanofiltration combined with ozone-based advanced oxidation processes. Chemical Engineering Journal 240, 211–220.

Liu, Y., Yan, J., Yuan, D., Li, Q., Wu, X., 2013. The study of lead removal from aqueous solution using an electrochemical method with a stainless steel net electrode coated with single wall carbon nanotubes. Chemical Engineering Journal 218, 81–88.

López Fernández, R., Coleman, H.M., Le-Clech, P., 2014. Impact of operating conditions on the removal of endocrine disrupting chemicals by membrane photocatalytic reactor. Environmental Technology 35, 2068–2074.

Ma, Y.S., Sung, C.F., 2010. Investigation of carbofuran decomposition by a combination of ultrasound and Fenton process. Sustainable Environment Research 20, 213–219.

Maher, E.K., O'Malley, K.N., Heffron, J., Huo, J., Mayer, B.K., Wang, Y., McNamara, P.J., 2019. Analysis of operational parameters, reactor kinetics, and floc characterization for the removal of estrogens via electrocoagulation. Chemosphere 220, 1141–1149.

Mahmoodi, P., Borazjani, H.H., Farhadian, M., Nazar, A.S., 2015. Remediation of contaminated water from nitrate and diazinon by nanofiltration process. Desalination and Water Treatment 53, 2948–2953.

Mehrjouei, M., Mueller, S., Moeller, D., 2012. Synergistic effect of the combination of immobilized TiO_2, UVA and ozone on the decomposition of dichloroacetic acid. Journal of Environmental Science and Health – Part A: Toxic/Hazardous Substances & Environmental Engineering 47, 1073–1081.

Mehrjouei, M., Mueller, S., Moeller, D., 2014. Energy consumption of three different advanced oxidation methods for water treatment: a cost-effectiveness study. Journal of Cleaner Production 65, 178–183.

Mehrjouei, M., Mueller, S., Moeller, D., 2015. A review on photocatalytic ozonation used for the treatment of water and wastewater. Chemical Engineering Journal 263, 209–219.

Mena, E., Rey, A., Acedo, B., Beltran, F.J., Malato, S., 2012. On ozone-photocatalysis synergism in black-light induced reactions: oxidizing species production in photocatalytic ozonation versus heterogeneous photocatalysis. Chemical Engineering Journal 204, 131–140.

Méndez-Arriaga, F., Torres-Palma, R.A., Pétrier, C., Esplugas, S., Gimenez, J., Pulgarin, C., 2009. Mineralization enhancement of a recalcitrant pharmaceutical pollutant in water by advanced oxidation hybrid processes. Water Research 43, 3984–3991.

Miklos, D.B., Remy, C., Jekel, M., Linden, K.G., Drewes, J.E., Hubner, U., 2018. Evaluation of advanced oxidation processes for water and wastewater treatment – a critical review. Water Research 139, 118–131.

Mohammad, A.W., Othaman, R., Hilal, N., 2004. Potential use of nanofiltration membranes in treatment of industrial wastewater from Ni-P electroless plating. Desalination 168, 241–252.

Moreno-Piraján, J.C., Garcia-Cuello, V.S., Giraldo, L., 2010. The removal and kinetic study of Mn, Fe, Ni and Cu ions from wastewater onto activated carbon from coconut shells. Adsorption 17 (3), 505–514.

Neppolian, B., Park, J.-S., Choi, H., 2004. Effect of Fenton-like oxidation on enhanced oxidative degradation of para-chlorobenzoic acid by ultrasonic irradiation. Ultrasonics Sonochemistry 11, 273–279.

Niu, J., Li, Y., Wang, W., 2013. Light-source-dependent role of nitrate and humic acid in tetracycline photolysis: kinetics and mechanism. Chemosphere 92, 1423–1429.

Ochiai, T., Nanba, H., Nakagawa, T., Masuko, K., Nakata, K., Murakami, T., Nakano, R., Hara, M., Koide, Y., Suzuki, T., Ikekita, M., Morito, Y., Fujishima, A., 2012. Development of an O-3-assisted photocatalytic water-purification unit by using a TiO_2 modified titanium mesh filter. Catalysis Science & Technology 2, 76–78.

Oyama, T., Otsu, T., Hidano, Y., Koike, T., Serpone, N., Hidaka, H., 2011. Enhanced remediation of simulated wastewaters contaminated with 2-chlorophenol and other aquatic

pollutants by TiO_2-photoassisted ozonation in a sunlight-driven pilot-plant scale photoreactor. Solar Energy 85, 938–944.
Oyama, T., Yanagisawa, I., Takeuchi, M., Koike, T., Serpone, N., Hidaka, H., 2009. Remediation of simulated aquatic sites contaminated with recalcitrant substrates by TiO_2/ozonation under natural sunlight. Applied Catalysis B: Environmental 91, 242–246.
Pan, Y., Zhang, Y., Zhou, M., Cai, J., Li, X., Tian, Y., 2018. Synergistic degradation of antibiotic sulfamethazine by novel pre-magnetized FeO/PS process enhanced by ultrasound. Chemical Engineering Journal 354, 777–789.
Pei, Z., Li, L., Sun, L., Zhang, S., Shan, X.-q., Yang, S., Wen, B., 2013. Adsorption characteristics of 1,2,4-trichlorobenzene, 2,4,6-trichlorophenol, 2-naphthol and naphthalene on graphene and graphene oxide. Carbon 51, 156–163.
Pichat, P., Cermenati, L., Albini, A., Mas, D., Delprat, H., Guillard, C., 2000. Degradation processes of organic compounds over UV-irradiated TiO_2. Effect of ozone. Research on Chemical Intermediates 26, 161–170.
Pignatello, J.J., Oliveros, E., MacKay, A., 2006. Advanced oxidation processes for organic contaminant destruction based on the Fenton reaction and related chemistry. Critical Reviews in Environmental Science and Technology 36, 1–84.
Pradhan, A.A., Gogate, P.R., 2010. Degradation of p-nitrophenol using acoustic cavitation and Fenton chemistry. Journal of Hazardous Materials 173, 517–522.
Ranjit, P.J.D., Palanivelu, K., Lee, C.S., 2008. Degradation of 2,4-dichlorophenol in aqueous solution by sono-Fenton method. Korean Journal of Chemical Engineering 25, 112–117.
Ren, M.J., Ning, P., Xu, J., Qu, G.F., Xie, R.S., 2018. Concentration and treatment of ceric ammonium nitrate wastewater by integrated electrodialysis-vacuum membrane distillation process. Chemical Engineering Journal 351, 721–731.
Robberson, K.A., Waghe, A.B., Sabatini, D.A., Butler, E.C., 2006. Adsorption of the quinolone antibiotic nalidixic acid onto anion-exchange and neutral polymers. Chemosphere 63 (6), 934–941.
Rodriguez, E.M., Marquez, G., Leon, E.A., Alvarez, P.M., Amat, A.M., Beltran, F.J., 2013. Mechanism considerations for photocatalytic oxidation, ozonation and photocatalytic ozonation of some pharmaceutical compounds in water. Journal of Environmental Management 127, 114–124.
Rodriguez-Mozaz, S., Ricart, M., Kock-Schulmeyer, M., Guasch, H., Bonnineau, C., Proia, L., de Alda, M.L., Sabater, S., Barcelo, D., 2015. Pharmaceuticals and pesticides in reclaimed water: Efficiency assessment of a microfiltration-reverse osmosis (MF-RO) pilot plant. Journal of Hazardous Materials 282, 165–173.
Rubio-Clemente, A., Torres-Palma, R.A., Penuela, G.A., 2014. Removal of polycyclic aromatic hydrocarbons in aqueous environment by chemical treatments: a review. The Science of the Total Environment 478, 201–225.
Salihoglu, N.K., Karaca, G., Salihoglu, G., Tasdemir, Y., 2012. Removal of polycyclic aromatic hydrocarbons from municipal sludge using UV light. Desalination and Water Treatment 44, 324–333.
Salman, J.M., Njoku, V.O., Hameed, B.H., 2011. Adsorption of pesticides from aqueous solution onto banana stalk activated carbon. Chemical Engineering Journal 174 (1), 41–48.
Sanchez, L., Peral, J., Domenech, X., 1998. Aniline degradation by combined photocatalysis and ozonation. Applied Catalysis B: Environmental 19, 59–65.
Segura, Y., Molina, R., Martínez, F., Melero, J.A., 2009. Integrated heterogeneous sono–photo Fenton processes for the degradation of phenolic aqueous solutions. Ultrasonics Sonochemistry 16, 417–424.

Senthilnathan, J., Philip, L., 2009. Removal of mixed pesticides from drinking water system by photodegradation using suspended and immobilized TiO_2. Journal of Environmental Science and Health - Part B: Pesticides, Food Contaminants, and Agricultural Wastes 44, 262–270.

Sharma, V.K., 2008. Oxidative transformations of environmental pharmaceuticals by Cl_2, ClO_2, O_3, and Fe(VI): kinetics assessment. Chemosphere 73, 1379–1386.

Shi, Y., Zhang, T., Ren, H., Kruse, A., Cui, R., 2017. Polyethylene imine modified hydrochar adsorption for chromium (VI) and nickel (II) removal from aqueous solution. Bioresource Technology 247, 370–379.

Sichel, C., Garcia, C., Andre, K., 2011. Feasibility studies: UV/chlorine advanced oxidation treatment for the removal of emerging contaminants. Water Research 45, 6371–6380.

Siedlecka, E.M., Stepnowski, P., 2005. Phenols degradation by Fenton reaction in the presence of chlorides and sulfates. Polish Journal of Environmental Studies 14, 823–828.

Sirés, I., Arias, C., Cabot, P.L., Centellas, F., Garrido, J.A., Rodríguez, R.M., Brillas, E., 2007. Degradation of clofibric acid in acidic aqueous medium by electro-Fenton and photoelectro-Fenton. Chemosphere 66, 1660–1669.

Snyder, S.A., Wert, E.C., Rexing, D.J., Zegers, R.E., Drury, D.D., 2006. Ozone oxidation of endocrine disruptors and pharmaceuticals in surface water and wastewater. Ozone: Science & Engineering 28, 445–460.

Sun, Y., Fenster, M., Yu, A., Berry, R.M., Argyropoulos, D.S., 1999. The effect of metal ions on the reaction of hydrogen peroxide with Kraft lignin model compounds. Canadian Journal of Chemistry-Revue Canadienne De Chimie 77, 667–675.

Ting, W., Lu, M., Huang, Y., 2008. The reactor design and comparison of Fenton, electro-Fenton and photoelectro-Fenton processes for mineralization of benzene sulfonic acid (BSA). Journal of Hazardous Materials 156, 421–427.

Tokumura, M., Sekine, M., Yoshinari, M., Znad, H.T., Kawase, Y., 2007. Photo-Fenton process for excess sludge disintegration. Process Biochemistry 42, 627–633.

Tokumura, M., Katoh, H., Katoh, T., Znad, H.T., Kawase, Y., 2009. Solubilization of excess sludge in activated sludge process using the solar photo-Fenton reaction. Journal of Hazardous Materials 162, 1390–1396.

Vaishnave, P., Kumar, A., Ameta, R., Punjabi, P.B., Ameta, S.C., 2014. Photo oxidative degradation of azure-B by sono-photo-Fenton and photo-Fenton reagents. Arabian Journal of Chemistry 7, 981–985.

Wang, A., Li, Y., Estrada, A.L., 2011. Mineralization of antibiotic sulfamethoxazole by photoelectro-Fenton treatment using activated carbon fiber cathode and under UVA irradiation. Applied Catalysis B: Environmental 102, 378–386.

Wang, D., Bolton, J.R., Hofmann, R., 2012. Medium pressure UV combined with chlorine advanced oxidation for trichloroethylene destruction in a model water. Water Research 46, 4677–4686.

Wang, S., Shiraishi, F., Nakano, K., 2002. A synergistic effect of photocatalysis and ozonation on decomposition of formic acid in an aqueous solution. Chemical Engineering Journal 87, 261–271.

Wang, Z., Han, L., Sun, K., Jin, J., Ro, K.S., Libra, J.A., Liu, X., Xing, B., 2016. Sorption of four hydrophobic organic contaminants by biochars derived from maize straw, wood dust and swine manure at different pyrolytic temperatures. Chemosphere 144, 285–291.

Wilcoxon, J.P., Thurston, T.R., 1999. In: Lednor, P.W., Nagaki, D.A., Thompson, L.T. (Eds.), Advanced Catalytic Materials-1998, pp. 119–124.

Wu, C., Liu, X., Wei, D., Fan, J., Wang, L., 2001. Photosonochemical degradation of Phenol in water. Water Research 35, 3927–3933.
Wu, D.S., Chen, G.Q., Hu, B.S., Deng, H.N., 2019. Feasibility and energy consumption analysis of phenol removal from salty wastewater by electro-electrodialysis. Separation and Purification Technology 215, 44–50.
Yang, S.T., Chang, Y., Wang, H., Liu, G., Chen, S., Wang, Y., Liu, Y., Cao, A., 2010. Folding/aggregation of graphene oxide and its application in Cu^{2+} removal. Journal of Colloid and Interface Science 351 (1), 122–127.
Yang, X., Flowers, R.C., Weinberg, H.S., Singer, P.C., 2011. Occurrence and removal of pharmaceuticals and personal care products (PPCPs) in an advanced wastewater reclamation plant. Water Research 45, 5218–5228.
Yap, H.C., Pang, Y.L., Lim, S., Abdullah, A.Z., Ong, H.C., Wu, C.H., 2018. A comprehensive review on state-of-the-art photo-, sono-, and sonophotocatalytic treatments to degrade emerging contaminants. International Journal of Environmental Science and Technology 16 (1), 601–628.
Yildirim, A.O., Gul, S., Eren, O., Kusvuran, E., 2011. A Comparative Study of ozonation, homogeneous catalytic ozonation, and photocatalytic ozonation for CI reactive red 194 azo dye degradation. Clean - Soil, Air, Water 39, 795–805.
Yuan, H., Zhou, X., Zhang, Y.-L., 2013. Degradation of acid pharmaceuticals in the UV/H_2O_2 process: effects of humic acid and inorganic salts. Clean – Soil, Air, Water 41, 43–50.
Yuan, S., Tian, M., Cui, Y., Lin, L., Lu, X., 2006. Treatment of nitrophenols by cathode reduction and electro-fenton methods. Journal of Hazardous Materials 137, 573–580.
Zaleska, A., Hupka, J., Wiergowski, M., Biziuk, M., 2000. Photocatalytic degradation of lindane, p,p '-DDT and methoxychlor in an aqueous environment. Journal of Photochemistry and Photobiology A-Chemistry 135, 213–220.
Zhang, G., Ren, Z., Zhang, X., Chen, J., 2013. Nanostructured iron(III)-copper(II) binary oxide: a novel adsorbent for enhanced arsenic removal from aqueous solutions. Water Research 47 (12), 4022–4031.
Zhang, X.F., Wang, B., Yu, J., Wu, X.N., Zang, Y.H., Gao, H.C., Su, P.C., Hao, S.Q., 2018. Three-dimensional honeycomb-like porous carbon derived from corncob for the removal of heavy metals from water by capacitive deionization. RSC Advances 8 (3), 1159–1167.
Zhang, Y., Zhou, J.L., Ning, B., 2007. Photodegradation of estrone and 17 beta-estradiol in water. Water Research 41, 19–26.
Zhao, P., Gao, B.Y., Yue, Q.Y., Liu, S.C., Shon, H.K., 2016. The performance of forward osmosis in treating high-salinity wastewater containing heavy metal Ni^{2+}. Chemical Engineering Journal 288, 569–576.
Zou, L., Zhu, B., 2008. The synergistic effect of ozonation and photocatalysis on color removal from reused water. Journal of Photochemistry and Photobiology A-Chemistry 196, 24–32.

Further Reading

Akita, S., Castillo, L.P., Nii, S., Takahashi, K., Takeuchi, H., 1999. Separation of Co(II)/Ni(II) via micellar-enhanced ultrafiltration using organophosphorus acid extractant solubilized by nonionic surfactant. Journal of Membrane Science 162, 111–117.
Bhattacharjee, C., Saxena, V.K., Dutta, S., 2017. Fruit juice processing using membrane technology: a review. Innovative Food Science & Emerging Technologies 43, 136–153.

Dahdouh, L., Wisniewski, C., Ricci, J., Vachoud, L., Dornier, M., Delalonde, M., 2016. Rheological study of orange juices for a better knowledge of their suspended solids interactions at low and high concentration. Journal of Food Engineering 174, 15–20.

Efome, J.E., Rana, D., Matsuura, T., Lan, C.Q., 2019. Effects of operating parameters and coexisting ions on the efficiency of heavy metal ions removal by nano-fibrous metal-organic framework membrane filtration process. The Science of the Total Environment 674, 355–362.

Fan, X., Su, Y., Zhao, X., Li, Y., Zhang, R., Zhao, J., Jiang, Z., Zhu, J., Ma, Y., Liu, Y., 2014. Fabrication of polyvinyl chloride ultrafiltration membranes with stable antifouling property by exploring the pore formation and surface modification capabilities of polyvinyl formal. Journal of Membrane Science 464, 100–109.

Gu, Y., Wei, Y., Xiang, Q., Zhao, K., Yu, X., Zhang, X., Li, C., Chen, Q., Xiao, H., Zhang, X., 2019. C:N ratio shaped both taxonomic and functional structure of microbial communities in livestock and poultry breeding wastewater treatment reactor. The Science of the Total Environment 651, 625–633.

Haddad, M., Mikhaylin, S., Bazinet, L., Savadogo, O., Paris, J., 2017. Electrochemical acidification of Kraft black liquor by electrodialysis with bipolar membrane: ion exchange membrane fouling identification and mechanisms. Journal of Colloid and Interface Science 488, 39–47.

Hu, C., Wang, M.S., Chen, C.H., Chen, Y.R., Huang, P.H., Tung, K.L., 2019. Phosphorus-doped g-C_3N_4 integrated photocatalytic membrane reactor for wastewater treatment. Journal of Membrane Science 580, 1–11.

Oliveira, R.C., Docê, R.C., de Barros, S.T.D., 2012. Clarification of passion fruit juice by microfiltration: analyses of operating parameters, study of membrane fouling and juice quality. Journal of Food Engineering 111, 432–439.

Persico, M., Mikhaylin, S., Doyen, A., Firdaous, L., Nikonenko, V., Pismenskaya, N., Bazinet, L., 2017. Prevention of peptide fouling on ion-exchange membranes during electrodialysis in overlimiting conditions. Journal of Membrane Science 543, 212–221.

Rouquié, C., Dahdouh, L., Ricci, J., Wisniewski, C., Delalonde, M., 2019. Immersed membranes configuration for the microfiltration of fruit-based suspensions. Separation and Purification Technology 216, 25–33.

Ujang, Z., Anderson, G.K., 1996. Application of low-pressure reverse osmosis membrane for Zn^{2+} and Cu^{2+} removal from wastewater. Water Science and Technology 34, 247–253.

Valero, D., García-García, V., Expósito, E., Aldaz, A., Montiel, V., 2015. Application of electrodialysis for the treatment of almond industry wastewater. Journal of Membrane Science 476, 580–589.

Van der Bruggen, B., Mänttäri, M., Nyström, M., 2008. Drawbacks of applying nanofiltration and how to avoid them: a review. Separation and Purification Technology 63, 251–263.

Wang, Y.-H., Wu, Y.-H., Tong, X., Yu, T., Peng, L., Bai, Y., Zhao, X.-H., Huo, Z.-Y., Ikuno, N., Hu, H.-Y., 2019. Chlorine disinfection significantly aggravated the biofouling of reverse osmosis membrane used for municipal wastewater reclamation. Water Research 154, 246–257.

Ward, A.J., Arola, K., Thompson Brewster, E., Mehta, C.M., Batstone, D.J., 2018. Nutrient recovery from wastewater through pilot scale electrodialysis. Water Research 135, 57–65.

Yao, W., Wang, X., Yang, H., Yu, G., Deng, S., Huang, J., Wang, B., Wang, Y., 2016. Removal of pharmaceuticals from secondary effluents by an electro-peroxone process. Water Research 88, 826–835.

Yong, M., Zhang, Y., Sun, S., Liu, W., 2019. Properties of polyvinyl chloride (PVC) ultrafiltration membrane improved by lignin: hydrophilicity and antifouling. Journal of Membrane Science 575, 50–59.

CHAPTER

Biological technologies for cHRPs and risk control

9

Hui Huang, PhD, Lin Ye, PhD

State Key Laboratory of Pollution Control and Resource Reuse, School of the Environment, Nanjing University, Nanjing, China

Chapter outline

High-Risk Pollutants in Wastewater. https://doi.org/10.1016/B978-0-12-816448-8.00009-5

Biological wastewater treatment technologies, which possess the advantages of being highly efficient and cost-effective, are widely used in the wastewater treatment plants (WWTPs) and playing a critical role in the removal of chemical high-risk pollutants (cHRPs) from wastewater (Andersen et al., 2003; Braga et al., 2005). In the past several decades, a number of biological technologies, including conventional biological technology, biofiltration technology, membrane biotechnology, constructed wetland, and bioaugmentation technology, were invented and applied in engineering practice for cHRP removal and risk control (Alvarino et al., 2014; Doulia et al., 2016; Ternes et al., 2004; Kennedy et al., 2015). Microorganisms could degrade cHRPs via the cometabolism processes in biological treatment facilities (Margot et al., 2016). However, biological transformation processes of cHRPs varied among different technologies. The removal or degradation efficiency of cHRPs is not only highly depended on the physicochemical properties of cHRPs, but also associated with the technologies used and operational conditions. Investigation and review of these technologies and the affecting factors are of great importance to provide a comprehensive overview of cHRP risk control.

Nowadays, most municipal and industrial wastewater contains a variety of toxic organic matters with different concentrations (Dai et al., 2014). Conventional biological processes do not always provide satisfactory treatment results, especially for industrial wastewater. Because many of the organic substances in industrial wastewater are toxic or resistant to biological degradation (Garcıa et al., 2001; Lapertot et al., 2006; Muñoz and Guieysse, 2006). To improve the removal efficiency of toxic organic matter in wastewater, a series of biological filtration, coagulation sedimentation, and advanced oxidation processes (AOPs) are incorporated into the conventional WWTPs to enhance the removal of toxic organic matters (Petrasek et al., 1983). In recent years, many new and integrated wastewater treatment processes have been developed and higher removal efficiency has been achieved. On the other hand, both treatment efficiency and cost should be considered and technologies with high efficiency and low operational cost are needed to be further developed in the future.

In this chapter, we systematically introduced the biological treatment technologies and the biological transformation processes of various cHRPs. Furthermore, the characteristics of these technologies and their applications in the treatment and degradation of cHRPs were also presented. Meanwhile, a series of integrated technologies and their advantages and disadvantages were reviewed as well. Such information is essential for the development of more advanced cHRP control technologies in the future.

9.1 Biological transformation of cHRPs in wastewater

9.1.1 Brief introduction to biological transformation of cHRPs

Biological processes are wildly applied in wastewater treatment and it is of great importance to fully understand the biological transformation of cHRPs. Generally, cHRPs are converted into some intermediate products at first, and then, final products of small molecules, such as water and carbon dioxide, are generated under the degradation of microorganisms.

The biodegradation processes of cHRPs involve enzymatic reactions, oxidative reactions, reductive reactions, cometabolism decarboxylation, and deacetylation reactions. Microbial transformation of organics is generally driven by enzymatic reactions such as oxidation (e.g., hydroxylation, N- and S-oxidation, and dealkylation) and reduction reactions (e.g., dehalogenation, nitro reduction, and hydrolysis of amides and carboxyl esters). Other oxidative reactions including ring cleavage, oxidative deamination, and oxidative dichlorination also contribute to substrate degradation (Helbling et al. 2010a, 2010b; Murugesan et al., 2010; Terzic et al., 2011). Some refractory organic compounds are not available as carbon sources and energy directly. Therefore, microorganisms should obtain most or all of the carbon sources and energy from other substrates (cometabolism) and alter their chemical composition.

The nitro reduction of N,N-diethyl 1-4-nitrobenzamide was evident under aerobic conditions (Helbling et al., 2010a) and 53 kinds of cHRPs were observed after microbial transformation of 30 xenobiotic compounds with amide groups as products of amide hydrolysis, N-dealkylation, hydroxylation, oxidation, dehalogenation, glutathione conjugation, and many other pathways (Helbling et al., 2010a). Reduction reactions predominantly take place under anaerobic conditions, such as reductive dechlorination of 5-chlorobenzotriazole and chlorpromazine, and dehydration, and hydration of testosterone (Fahrbach et al., 2010; Trautwein and Kümmerer, 2012). The biodegradation processes of cHRPs also involve bacteria, algae, fungi, and other microorganisms (Garcia-Rodríguez et al., 2014). In conventional biodegradation process, microorganisms use organic compounds as primary substrates for their cell growth and induce enzymes for their assimilation (Tran et al., 2013).

9.1.2 Advances in biological transformation of cHRPs in wastewater

In recent years, studies on biotransformation of cHRPs mainly focused on revealing the in-depth degradation mechanism and functional bacteria involved.

As to the mechanism of biological transformation, new reaction mechanisms have been discovered. Microorganisms, such as ammonia-oxidizing bacteria, could oxidize cHRPs cometabolically due to the presence of ammonia monooxygenase, thus the removal of organic HRPs was improved (Margot et al., 2016). Sulfonamides resistance genes (*sul1* and *sul2*) were the most prevalent genes with the detection frequency of 100% in water samples. Domestic wastewater contained more types of resistance genes than other kinds of sewage and the abundance of resistance genes generally differed among WWTPs (Du et al., 2015). It was reported that *Hydrogenophaga, Aeromonas, Pseudomonas,* and *Methyloversatilis* dominated in the moving bed biofilm reactors treating tetracycline wastewater (Peng et al., 2018). In another report, four bacteria discovered from soil and sediments, which were identified as *Achromobacter* sp. *NP03*, *Ochrobactrum* sp. *NP04*, *Lysinibacillus* sp. *NP05*, and *Pseudomonas* sp. *NP06*, were applied to investigate their PCB degradation efficiencies (Pathiraja et al., 2019b).

The removal or degradation capacity of cHRPs is highly depended on the chemical and biological persistence together with the physicochemical properties of cHRPs, and the technologies used and operational conditions are also significant influencing factors. There are still some deficiencies in the existing studies of biotransformation mechanisms in biological processes, and further studies are required to explore the principles.

9.2 Conventional biological technology

9.2.1 Types of conventional biological technology

Biological technology generally means to create an environment conducive to the growth and reproduction of microorganisms by certain artificial measures, so that a great number of microorganisms can proliferate and achieve oxidation and decomposition of organic pollutants, which are finally converted into harmless substances.

Conventional biological technologies can be divided into two types, aerobic biological treatment and anaerobic biological treatment, depending on whether or not oxygen is involved. The former is of higher treatment efficiency and has been widely used, which is also the main method applied in biological treatment.

Aerobic biological treatment technology uses aerobic microorganisms including facultative microorganisms, etc. to conduct biological metabolism in the presence of oxygen to degrade organic matters and make it stable and harmless. The advantages of aerobic biological treatment technology include high reaction rate, short residence time, small volume of treatment structures and less odor generation in the process of treatment. The disadvantages include the low removal rate of refractory organics, more sludge than anaerobic treatment and higher operating costs. Anaerobic biological treatment processes include upflow anaerobic sludge blanket (UASB), expanded granular sludge bed, and internal circulation (IC) reactor. Anaerobic decomposition of complex organic compounds in wastewater can be divided into four stages: hydrolysis, acidification, acetic acid production, and methane production, through which a high removal rate of organic matter, less sludge production, and less operating costs can be achieved in anaerobic treatment. Disadvantages include long wastewater residence time, incomplete decomposition of organic matter, and more odor generation.

Conventional biological technologies can be divided into activated sludge process and biofilm processes according to the living conditions of microorganisms. Activated sludge process uses the biological coagulation, adsorption, and oxidation of activated sludge to decompose and remove organic pollutants in sewage. Typical activated sludge method includes oxidation ditch, multipoint water intake method, adsorption and regeneration method, delayed aeration method, anaerobic/aerobic

(A/O) process, anaerobic/anoxic/aerobic (A^2/O) process, adsorption biodegradation (AB) process, and sequencing batch reactor activated sludge process. Biofilm process is a method of wastewater treatment that uses microorganisms (biofilms) attached to the surface of certain solids and it generally includes biological filter, biological rotary plate, biological contact oxidation, biological fluidized bed, and so on. The biofilm method has strong adaptability to the fluctuation of water quality and temperature with the advantages of good treatment effect and nitrification function, low sludge production, easy solid—liquid separation, and low power cost.

9.2.2 Removal of cHRPs by conventional biological technologies

Among all conventional wastewater treatment processes, the activated sludge process is the most widely applied around the world. On the removal of cHRPs, the proportion removed by primary setting, chemical precipitation, aerating volatilization, and sludge absorption is small, while most cHRPs in wastewater are removed by biodegradation (Andersen et al., 2003; Braga et al., 2005; Liu et al., 2009). It has been reported that traditional biological treatment via activated sludge is not very effective for the removal of such contaminants. For cHRPs, the removal of these molecules occurs due to the combined effect of biodegradation and sorption. Hence, conventional activated sludge systems keep a wide range of removal efficiencies regarding cHRPs (Verlicchi et al., 2012). It is worth mentioning that the degradation of some cHRPs such as CFN, ACE, and metformin that are of high degradability, is slowed or stopped at trace but notable at high concentrations within an activated sludge system. A degradation plateau has been observed with these molecules: 40 ng L^{-1} for CFN, 90 ng L^{-1} for ACE, and 1000 ng L^{-1} for metformin (Blair et al., 2015). Another potential theory is that the undetected pharmaceutical and personal care products (PPCPs) metabolites are further transformed back into the parent compounds through microbial activity (Verlicchi et al., 2012). The deconjugation of conjugates by hydrolysis during treatment and yielding the parent compound can lead to an additional source of contaminant load (Suárez et al., 2008). Diclofenac (DCF) removal efficiencies range from 0% to 70%, which have been reported according to the biological composition of the sludge used (Bernhard et al., 2006; Clara et al., 2005; De Wever et al., 2007; Kimura et al., 2007; Radjenović et al., 2009).

Among the conventional biological technologies, the removal of PPCPs by traditional biotechnology is the most studied. Table 9.1 presented the removal efficiency of PPCPs by different conventional technologies. Different treatment techniques have different efficiencies in removing different pollutants. The removal rate of pollutants by these treatment technologies generally ranges from 0% to 98.5%. The conventional activated sludge method has a certain removal effect on cHRPs but it has no removal ability for some substances, of which even with an increase in the concentration after treatment, such as ciprofloxacin.

Traditional biological treatment technology has limited the removal efficiency of cHRPs, so it is of great necessity to improve traditional technology or combine it

Table 9.1 cHRPs removal of conventional technologies.

Categories	cHRPs	Technologies	Removal efficiencies (%)	References
PPCPs	Celestolide	CAS	85	Alvarino et al. (2014)
	Ethinylestradiol (EE2)	CAS	63	Alvarino et al. (2014)
	Diclofenac (DCF)	CAS	8	Alvarino et al. (2014)
		Aerobic activated sludge process	11.3	Kosma et al. (2010)
		A/O	14.8	Kosma et al. (2014)
	Sulfamethoxazole (SMX)	CAS	42	Alvarino et al. (2014)
		A^2/O	13.1	Kosma et al. (2010)
		A/O	82.1	Kosma et al. (2014)
	Naproxen (NPX)	CAS	92	Alvarino et al. (2014)
		A/O	62.1	Kosma et al. (2014)
		A^2/O	46.9	Kosma et al. (2010)
		Aerobic activated sludge process	48.8	Kosma et al. (2010)
	Ibuprofen (IBP)	CAS	94	Alvarino et al. (2014)
		A^2/O	78.6	Kosma et al. (2010)
		Aerobic activated sludge process	58.7	Kosma et al. (2010)
	Fluoxetine (FLX)	CAS	82	Alvarino et al. (2014)
	Erythromycin (ERY)	CAS	72	Alvarino et al. (2014)
	Carbamazepine (CBZ)	CAS	0	Alvarino et al. (2014)
		A^2/O	8.9	Kosma et al. (2010)
			75.6	Kosma et al. (2010)

Table 9.1 cHRPs removal of conventional technologies.—*cont'd*

Categories	cHRPs	Technologies	Removal efficiencies (%)	References
		Aerobic activated sludge process		
	Caffeine	A^2/O	79.5	Kosma et al. (2010)
		A/O	85.7	Kosma et al. (2014)
	Paracetamol	Aerobic activated sludge process	87.7	Kosma et al. (2010)
		A/O	98.5	Kosma et al. (2014)
	Trimethoprim (TMP)	A/O	42.9	Kosma et al. (2014)
	Salicylic acid	A/O	73.2	Kosma et al. (2014)
Polychlorinated biphenyls	Total PCB	Alternating anaerobic aerobic treatments (a facultative anaerobic bacterial consortium)	55.8	Pathiraja et al. (2019a)
	Total PCB	A/O (a facultative anaerobic bacterial consortium)	47.6	

with other technologies to achieve more efficient removal of cHRPs. The ability of traditional biotechnologies to remove other cHRPs (such as organic pesticides, and polychlorinated biphenyls) should be further investigated.

9.3 Biofiltration technology

9.3.1 Definition of biofiltration

Biofiltration relies on autochthonous microbial communities attached to granular media in depth filtration processes to biotransform organic chemicals, nitrogen, phosphorus, and dissolved metal constituents (Bouwer and Crowe, 1988).

Biofiltration is an alternative drinking water treatment process that has the potential for broad cHRPs removal without the need for significant inputs of energy or chemical reagents. Due to the rationality of the process, the biological filter can achieve better treatment effect without secondary pollution and avoid the phenomenon of microbial loss or turbidity caused by water flow.

The biofiltration processes include biological aerated filter (BAF), nitrifying biological filter, and denitrification biological filter. BAF, compared with normal activated sludge process with high organic load, covers advantages of smaller area occupied, less investment, less sludge production, higher oxygen transfer efficiency, and better effluent quality. However, it is restricted by SS (suspended solids) in the influent, so the preprocessing of wastewater is needed. Meanwhile, the backwashing volume and head loss are large. With the function of simultaneous biological oxidation and filtration of SS, no subsequent sedimentation tank (e.g., secondary sedimentation tank) was required after BAF. It also has advantages such as large hydraulic load, short hydraulic retention time, low energy consumption, and low operating costs.

Denitrification biological filter is more efficient than the conventional activated sludge process with obvious improvement. Microorganisms are attached to the packing materials and are less likely to be broken up by wastewater disturbance, making the reactor more stable. In addition, the higher efficiency and stable operation lead to the great decrease in the volume of denitrifying biological filter.

9.3.2 Removal of cHRPs by biofiltration technology

There are a few reports concerning the removal of cHRPs by biofilters (Hallé et al., 2015; Kennedy et al., 2015; L Zearley and Scott Summers, 2012; Snyder et al., 2007). Conventional sand or anthracite-sand biofilters can achieve removal efficiencies of 50%–80% for biodegradable compounds but little or no removal for compounds that are typically persistent in the granular activated carbon (GAC) biofilters. However, GAC biofilters were reported to provide more than 90% removal rate of atrazine and carbamazepine over the 200 days of operation (Kennedy et al., 2015).

There is an increasing need for the application of biofilm process in the upcycling of wastewater treatment plants all around the world in recent years, yet there are few literatures on summarizing wastewater biofilm during the life cycle. In considering this, the authors have systematically investigated the microcosmic mechanism (Huang et al. 2015, 2018a) and enhancement methods (Peng et al. 2018a, 2018b, 2019) of biofilm formation, biofilm activity (Huang et al. 2014, 2019b), *in situ* and *ex situ* activity recovery of aging biofilm on surfaces of organic and inorganic filter medias (Huang et al. 2018b, 2019a; Yu et al., 2016) in recent years. However, there is still a vacancy on characterization at various stages of biofilm and its regulation (Huang et al., 2019). Further studies are needed on the formation of biofilms and the mechanism, prevention of biofilm aging, and application in removing cHRPs to improve the performance of the biological filter.

9.4 Membrane biotechnology

9.4.1 Presentation of membrane bioreactor

Membrane bioreactor (MBR) is the combination of biological treatment with microfiltration (MF) or ultrafiltration (UF) membranes and is widely used in water reuse systems. MBR provides an alternative to conventional secondary treatment with activated sludge and could enhance the organic compounds and suspended solid removal.

MBR possess the following advantages over conventional wastewater treatment: high effluent quality, excellent microbial separation ability, absolute control of sludge, long sludge retention time (SRT) and short hydraulic retention time (HRT), high biomass content and less sludge bulking problem, low sludge production, small footprint requirement, and possibilities for a flexible and phased extension of existing WWTPs.

9.4.2 Removal of cHRPs in MBR

The presence of membranes improves the removal of cHRPs in MBR because of the intrinsic effect of the membrane and the increase of SRT. Higher SRT leads to the diversification of microorganisms, including some slow-growing bacteria such as nitrifying bacteria, which improves the removal of cHRPs. Sorption to biomass and infiltration into the membrane biofilm are also important mechanisms of cHRPs' removal. Regarding cHRPs in municipal wastewater, MBR does not efficiently remove them. MF and UF membranes are also widely used as a pretreatment step to help prevent fouling of the less permeable nanofiltration (NF) and reverse osmosis (RO) membranes.

The speciation of cHRPs may result in a significant change of membranes with much greater retention for ionized, negatively charged compounds. For uncharged cHRPs, intrinsic physicochemical properties of the molecules play a role in their retention. UF and NF exhibited much higher removals for less polar cHRPs (Yoon et al., 2006). An increase in cation concentration leads to a decrease in the removal of neutral compounds probably by decreasing the availability of HS interaction sites (Sadmani et al., 2014).

MBR is regarded as being a state-of-the-art technology for municipal and industrial wastewater treatment due to its high effluent quality achieved with respect to many cHRPs (Abegglen et al., 2009; Li et al., 2014; Luo et al., 2014a; Meng et al., 2012). MBR is able to effectively remove a wide range of cHRPs including compounds that are resistant to activated sludge process and constructed wetland (Luo et al., 2014b; Radjenović et al., 2009; Westerhoff et al., 2005).

Table 9.2 presents the removal efficiency of PPCPs and organic pesticides by MBR. MBR systems seem to enhance the removal of cHRPs compared to CAS. MF is often used to remove organic pesticides with a removal rate higher than 40% and UF is used to remove PPCPs. The removal of cHRPs in MBR system can be affected by sludge age, mixed liquor suspended solids concentration, the

Table 9.2 cHRPs removal efficiency by MBR-based systems.

Categories	cHRPs	Technologies	Influent	Removal efficiencies (%)	References
Organic pesticides	Aldrin	MF pore size 0.45 μm Cellulose acetate	0.1 mg/L	63.4 ± 1.2	Doulia et al. (2016)
	Lindane	MF pore size 0.45 μm Cellulose acetate	0.1 mg/L	40.7 ± 7.7	Doulia et al. (2016)
	Tetradifon	MF pore size 0.45 μm Cellulose acetate	0.1 mg/L	88.9 ± 2.3	Doulia et al. (2016)
	Bromophos	MF	0.1 mg/L	96.6 ± 0.5	Doulia et al. (2016)
	Chlorpyrifos	MF	0.1 mg/L	91.6 ± 3.4	Doulia et al. (2016)
PPCPs	CBZ	Side-stream UF-modules (tubular PVDF membranes)	177 ± 38 ng/L	2 ± 4	Echevarría et al. (2019)
		PAC-UF	177 ± 38 ng/L	67 ± 16	Echevarría et al. (2019)
	DIU	UF	140 ± 48 ng/L	13 ± 21	Echevarría et al. (2019)
		PAC-UF	140 ± 48 ng/L	93 ± 5	
	ERY	UF	152 ± 95 ng/L	73 ± 24	Echevarría et al. (2019)
		PAC-UF	152 ± 95 ng/L	80 ± 9	
	DCF	UF	614 ± 258 ng/L	72 ± 19	Echevarría et al. (2019)
		PAC-UF	614 ± 258 ng/L	75 ± 14	
	SMX	UF	313 ± 112 ng/L	60 ± 13	Echevarría et al. (2019)
		PAC-UF	313 ± 112 ng/L	54 ± 28	

existence of anoxic and anaerobic compartments, the composition of wastewater, operating temperature, pH, and conductivity (Luo et al., 2014b).

MBR can realize a high removal rate for cHRPs and further studies on the mechanism of MBR degradation toward cHRPs are needed, to make MBR a process with a higher removal efficiency of cHRPs. Similarly, it is also necessary to combine MBR with other treatment technologies to further improve the removal rate of cHRPs.

9.5 Constructed wetland (CW) systems

9.5.1 Definition and characteristics of CWs

Wetlands are areas of marsh, fen, peatland or water, whether natural or artificial, permanent, or temporary, with water that is static or flowing, fresh, brackish, or salt, including areas of marine water the depth of which at low tide does not exceed 6 meters (Goodwin, 2017). It is widely accepted that wetlands have the following three characteristics (Scholz and Lee, 2005): (1) the presence of water; (2) unique soils that differ from upland soils; (3) the presence of vegetation adapted to saturated conditions. Given the special features, wetlands are not only of great ecological importance but also have purification functions to remove pollutants from water or wastewater.

Wetlands can be classified as either natural wetlands or CWs. Natural wetlands usually improve the quality of natural water passing through them and act as natural ecosystem filters; whereas, constructed wetlands are artificially created wetlands used to treat wastewater or polluted water (Scholz and Lee, 2005). CWs have the purpose of removing bacteria, enteric viruses, suspended solids, biological oxygen demand (BOD), nitrogen (as ammonia and nitrate), metals, phosphorus, etc (Pinney et al., 2000). Two general forms of CWs are used in practice: surface flow (horizontal flow) and subsurface flow (vertical flow). Surface-flow CWs most closely mimic natural environments and are usually more suitable for wetland because of the permanent standing water. In subsurface-flow wetlands, water passes laterally through a porous medium (usually sand and gravel) with a limited number of macrophyte (Scholz and Lee, 2005), and subsurface-flow CWs are one of the most common types of extensive wastewater systems used throughout the world (Garcia et al., 2010). Overall, as a green treatment technology by simulating natural wetlands, CWs have been widely used to treat various kinds of wastewater such as domestic sewage, agricultural wastewater, industrial effluent, mine drainage, landfill leachate, stormwater, polluted river water, and urban runoff in the last few decades (Harrington and Scholz, 2010; Saeed and Sun, 2012; Wu et al., 2015; Yalcuk and Ugurlu, 2009).

9.5.2 Removal of pesticides and herbicides

The use of agricultural pesticides results in increased crop yields, but their effects are undesirable when they leave agricultural ecosystems, especially by entering waterways. Widespread use of pesticides in modern agriculture contributes to agricultural nonpoint source pollution in rivers and streams across the world threating drinking water resources and aquatic ecosystems (Vymazal and Březinová, 2015).

Researchers have conducted extensive studies on the removal of pesticides from CWs and natural wetlands. In the Mississippi Delta region of the United States, Moore et al. used artificial wetlands to remove the insecticide diazinon (Moore et al., 2007), which showed that plants absorbed 43% of the diazinon; whereas, 23% and 34% were still present in the sediment and water, respectively. Budd et al. studied the removal of pesticides from agricultural irrigation tailwater during the peak period of agricultural irrigation (Budd et al., 2009). Their results showed that the artificial wetland had proper removal of pesticides if a reasonable design was carried out. The concentration of pesticides in the wetland effluent could be much lower than that in the influent. Researchers have also conducted studies on the migration and transformation mechanisms of pollutants entering the wetland. Their results showed that in the process of removing pollutants from CWs, the matrix, the plant, and the microbes were interrelated and mutually benefited to form a symbiotic system, Moreover, pollutants, including pesticides and herbicides, were removed through filtration, adsorption, precipitation, coprecipitation, ion exchange, plants absorption, and microbial degradation.

Many factors affect the removal of pesticides in CWs. Plants are an essential part of CWs and an essential factor affecting the removal efficiency of pesticides. Plants can provide adsorption and growth sites for microorganisms, physically intercept pollutants, and promote the stabilization of the sediments. Microorganisms play an essential role in the degradation of many organic pollutants. Slemp et al. (2004) determined the preferred electron acceptor for the microbial degradation of simazine in horizontal subsurface-flow CWs. They found that denitrification and sulfate reduction did not have a significant effect on simazine removal. However, the aerobic microbial degradation in the plant rhizosphere might be the main biodegradation mechanism (Garcia et al., 2010). The microbial communities in wetlands are incredibly complex, and the mechanism of microbial removal in wetlands is very complicated. Therefore, further intensive research on the community structure and functions of the microbial community in wetlands is required. In addition, hydraulic retention time and temperature are also important factors affecting the removal of pesticides from CWs.

9.5.3 Removal of PPCPs

The U.S. EPA defines PPCPs as "any product used by individuals for personal health or cosmetic reasons or used by agribusiness to enhance growth or health of livestock." PPCPs are a unique group of emerging environmental contaminants, due to their inherent ability to induce physiological effects in human at low doses. Although most of these contaminants are readily degradable, some compounds are refractory to biodegradation and cannot be removed in conventional activated sludge of WWTPs (Ternes et al., 2004). A study has confirmed that some types of

PPCPs could be effectively removed in a WWTP using an oxidation pond constructed wetland wastewater treatment system in Louisiana, USA (Park et al., 2009). Conkle et al. (2008) investigated 15 pharmaceutically active compounds in and out of water, and found that thirteen of them could be detected in the influent, while only nine could be detected in the effluent. Moreover, most compounds have a removal rate of over 90%.

Adsorption is a meaningful way to remove PPCPs from CWs. In CWs, packed bed matrix, soil, and sediments play major roles in adsorption of PPCPs. Besides, plant root biofilm, can also adsorb soluble organic matter. By studying the adsorption and desorption behavior of ciprofloxacin, ofloxacin, and norfloxacin in CWs, Conkle et al. concluded that adsorption is an effective and long-term removal potential for removing these antibiotics from sewage (Conkle et al., 2010).

Aquatic plants can use roots, stems, and leaves to adsorb, enrich, and degrade pollutants in water bodies and sediments. In CWs, plants can remove PPCPs by absorption. This can be verified by testing the linear relationship between transpiration and mass removal rate. Hijosa-Valsero et al. (2010) found that the quality removal rate of carbenicillin in winter has a significant linear relationship with transpiration. The mass removal rate of methyl dihydrojasmonate in summer has a significant linear relationship with transpiration.

Microbial degradation of antibiotics refers to the action of microorganisms. The microorganisms degrade the antibiotic from the macromolecular compounds to small molecules, and finally convert them into water and carbon dioxide. Moreover, during this process, the resistant bacteria play an essential role in the CW system and the biodegradation usually occurs at the biofilm and the interface between the soil and the roots (Hussain et al., 2012).

9.6 Bioaugmentation technology

9.6.1 Introduction to bioaugmentation technology

Bioaugmentation (the process of adding selected strains/mixed cultures to wastewater reactors to improve the catabolism of specific compounds, for example, refractory organics, or overall COD) is a promising technique to solve practical problems in WWTPs and enhance removal efficiency (Herrero and Stuckey, 2015). Bioaugmentation generally falls in two main strategies: (1) bioaugmentation by enrichment with indigenous microorganisms; and (2) bioaugmentation by enrichment with nonindigenous microorganisms. The reinoculation of an environment with previously adapted indigenous microorganisms directly isolated from the site is often termed indigenous bioaugmentation. However, if the sites do not contain active pollutant-degrading microbes, the addition of exogenous microbial strains be a solution (Poi et al., 2017).

Bioaugmentation uses added microorganisms to "reinforce" biological waste treatment populations so that they can effectively reduce the contaminant load by transforming it into less dangerous compounds (Herrero and Stuckey, 2015). The most common way to apply bioaugmentation technology is to directly add microorganisms that have specific degradation capabilities to target contaminants. This needs to obtain one or more high-efficiency microbial strains with target contaminants as the primary carbon source and energy source. The added microorganisms may be attached to the carrier to form a highly efficient biofilm, and they may also exist in a free state. Although many pure bacteria have an excellent degradation effect on specific target pollutants, in the actual wastewater treatment, people tend to use the mixed flora for bioenhancement and bioaugmentation. Compared with pure bacteria, the mixed bacteria generally have a higher degradation ability and higher adaptability.

9.6.2 Selection of high-performance bacteria

Under certain conditions, bacteria strains with relatively high degradability can be obtained by selective enrichment, which is not necessarily typical or representative microorganisms of indigenous environments. Moreover, the enrichment process is unlikely to increase the competitiveness of the strain in the real environment with indigenous microorganisms. Therefore, due to the lack of understanding of the original microbial flora and microbial population composition and dynamic changes in the target contaminated land, it is easy to make the selection of dominant strains fail.

In addition, some strains screened based on the degradation ability as the single standard may be pathogenic bacteria, which may cause negative impacts when applied to environmental engineering. For example, *Pseudomonas aeruginosa* strain is a multifunctional gram-negative bacterium, but may cause pneumonia and severe respiratory diseases and should not be used in engineering practice.

9.6.3 Removal of cHRPs with bioaugmentation technology

Researchers use a variety of methods to isolate microorganisms with specific degradation properties and enhance the effectiveness of microorganisms through biofortification to improve the treatment of biological systems.

Wang et al. (2002) used quinoline as the target pollutant and intensified the treatment of coking wastewater by the quinoline degrading bacteria *Burkholderia pickettii* in the A/A/O system. The results showed that the contribution of anaerobic, anoxic, and oxic reactors to COD removal rate was 25%, 16%, and 59%, respectively. A reactor for treating the mixture of domestic sewage and polychlorophenol with 2,4-DCP degrading bacteria as bioenhanced bacteria was investigated by Quan et al. (2004). Their results showed that the 2,4-DCP-degrading bacteria not only improved the treatment efficiency of 2,4-DCP, but also improved the removal rates of other polychlorophenols such as 4-chlorophenol and 2,4,5-trichlorophenol. The bacterial strain *Paracoccus denitrificans* W12, which can utilize pyridine as its sole source of carbon and nitrogen, was added into an MBR to enhance the treatment of pharmaceutical wastewater (Wen et al., 2013). When the hydraulic retention time was 60 h and the influent concentration of pyridine was 250–500 mg/L, the mean effluent concentration of pyridine without adding W12 was 57.2 mg/L, while the pyridine was degraded to an average of 10.2 mg/L with addition of W12.

In the biological treatment of papermaking wastewater, pharmaceutical wastewater, dye wastewater, and pesticide wastewater containing refractory compounds, high-efficiency microbial enhancement has been achieved (Nicolella et al., 2005). As the number of pollutants entering the water environment increases, many are complex compounds or a mixture of compounds, and only particular strains or microorganisms with specific metabolic properties can degrade these complex compounds. As these substances are toxic to microorganisms, the number of microorganisms with corresponding metabolic functions is limited. Therefore, it is difficult to achieve the desired effect with conventional biological treatment methods, and bioenhancement technology has become an effective approach.

9.7 Integrated technologies for toxic organic wastewater treatment

9.7.1 Interpretation of integrated technologies

With the rapid growth of the world's economy, a large amount of discharged wastewater has caused more and more serious water pollution problems. Most wastewater including municipal wastewater and industrial wastewater contains toxic organics. Among them, municipal wastewater contains a variety of cHRPs with low concentrations (Dai et al., 2014). At present, the treatment methods for high-concentration organic wastewater mainly include physical, chemical, and biological treatment. The principle of physicochemical treatment is converting pollutants in wastewater into harmless substances, such as coagulation, oxidation-adsorption, incineration, extraction, wet catalytic oxidation, electrochemical, and membrane separation methods. Biological treatment is a method that microorganisms degrade pollutants in water and used them as their nutrients and energy, and simultaneously purify the wastewater. In this context, conventional biological processes do not always provide satisfactory results, especially for industrial wastewater treatment, as many of the organic substances produced by the chemical industry are toxic or resistant to biological treatment (Garcia et al., 2001; Lapertot et al., 2006; Muñoz and Guieysse, 2006). To improve the removal efficiency of toxic organic matter in wastewater, different integrated technologies have been developed (Table 9.3), and a series of biological filter, coagulation sedimentation, and AOPs are added to enhance the removal of cHRPs (Petrasek et al., 1983). The most significant challenge in the treatment of toxic organics is to optimize the use of biological, physical, and chemical wastewater treatment processes. The selection of the processes and the integration of the processes depend on the characteristics of the wastewater, the concentration of toxic organics matter, and the required efficiency.

Table 9.3 Comparison of different integrated technologies.

Integrated technologies		Characteristics
Chemical oxidation and biological method	Ultrasonic irradiation and activated sludge	Low in price; no secondary pollution
	Ultrasound oxidation with membrane bioreactor	Saving footprints; reducing HRT; migrating the membrane fouling
	Electrochemical oxidation and anaerobic process	Higher efficiency; lower energy consumption
Adsorption and biological treatment	GAC and biological treatment	Enriches substrates, nutrients and oxygen concentrations; readily adsorbed natural organic matter
	MBR combined with GAC adsorption	Prevent fouling of membranes; trace organics removal; ccolor removal
Constructed wetland and biological treatment	Constructed wetland and microbial fuel cells	Generating electricity
	Constructed wetland and UASB reactor	Simple construction; low energy consumption; low cost in maintenance and operation
Membrane bioreactor (MBR) and physical-chemical treatment	MBR and RO treatment	Reduced bacteria and viruses
	MBR and fenton	Minimizing production of excess sludge; reducing membrane fouling

9.7.2 Integrated technologies and their characteristics

9.7.2.1 Chemical oxidation and biological methods

To reduce operating costs, a combination of chemical oxidation as pretreatment and biological treatment as a posttreatment has received increasing attention.

The method of ultrasonic catalytic oxidation treatment of refractory organic wastewater is based on the effects of free radical oxidation, high-temperature pyrolysis, and supercritical water oxidation generated by ultrasonic cavitation process. The refractory organic matters are directly converted into carbon dioxide, water, or small molecular organic matter; meanwhile, the biological toxicity in the wastewater was reduced. It can also significantly reduce the load of the microorganisms, and ensure the stable and efficient operation of the subsequent biological treatment. The catalyst used is cheap and does not cause secondary pollution. It has excellent degradation and removal efficiency for high, medium, and low concentrations of refractory organics and has wide adaptability.

Most of the organic components in landfill leachate are organic compounds that are difficult to biodegrade, such as phenols, heterocyclics, heterocyclic aromatic hydrocarbons, and polycyclic aromatic hydrocarbons, accounting for more than 70% of the organic components in the leachate. Degradation characteristics of typical organic compounds in landfill leachate were investigated by electrochemical oxidation, and anaerobic process combined treatment system by Li et al. (2004). After the electrochemical oxidation process, the content of volatile fatty acids markedly increased from 0.68% of raw leachate to 16–18% of the effluent of the electrochemical reactor. The results indicated that synergetic and antagonistic effects of complex components on biotoxicity in leachate were notably decreased in such combined treatment system, and biodegradability of effluent was improved.

The combined process of electrochemical oxidation method and biological fluidized bed method was applied to treat phenolic resin wastewater. A novel bioelectrochemical oxidation (BEO) process was proposed for the treatment of highly toxic biodegradable pesticide wastewater (Liu et al., 2010). Its biodegradability in BOD5/total organic carbon (TOC) is only 0.0057, indicating its highly refractory property. In a BEO system with a boron-doped diamond electrode, the value of BOD5/TOC was increased to 1.17, and the toxicity was reduced by 40% within 3 h, thereby maintaining the activity of the microorganisms on the biofilter at an appropriate level. The intermediates (mainly small molecule carboxylic acids) produced in the electrochemical process are then biodegraded as microbial nutrients, and the biodegradation of these intermediates further enhances the electrochemical oxidation of the original contaminants. The synergy of electrochemical and biological processes improves the removal of toxic organic pollutants with higher efficiency and lower energy consumption. For the same TOC removal, the MCE in BEO is at least 40% higher than in single electrochemical oxidation (EO), and the energy consumption in BEO is only 53.3 (kW h)/m^3 in the BEO in 5 h, which is 44.5% lower than that in the EO. Therefore, BEO technology is an effective method for treating high concentration and biodegradable wastewater.

9.7.2.2 Integration of adsorption and biological treatment

When the wastewater contains substances that are toxic or inhibitory to the microorganisms, the GAC adsorption is often applied pior to biological treatment. However, in general, GAC adsorption is placed in tertiary wastewater treatment when the secondary effluent contains nonbiodegradable and toxic organics that are readily absorbed.

In WWTPs, tertiary GAC adsorption is primarily used to remove trace contaminants such as drugs and hormones that are not adequately eliminated in secondary treatment. Reducing discharge of these compounds helps protect the aquatic environment and enables reuse of treated wastewater. However, the primary evidence for biological activity in GAC filters is that attained effluent levels can not be achieved with adsorption alone. In the 1970s, Weber and his colleagues carried out experiments and convert GAC into biological activated carbon (BAC) (Weber et al., 1978).

Full-scale plants applying MBR technologies in municipal and industrial wastewater treatment are already in operation in several countries. The MBR can be introduced at different locations in the treatment processes, for example, after primary treatment of wastewater for secondary treatment purposes. Then, adsorption in GAC columns is introduced as a tertiary treatment step (Çeçen and Aktas, 2011). However, this combination would not prevent fouling of membranes. When MBR is combined with GAC adsorption, biofouling can be prevented by immobilizing the bacteria on activated carbon instead of the membrane surface because the attached bacteria cannot reach the membrane. Membrane-assisted BAC is used to remove preferential contaminants from secondary wastewater and drinking water. When coupling MBR to GAC adsorption, one of the primary purposes is usually to enhance the removal of microcontaminants. For example, the patented BioMAC technique uses a combination of BAC filtration and micro/ultrafiltration to enhance the treatment of the WWTP effluent or to treat the concentrated stream. The removal of antibiotics, drugs, pathogens, and endocrine disrupting compounds from wastewater is effective. The BioMAC reactor treats the secondary wastewater with a very low level of organic matter, nutrients, odors and indicator organisms that can be used for direct industrial reuse or groundwater recharge (Van Hege et al., 2002).

9.7.2.3 Constructed wetland and microbial fuel cells

CWs and microbial fuel cells (MFCs) are compatible technologies because they rely on the action of bacteria to remove contaminants from wastewater. The anodes of the MFCs need to be kept anaerobic and the cathodes are exposed to oxygen. Moreover, these redox conditions can be naturally developed in CWs. Therefore, in recent years, research combining these two technologies (called as CW-MFC) has appeared to improve the wastewater treatment capacity of wetlands while generating electricity at the same time.

If we consider that the electrogenic bacteria in MFCs are mainly supplied by the rhizosphere sediments and exudates of plants, the purpose is to produce electricity (Strik et al., 2008). The first CW-MFC study was carried out by Yadav et al. (2012). A vertical flow CW was combined with MFC to treat the azo dye synthesis wastewater. The compatibility and subsequent combination of CW and MFC are based on the fact that they are all biological systems involved in the degradation of organic matter. In addition, MFCs require redox gradients—an anaerobic anode and aerobic cathodes—which can be naturally found in CWs depending on the flow direction and wet depth (Corbella et al., 2014; Logan et al., 2006; Yadav et al., 2012).

To treat different kinds of wastewater, a variety of synthetic wetland processes have been studied. Zhao et al. (2013) and Oon et al. (2015) combined the CW and MFC to treat wastewater and generate electricity. The results indicated that the CW-MFC system had a 75% removal efficiency of COD (with an initial concentration of 3190–7080 mg/L) and produced a peak power density of 12.83 $\mu W/m^2$. In Oon's study, an innovative design of upflow constructed wetland—microbial fuel cell (UFCW-MFC) planted with cattailwas used for simultaneous wastewater treatment and electricity generation. Biodegradation of organic matter, nitrification, and denitrification were investigated and the removal efficiencies of COD, NO_3^-, and NH_4^+ were 100%, 40%, and 91%, respectively. Abdel-Shafy et al. (2009) treated the blackwater and graywater via UASB and CW. According to this study, two separate UASB reactors were used as a primary treatment step, followed by CW for the treatment of blackwater and graywater separately. The removal efficiency of COD in the UASB reactor was about 60% for graywater and 68% for blackwater. Further improvement of the quality of the treated wastewater was obtained after the application of the horizontal subsurface flow CW. The overall results indicated that the integration between the UASB and the CW was useful for the treatment of blackwater. The overall removal of the critical constituents represented by COD, BOD, and TSS in the final effluent was 87.7%, 89.5%, and 94% for graywater and 94.2%, 95.6%, and 94.9% for blackwater.

9.7.2.4 MBR and physical–chemical treatment

MBR, a combination of conventional activated sludge process and membrane separation, can produce consistently high quality of effluent from municipal wastewater. As for the toxic organic wastewater, it is highly toxic and difficult to be degraded, which may lead to sludge bulking and other problems in the general biological treatment process. MBR process can effectively avoid these problems, and membrane filtration can also improve the removal efficiency of toxic organic matter.

Based on the development of MBR, a new concept with the addition of RO after MBR recently has been developed to reclaim municipal wastewater. Dolar et al. (2012) showed that the combination of MBR and RO treatment could be used to remove emerging contaminants from municipal wastewater. MBR-RO system showed excellent overall removal of target emerging contaminants with removal rates above 99% for all of them. A biofilm-MBR process was developed to process shipboard wastewater and recycling effluent (Sun et al., 2010). The treatment capacity of the biofilm-MBR could be increased up to 160 mg/L influent fuel oil concentration and even higher (400 mg/L influent oil in shock load test) with an HRT of 8 h. Yuan and He (2015) integrated membrane filtration into bioelectrochemical systems as next-generation energy-efficient wastewater treatment technologies for water recovery.

Meanwhile, MBR is particularly suitable for advanced biological treatment of wastewater containing recalcitrant compounds and shows several advantages that make it an excellent alternative to be coupled with photo-Fenton, especially for water reclamation (López et al., 2010). Zhang et al. (2009) used anoxic/aerobic/MBR membrane systems to replace the conventional A/A/O method in paper mill wastewater treatment. The effluent from the first sedimentation tank was treated with the anoxic/aerobic/MBR membrane system to eliminate NH_4^+-N and dissolved or undissolved organic compounds. Laera et al. (2011) demonstrated that an integrated MBR-TiO_2 photocatalysis process was effective for the removal of carbamazepine from simulated pharmaceutical industrial effluent. MBR as a combination of the activated sludge process with micro- and ultrafiltration is widely regarded as an effective tool for industrial water treatment and water reuse due to its high product water quality and low footprint (Hoinkis et al., 2012).

9.7.2.5 Advanced oxidation processes and biological treatment

One feasible option for treatment of refractory wastewater is the use of technologies based on chemical oxidation, such as the AOPs, widely recognized as highly efficient methods for recalcitrant wastewater. AOPs are considered to be a highly competitive water treatment technology for the removal of cHRPs that cannot be processed by conventional techniques due to their high chemical stability and low biodegradability. AOPs are used to improve the biodegradability of wastewater, which lays a foundation for subsequent biological treatment. AOPs processes degrade organic pollutants by forming hydroxyl radicals (Balcioglu et al., 2001; García-Montaño et al., 2006; Neyens and Baeyens, 2003). In general, most studies in this field employ conventional bioassays, such as biological oxygen demand to determine enhancement of the biodegradation rate after pretreatment of recalcitrant wastewater by AOPs (Bacardit et al., 2006; Méndez-Arriaga et al., 2008; Suh and Mohseni, 2004).

Scott and Ollis (1995) have reported that four types of wastewater can be treated by combining AOP/biodegradation: (1) wastewater containing biological calcium-lowering compounds, such as soluble polymers, due to their large molecular size and lack of active biodegradable active center (Kiwi et al., 1993; Somich et al., 1990); (2) high concentration of biodegradable industrial wastewater still requiring chemical posttreatment because of the presence of low concentrations of stubborn compounds (Berge et al., 1994; Sierka and Bryant, 1994); (3) wastewater containing inhibitory compounds (Manilal et al., 1992); (4) wastewater containing inert intermediates such as specific metabolites accumulating in the medium and inhibiting the growth of microorganisms (Hapeman et al., 1994).

9.8 Summary

Removal of the cHRPs from wastewater is of great importance for the protection of natural environments and human health. Biological treatment processes are the most widely used methods for cHRP removal due to the high efficiency and low cost. As the traditional wastewater treatment process has low removal efficiency of toxic organic substances in wastewater, the comprehensive and integrated wastewater treatment processes have been extensively studied by researchers. Meanwhile, the mechanisms behind the biological transformation process of cHRPs were also revealed over the last decades. In recent years, many new and integrated wastewater treatment processes have been developed, and by applying these new processes, higher cHRP removal efficiency has been achieved. On the other hand, although the application of the combined processes has improved the removal of cHRPs from wastewater, it also increases the cost of wastewater treatment processes compared to the conventional approaches. Therefore, the selection of wastewater combination process is particularly important, and both treatment efficiency and cost should be considered. Therefore, novel technologies with both high efficiency and low operational cost are needed to be developed in the future.

References

Abdel-Shafy, H.I., El-Khateeb, M.A., Regelsberger, M., El-Sheikh, R., Shehata, M., 2009. Integrated system for the treatment of blackwater and greywater via uasb and constructed wetland in Egypt. Desalination and Water Treatment 8 (1–3), 272–278.

Abegglen, C., Joss, A., McArdell, C.S., Fink, G., Schlüsener, M.P., Ternes, T.A., Siegrist, H., 2009. The fate of selected micropollutants in a single-house mbr. Water Research 43 (7), 2036–2046.

Alvarino, T., Suarez, S., Lema, J.M., Omil, F., 2014. Understanding the removal mechanisms of ppcps and the influence of main technological parameters in anaerobic uasb and aerobic cas reactors. Journal of Hazardous Materials 278, 506–513.

Andersen, H., Siegrist, H., Halling-Sørensen, B., Ternes, T., 2003. Fate of estrogens in a municipal sewage treatment plant. Environmental Science and Technology 37 (18), 4021–4026.

Bacardit, J., Hultgren, A., García-Molina, V., Esplugas, S., 2006. Biodegradability Enhancement of Wastewater Containing 4-chlorophenol by Means of Photo-Fenton, p. 27.

Balcioglu, I.A., Arslan, I., Sacan, M.T., 2001. Homogenous and heterogenous advanced oxidation of two commercial reactive dyes. Environmental Technology 22 (7), 813–822.

Berge, D., Ratnaweera, H., Efraimsen, H., 1994. Degradation of recalcitrant chlorinated organics by radiochemical and biochemical oxidation. Water Science and Technology 29 (5–6), 219–228.

Bernhard, M., Müller, J., Knepper, T.P., 2006. Biodegradation of persistent polar pollutants in wastewater: comparison of an optimised lab-scale membrane bioreactor and activated sludge treatment. Water Research 40 (18), 3419–3428.

Blair, B., Nikolaus, A., Hedman, C., Klaper, R., Grundl, T., 2015. Evaluating the degradation, sorption, and negative mass balances of pharmaceuticals and personal care products during wastewater treatment. Chemosphere 134, 395–401.

Braga, O., Smythe, G., Schaefer, A., Feitz, A., 2005. Fate of steroid estrogens in Australian inland and coastal wastewater treatment plants. Environmental Science and Technology 39 (9), 3351–3358.

Bouwer, E.J., Crowe, P., 1988. Biological processes in drinking water treatment. JAWWA 80 (9), 82–93.

Budd, R., O'Geen, A., Goh, K.S., Bondarenko, S., Gan, J., 2009. Efficacy of constructed wetlands in pesticide removal from tailwaters in the central valley, California. Environmental Science and Technology 43 (8), 2925–2930.

Çeçen, F., Aktas, O., 2011. Activated Carbon for Water and Wastewater Treatment: Integration of Adsorption and Biological Treatment. John Wiley & Sons.

Clara, M., Strenn, B., Gans, O., Martinez, E., Kreuzinger, N., Kroiss, H., 2005. Removal of selected pharmaceuticals, fragrances and endocrine disrupting compounds in a membrane bioreactor and conventional wastewater treatment plants. Water Research 39 (19), 4797–4807.

Conkle, J.L., Lattao, C., White, J.R., Cook, R.L., 2010. Competitive sorption and desorption behavior for three fluoroquinolone antibiotics in a wastewater treatment wetland soil. Chemosphere 80 (11), 1353–1359.

Conkle, J.L., White, J.R., Metcalfe, C.D., 2008. Reduction of pharmaceutically active compounds by a lagoon wetland wastewater treatment system in southeast Louisiana. Chemosphere 73 (11), 1741–1748.

Corbella, C., Garfí, M., Puigagut, J., 2014. Vertical redox profiles in treatment wetlands as function of hydraulic regime and macrophytes presence: surveying the optimal scenario for microbial fuel cell implementation. The Science of the Total Environment 470–471, 754–758.

Dai, G., Huang, J., Chen, W., Wang, B., Yu, G., Deng, S., 2014. Major pharmaceuticals and personal care products (ppcps) in wastewater treatment plant and receiving water in beijing, China, and associated ecological risks. Bulletin of Environmental Contamination and Toxicology 92 (6), 655–661.

De Wever, H., Weiss, S., Reemtsma, T., Vereecken, J., Müller, J., Knepper, T., Rörden, O., Gonzalez, S., Barcelo, D., Dolores Hernando, M., 2007. Comparison of sulfonated and other micropollutants removal in membrane bioreactor and conventional wastewater treatment. Water Research 41 (4), 935–945.

Dolar, D., Gros, M., Rodriguez-Mozaz, S., Moreno, J., Comas, J., Rodriguez-Roda, I., Barceló, D., 2012. Removal of emerging contaminants from municipal wastewater with an integrated membrane system, mbr–ro. Journal of Hazardous Materials 239–240, 64–69.

Doulia, D.S., Anagnos, E.K., Liapis, K.S., Klimentzos, D.A., 2016. Removal of pesticides from white and red wines by microfiltration. Journal of Hazardous Materials 317, 135–146.

Du, J., Geng, J., Ren, H., Ding, L., Xu, K., Zhang, Y., 2015. Variation of antibiotic resistance genes in municipal wastewater treatment plant with A^2O-MBR system. Environmental Science and Pollution Research 22 (5), 3715–3726.

Echevarría, C., Valderrama, C., Cortina, J.L., Martín, I., Arnaldos, M., Bernat, X., De la Cal, A., Boleda, M.R., Vega, A., Teuler, A., Castellví, E., 2019. Techno-economic evaluation and comparison of pac-mbr and ozonation-uv revamping for organic micropollutants removal from urban reclaimed wastewater. The Science of the Total Environment 671, 288–298.

Fahrbach, M., Krauss, M., Preiss, A., Kohler, H.-P.E., Hollender, J., 2010. Anaerobic testosterone degradation in steroidobacter denitrificans – identification of transformation products. Environmental Pollution 158 (8), 2572–2581.

García-Montaño, J., Ruiz, N., Muñoz, I., Domènech, X., García-Hortal, J.A., Torrades, F., Peral, J., 2006. Environmental assessment of different photo-fenton approaches for commercial reactive dye removal. Journal of Hazardous Materials 138 (2), 218–225.

Garcia-Rodríguez, A., Matamoros, V., Fontàs, C., Salvadó, V., 2014. The ability of biologically based wastewater treatment systems to remove emerging organic contaminants—a review. Environmental Science and Pollution Research 21 (20), 11708–11728.

Garcia, J., Rousseau, D.P., Morato, J., Lesage, E., Matamoros, V., Bayona, J.M., 2010. Contaminant removal processes in subsurface-flow constructed wetlands: a review. Critical Reviews in Environmental Science and Technology 40 (7), 561–661.

Garcıa, M.T., Ribosa, I., Guindulain, T., Sánchez-Leal, J., Vives-Rego, J., 2001. Fate and effect of monoalkyl quaternary ammonium surfactants in the aquatic environment. Environmental Pollution 111 (1), 169–175.

Goodwin, E.J., 2017. Elgar Encyclopedia of Environmental Law. Edward Elgar Publishing Limited, pp. 101–108.

Hallé, C., Huck, P., Peldszus, S., 2015. Emerging contaminant removal by biofiltration: temperature, concentration and ebct impacts. Journal of the American Water Works Association 107 (7), 364–379.

Hapeman, C., Shelton, D., Peyton, G., Bell, O., LeFaivre, M., 1994. Oxidation and Microbial Mineralization to Remediate Pesticide Contaminated Waters—Overcoming the Technical Challenges.

Harrington, C., Scholz, M., 2010. Assessment of pre-digested piggery wastewater treatment operations with surface flow integrated constructed wetland systems. Bioresource Technology 101 (20), 7713–7723.

Helbling, D.E., Hollender, J., Kohler, H.-P.E., Fenner, K., 2010a. Structure-based interpretation of biotransformation pathways of amide-containing compounds in sludge-seeded bioreactors. Environmental Science and Technology 44 (17), 6628–6635.

Helbling, D.E., Hollender, J., Kohler, H.-P.E., Singer, H., Fenner, K., 2010b. High-throughput identification of microbial transformation products of organic micropollutants. Environmental Science and Technology 44 (17), 6621–6627.

Herrero, M., Stuckey, D., 2015. Bioaugmentation and its application in wastewater treatment: a review. Chemosphere 140, 119–128.

Hijosa-Valsero, M., Matamoros, V., Sidrach-Cardona, R., Martín-Villacorta, J., Bécares, E., Bayona, J.M., 2010. Comprehensive assessment of the design configuration of constructed wetlands for the removal of pharmaceuticals and personal care products from urban wastewaters. Water Research 44 (12), 3669–3678.

Hoinkis, J., Deowan, S.A., Panten, V., Figoli, A., Huang, R.R., Drioli, E., 2012. Membrane bioreactor (mbr) technology – a promising approach for industrial water reuse. Procedia Engineering 33, 234–241.

Huang, H., Ding, L.L., Ren, H.Q., Geng, J.J., Xu, K., Zhang, Y., 2015. Preconditioning of model biocarriers by soluble pollutants: a QCM-D study. ACS Applied Materials and Interfaces 7 (13), 7222–7230.

Huang, H., Fan, X., Peng, C., Geng, J., Ding, L., Xu, K., Zhang, Y., Ren, H., 2019a. Biofilm aging in full-scale aerobic bioreactors from perspectives of metabolic activity and microbial community. Biochemical Engineering Journal 146, 69–78.

Huang, H., Fan, X., Peng, C., Geng, J., Ding, L., Zhang, X., Ren, H., 2019b. Linking microbial respiratory activity with phospholipid fatty acid of biofilm from full-scale bioreactors. Bioresource Technology 272, 599–605.

Huang, H., Lin, Y., Peng, P.C., Geng, J.J., Xu, K., Zhang, Y., Ding, L.L., Ren, H.Q., 2018a. Calcium ion- and rhamnolipid-mediated deposition of soluble matters on biocarriers. Water Research 133, 37–46.

Huang, H., Peng, C., Peng, P., Lin, Y., Zhang, X., Ren, H., 2019c. Towards the biofilm characterization and regulation in biological wastewater treatment. Applied Microbiology and Biotechnology 103 (3), 1115–1129.

Huang, H., Ren, H., Ding, L., Geng, J., Xu, K., Zhang, Y., 2014. Aging biofilm from a full-scale moving bed biofilm reactor: characterization and enzymatic treatment study. Bioresource Technology 154, 122–130.

Huang, H., Yu, Q.S., Ren, H.Q., Geng, J.J., Xu, K., Zhang, Y., Ding, L.L., 2018b. Towards physicochemical and biological effects on detachment and activity recovery of aging biofilm by enzyme and surfactant treatments. Bioresource Technology 247, 319–326.

Hussain, S.A., Prasher, S.O., Patel, R.M., 2012. Removal of ionophoric antibiotics in free water surface constructed wetlands. Ecological Engineering 41, 13–21.

Wang J., Quan X., Wu L., Qian Y., Hegemannc W., Bioaugmentation as a tool to enhance the removal of refractory compound in coke plant wastewater. Process Biochemistry 38 (5), 2002, 777–781.

Kennedy, A.M., Reinert, A.M., Knappe, D.R.U., Ferrer, I., Summers, R.S., 2015. Full- and pilot-scale gac adsorption of organic micropollutants. Water Research 68, 238–248.

Kimura, K., Hara-Yamamura, H., Watanabe, Y., 2007. Elimination of selected acidic pharmaceuticals from municipal wastewater by an activated sludge system and membrane bioreactors. Environmental Science and Technology 41 (10), 3708–3714.

Kiwi, J., Pulgarin, C., Peringer, P., Grätzel, M., 1993. Beneficial effects of homogeneous photo-fenton pretreatment upon the biodegradation of anthraquinone sulfonate in waste water treatment. Applied Catalysis B: Environmental 3 (1), 85–99.

Kosma, C.I., Lambropoulou, D.A., Albanis, T.A., 2014. Investigation of ppcps in wastewater treatment plants in Greece: occurrence, removal and environmental risk assessment. Science of the Total Environment, 466-467, 421-438.

Kosma, C.I., Lambropoulou, D.A., Albanis, T.A., 2010. Occurrence and removal of ppcps in municipal and hospital wastewaters in Greece. Journal of Hazardous Materials 179 (1), 804–817.

López, J.L.C., Reina, A.C., Gómez, E.O., Martín, M.M.B., Rodríguez, S.M., Pérez, J.A.S., 2010. Integration of solar photocatalysis and membrane bioreactor for pesticides degradation. Separation Science and Technology 45 (11), 1571–1578.

L Zearley, T., Scott Summers, R., 2012. Removal of trace organic micropollutants by drinking water biological filters. Environmental Science and Technology 46 (17), 9412–9419.

Laera, G., Chong, M.N., Jin, B., Lopez, A., 2011. An integrated mbr–tio2 photocatalysis process for the removal of carbamazepine from simulated pharmaceutical industrial effluent. Bioresource Technology 102 (13), 7012–7015.

Lapertot, M., Pulgarín, C., Fernández-Ibáñez, P., Maldonado, M.I., Pérez-Estrada, L., Oller, I., Gernjak, W., Malato, S., 2006. Enhancing biodegradability of priority substances (pesticides) by solar photo-fenton. Water Research 40 (5), 1086–1094.

Li, C., Cabassud, C., Guigui, C., 2014. Evaluation of membrane bioreactor on removal of pharmaceutical micropollutants: a review. Desalination and Water Treatment 55 (4), 845–858.

Li, T., Li, X., Chen, J., 2004. Biodegradation characteristics of organic compounds in leachate in electrochemical oxidation and anaerobic process combined treatment system. Environmental Sciences 25 (5), 172–176.

Liu, L., Zhao, G., Pang, Y., Lei, Y., Gao, J., Liu, M., 2010. Integrated biological and electrochemical oxidation treatment for high toxicity pesticide pollutant. Industrial and Engineering Chemistry Research 49 (12), 5496–5503.

Liu, Z.-h., Kanjo, Y., Mizutani, S., 2009. Removal mechanisms for endocrine disrupting compounds (edcs) in wastewater treatment — physical means, biodegradation, and chemical advanced oxidation: a review. The Science of the Total Environment 407 (2), 731–748.

Logan, B.E., Hamelers, B., Rozendal, R., Schröder, U., Keller, J., Freguia, S., Aelterman, P., Verstraete, W., Rabaey, K., 2006. Microbial fuel cells: Methodology and technology. Environmental Science and Technology 40 (17), 5181–5192.

Luo, W., Hai, F.I., Price, W.E., Guo, W., Ngo, H.H., Yamamoto, K., Nghiem, L.D., 2014a. High retention membrane bioreactors: challenges and opportunities. Bioresource Technology 167, 539–546.

Luo, Y., Guo, W., Ngo, H.H., Nghiem, L.D., Hai, F.I., Zhang, J., Liang, S., Wang, X.C., 2014b. A review on the occurrence of micropollutants in the aquatic environment and their fate and removal during wastewater treatment. The Science of the Total Environment 473–474, 619–641.

Méndez-Arriaga, F., Esplugas, S., Giménez, J., 2008. Photocatalytic degradation of non-steroidal anti-inflammatory drugs with TiO_2 and simulated solar irradiation. Water Research 42 (3), 585–594.

Manilal, V.B., Haridas, A., Alexander, R., Surender, G.D., 1992. Photocatalytic treatment of toxic organics in wastewater: toxicity of photodegradation products. Water Research 26 (8), 1035–1038.

Margot, J., Lochmatter, S., Andrew Barry, D., Holliger, C., 2016. Role of ammonia-oxidizing bacteria in micropollutant removal from wastewater with aerobic granular sludge. Water Science and Technology 73 (3), 564–575.

Meng, F., Chae, S.-R., Shin, H.-S., Yang, F., Zhou, Z., 2012. Recent advances in membrane bioreactors: configuration development, pollutant elimination, and sludge reduction. Environmental Engineering Science 29 (3), 139–160.

Moore, M., Cooper, C., Smith, S., Cullum, R., Knight, S., Locke, M., Bennett, E., 2007. Diazinon mitigation in constructed wetlands: influence of vegetation. Water, Air, and Soil Pollution 184 (1–4), 313–321.

Muñoz, R., Guieysse, B., 2006. Algal–bacterial processes for the treatment of hazardous contaminants: a review. Water Research 40 (15), 2799–2815.

Murugesan, K., Chang, Y.-Y., Kim, Y.-M., Jeon, J.-R., Kim, E.-J., Chang, Y.-S., 2010. Enhanced transformation of triclosan by laccase in the presence of redox mediators. Water Research 44 (1), 298–308.

Neyens, E., Baeyens, J., 2003. A review of classic fenton's peroxidation as an advanced oxidation technique. Journal of Hazardous Materials 98 (1), 33–50.

Nicolella, C., Zolezzi, M., Rabino, M., Furfaro, M., Rovatti, M., 2005. Development of particle-based biofilms for degradation of xenobiotic organic compounds. Water Research 39 (12), 2495–2504.

Oon, Y.-L., Ong, S.-A., Ho, L.-N., Wong, Y.-S., Oon, Y.-S., Lehl, H.K., Thung, W.-E., 2015. Hybrid system up-flow constructed wetland integrated with microbial fuel cell for simultaneous wastewater treatment and electricity generation. Bioresource Technology 186, 270–275.

Park, N., Vanderford, B.J., Snyder, S.A., Sarp, S., Kim, S.D., Cho, J., 2009. Effective controls of micropollutants included in wastewater effluent using constructed wetlands under anoxic condition. Ecological Engineering 35 (3), 418–423.

Pathiraja, G., Egodawatta, P., Goonetilleke, A., Te'o, V.S.J., 2019a. Effective degradation of polychlorinated biphenyls by a facultative anaerobic bacterial consortium using alternating anaerobic aerobic treatments. The Science of the Total Environment 659, 507–514.

Pathiraja, G., Egodawatta, P., Goonetilleke, A., Te'o, V.S.J., 2019b. Solubilization and degradation of polychlorinated biphenyls (PCBS) by naturally occurring facultative anaerobic bacteria. The Science of the Total Environment 651, 2197–2207.

Peng, C., Gao, Y., Fan, X., Peng, P., Huang, H., Zhang, X., Ren, H., 2019. Enhanced biofilm formation and denitrification in biofilters for advanced nitrogen removal by rhamnolipid addition. Bioresource Technology 287. https://doi.org/10.1016/j.biortech.2019.121387.

Peng, P., Huang, H., Ren, H., 2018a. Effect of adding low-concentration of rhamnolipid on reactor performances and microbial community evolution in MBBRs for low C/N ratio and antibiotic wastewater treatment. Bioresource Technology 256, 557–561.

Peng, P., Huang, H., Ren, H., Ma, H., Lin, Y., Geng, J., Xu, K., Zhang, Y., Ding, L., 2018b. Exogenous N-acyl homoserine lactones facilitate microbial adhesion of high ammonia nitrogen wastewater on biocarrier surfaces. The Science of the Total Environment 624, 1013–1022.

Petrasek, A.C., Kugelman, I.J., Austern, B.M., Pressley, T.A., Winslow, L.A., Wise, R.H., 1983. Fate of toxic organic compounds in wastewater treatment plants. Journal – Water Pollution Control Federation 55 (10), 1286–1296.

Pinney, M.L., Westerhoff, P.K., Baker, L., 2000. Transformations in dissolved organic carbon through constructed wetlands. Water Research 34 (6), 1897–1911.

Poi, G., Shahsavari, E., Aburto-Medina, A., Ball, A.S., 2017. Bioaugmentation: an effective commercial technology for the removal of phenols from wastewater. Microbiology Australia 38 (2), 82–84.

Quan, X., Shi, H., Liu, H., Wang, J., Qian, Y., 2004. Removal of 2, 4-dichlorophenol in a conventional activated sludge system through bioaugmentation. Process Biochemistry 39 (11), 1701–1707.

Radjenović, J., Petrović, M., Barceló, D., 2009. Fate and distribution of pharmaceuticals in wastewater and sewage sludge of the conventional activated sludge (cas) and advanced membrane bioreactor (MBR) treatment. Water Research 43 (3), 831–841.

Sadmani, A.H.M.A., Andrews, R.C., Bagley, D.M., 2014. Nanofiltration of pharmaceutically active and endocrine disrupting compounds as a function of compound interactions with dom fractions and cations in natural water. Separation and Purification Technology 122, 462–471.

Saeed, T., Sun, G., 2012. A review on nitrogen and organics removal mechanisms in subsurface flow constructed wetlands: dependency on environmental parameters, operating conditions and supporting media. Journal of Environmental Management 112, 429–448.

Scholz, M., Lee, B.h., 2005. Constructed wetlands: a review. International Journal of Environmental Studies 62 (4), 421–447.

Scott, J.P., Ollis, D.F., 1995. Integration of chemical and biological oxidation processes for water treatment: review and recommendations. Environmental Progress 14 (2), 88–103.

Sierka, R.A., Bryant, C.W., 1994. Enhancement of biotreatment effluent quality by illuminated titanium dioxide and membrane pretreatment of the kraft extraction waste stream and by increased chlorine dioxide substitution. Water Science and Technology 29 (5–6), 209–218.

Slemp, J.D., Norman, J.L., George, D.B., Stearman, G.K., Wells, M.J.M., 2004. In: Proceedings 9th International Conference on Wetland Systems for Water Pollution Control, Vol. 1, 359 Avignon, France.

Snyder, S.A., Adham, S., Redding, A.M., Cannon, F.S., DeCarolis, J., Oppenheimer, J., Wert, E.C., Yoon, Y., 2007. Role of membranes and activated carbon in the removal of endocrine disruptors and pharmaceuticals. Desalination 202 (1), 156–181.

Somich, C.J., Muldoon, M.T., Kearney, P.C., 1990. On-site treatment of pesticide waste and rinsate using ozone and biologically active soil. Environmental Science and Technology 24 (5), 745–749.

Strik, D.P.B.T.B., Hamelers, H.V.M., Snel, J.F.H., Buisman, C.J.N., 2008. Green electricity production with living plants and bacteria in a fuel cell. International Journal of Energy Research 32 (9), 870–876.

Suárez, S., Carballa, M., Omil, F., Lema, J.M., 2008. How are pharmaceutical and personal care products (ppcps) removed from urban wastewaters? Reviews in Environmental Science and Biotechnology 7 (2), 125–138.

Suh, J.H., Mohseni, M., 2004. A study on the relationship between biodegradability enhancement and oxidation of 1,4-dioxane using ozone and hydrogen peroxide. Water Research 38 (10), 2596–2604.

Sun, C., Leiknes, T., Weitzenböck, J., Thorstensen, B., 2010. Development of an integrated shipboard wastewater treatment system using biofilm-mbr. Separation and Purification Technology 75 (1), 22–31.

Ternes, T.A., Joss, A., Siegrist, H., 2004. Scrutinizing pharmaceuticals and personal care products in wastewater treatment. Environmental Science and Biotechnology 38 (20), 392a–399a.

Terzic, S., Senta, I., Matosic, M., Ahel, M., 2011. Identification of biotransformation products of macrolide and fluoroquinolone antimicrobials in membrane bioreactor treatment by ultrahigh-performance liquid chromatography/quadrupole time-of-flight mass spectrometry. Analytical and Bioanalytical Chemistry 401 (1), 353–363.

Tran, N.H., Urase, T., Ngo, H.H., Hu, J., Ong, S.L., 2013. Insight into metabolic and cometabolic activities of autotrophic and heterotrophic microorganisms in the biodegradation of emerging trace organic contaminants. Bioresource Technology 146, 721–731.

Trautwein, C., Kümmerer, K., 2012. Degradation of the tricyclic antipsychotic drug chlorpromazine under environmental conditions, identification of its main aquatic biotic and abiotic transformation products by lc–msn and their effects on environmental bacteria. Journal of Chromatography B 889–890, 24–38.

Van Hege, K., Dewettinck, T., Claeys, T., De Smedt, G., Verstraete, W., 2002. Reclamation of treated domestic wastewater using biological membrane assisted carbon filtration (biomac). Environmental Technology 23 (9), 971–980.

Verlicchi, P., Al Aukidy, M., Zambello, E., 2012. Occurrence of pharmaceutical compounds in urban wastewater: removal, mass load and environmental risk after a secondary treatment—a review. The Science of the Total Environment 429, 123–155.

Vymazal, J., Březinová, T., 2015. The use of constructed wetlands for removal of pesticides from agricultural runoff and drainage: a review. Environment International 75, 11–20.

Weber, J., Walter, Pirbazari, M., Melson, G., 1978. Biological growth on activated carbon: an investigation by scanning electron microscopy. Environmental Science and Technology 12 (7), 817–819.

Wen, D., Zhang, J., Xiong, R., Liu, R., Chen, L., 2013. Bioaugmentation with a pyridine-degrading bacterium in a membrane bioreactor treating pharmaceutical wastewater. Journal of Environmental Sciences 25 (11), 2265–2271.

Westerhoff, P., Yoon, Y., Snyder, S., Wert, E., 2005. Fate of endocrine-disruptor, pharmaceutical, and personal care product chemicals during simulated drinking water treatment processes. Environmental Science and Technology 39 (17), 6649–6663.

Wu, H., Zhang, J., Ngo, H.H., Guo, W., Hu, Z., Liang, S., Fan, J., Liu, H., 2015. A review on the sustainability of constructed wetlands for wastewater treatment: design and operation. Bioresource Technology 175, 594–601.

Yadav, A.K., Dash, P., Mohanty, A., Abbassi, R., Mishra, B.K., 2012. Performance assessment of innovative constructed wetland-microbial fuel cell for electricity production and dye removal. Ecological Engineering 47, 126–131.

Yalcuk, A., Ugurlu, A., 2009. Comparison of horizontal and vertical constructed wetland systems for landfill leachate treatment. Bioresource Technology 100 (9), 2521–2526.

Yoon, Y., Westerhoff, P., Snyder, S., C. Wert, E., 2006. Nanofiltration and ultrafiltration of endocrine disrupting compounds, pharmaceuticals and personal care products. Journal of Membrane Science 270 (1–2), 88–100.

Yuan, H., He, Z., 2015. Integrating membrane filtration into bioelectrochemical systems as next generation energy-efficient wastewater treatment technologies for water reclamation: a review. Bioresource Technology 195, 202–209.

Yu, Q., Huang, H., Ren, H., Ding, L., Geng, J., 2016. In situ activity recovery of aging biofilm in biological aerated filter: surfactants treatment and mechanisms study. Bioresource Technology 219, 403–410.

Zhang, Y., Ma, C., Ye, F., Kong, Y., Li, H., 2009. The treatment of wastewater of paper mill with integrated membrane process. Desalination 236 (1), 349–356.

Zhao, Y., Collum, S., Phelan, M., Goodbody, T., Doherty, L., Hu, Y., 2013. Preliminary investigation of constructed wetland incorporating microbial fuel cell: batch and continuous flow trials. Chemical Engineering Journal 229, 364–370.

CHAPTER

Technologies for bHRPs and risk control

10

Jinbao Yin, PhD, Xuxiang Zhang, PhD

State Key Laboratory of Pollution Control and Resource Reuse, School of the Environment, Nanjing University, Nanjing, China

Chapter outline

High-Risk Pollutants in Wastewater. https://doi.org/10.1016/B978-0-12-816448-8.00010-1

Wastewater treatment facilities have become necessary processes to remove the chemical pollutants to minimize the negative effects of the wastewater discharge on human health and environmental safety (Ahmed et al., 2017; Collivignarelli et al., 2017). Wastewater also is a major source of biological high-risk pollutants (bHRPs), including pathogenic bacteria, viruses, parasitic protozoa, antibiotic-resistant bacteria (ARB), and antibiotic resistance genes (ARGs), which pose serious risks to human beings and ecosystem. To minimize the risks, bHRPs must be removed before wastewater discharge or reclamation (Collivignarelli et al., 2017; Amin et al., 2013; Curtis, 2003). Diverse technologies have been used to remove the bHRPs in wastewater, including the conventional disinfection technologies specially designed and widely used for bHRPs removal (Amin et al., 2013). Some existing or new wastewater treatment technologies are also being studied or applied to control the bHRPs. However, the removal efficiency of bHRPs varies greatly and depends on the type of the treatment processes used (Ottoson et al., 2006; Lazarova et al., 1999; Zhang and Farahbakhsh, 2007).

The objective of this chapter is to introduce the technologies of controlling the bHRPs in wastewater, focusing on the application, action mechanisms of conventional disinfection technologies such as chlorination, chlorine dioxide, ozone, ultraviolet (UV), and peracetic acid (PAA). Meanwhile, the effects of other effective technologies including anaerobic and/or aerobic treatment reactors, membrane bioreactor (MBR), membrane filtration, advanced oxidation processes (AOPs), constructed wetland, wastewater stabilization pond, nanomaterials, biochar, and bacteriophage are also elucidated.

10.1 Conventional disinfection technologies

Conventional disinfection technologies have been specially designed and widely used to remove pathogenic organisms from wastewater, including pathogenic bacteria, virus, parasitic protozoa, or helminth (Collivignarelli et al., 2017; Amin et al.,

2013). Meanwhile, effective disinfection is a vital and regular process for disruption and inactivation of ARB and ARGs (Sharma et al., 2016). In fact, none of the tested disinfection technologies has been designed to specifically target ARB or ARGs, whereas their effect can result in a deactivation of bacterial cells, and this process is not necessarily associated with the concomitant deactivation of the ARGs at the community level (Barancheshme and Munir, 2018). They are the last barrier for direct discharge or reclamation of wastewater to protect ecosystem safety and human health. The widely applied disinfectants mainly include chlorine, chlorine dioxide, ozone, UV, and peracetic acid.

10.1.1 Chlorination

Chlorination is the most popular disinfection technology used to inactivate waterborne pathogenic organisms in wastewater treatment. Conventional disinfectant is gaseous chlorine (Cl_2), and chloramine (NH_2Cl) and sodium hypochlorite are also often used. Cl_2 is by far the most commonly used disinfection technology characterized by low cost, stable process, and reliable performance. Chlorine is a highly toxic dangerous product, which is necessary to store cylinders with high pressure capacity. It is necessary to build a chlorine storage and chlorination room according to safety regulations in wastewater treatment plants. Sodium hypochlorite is a relatively simple and cost-effective process and stored in a storage tank, which does not require extensive technical expertise (Collivignarelli et al., 2017).

Chlorination mechanism is that chlorine-containing disinfectants react to form hypochlorous acid (HOCl) when they are added into water. HOCl is a weak acid but a strong oxidizing agent, and a neutral small molecule, which can damage the genome- and protein-mediated functions (Wigginton and Kohn, 2012). HOCl has a pKa of 7.6 at room temperature, which means that HOCl exists in its neutral form, HOCl, or hypochlorite ion, OCl^-. Chlorine, HOCl and OCl^- are called free chlorine, which ensures continuous sterilization and prevents water from being repolluted.

Chlorine rapidly hydrolyzes to hydrochloric acid (HCl) and HOCl in water:

$$Cl_2 + H_2O \rightleftharpoons H^+ + Cl^- + HOCl$$

When sodium hypochlorite is dissolved in water, it mostly hydrolyzes to sodium hydroxide (NaOH):

$$NaClO + H_2O \rightarrow HClO + NaOH$$

Chlorine rapidly hydrolyzes HOCl, and then HOCl can react with ammonia to form NH_2Cl:

$$NH_2Cl + H_2O \rightleftharpoons NH_3 + HOCl$$

There is a dynamic equilibrium in this reaction. The antiseptic effect of chloramine disinfection is its slow release of HClO, when HClO is consumed.

10.1.1.1 Pathogenic bacteria

HOCl can spread to the surface of negatively charged bacteria, penetrate the cell wall of the bacteria, and reach the inside of the bacteria (Kim et al., 2002). The main actions of HOCl include inhibition of glucose oxidation, depletion of adenine nucleotides, inhibition of DNA replication, protein unfolding, and aggregation that are the basic physiological processes of bacteria, inactivating them and thus inhibiting their growth. HOCl is known to cause posttranslational modifications to proteins, the notable ones being cysteine and methionine oxidation. HOCl's bactericidal role is to be a potent inducer of protein aggregation. In general, when the free residual chlorine is above 0.3 mg/L, and the contact time is above 30 min, it can kill the most of pathogenic bacteria, such as *Escherichia coli*, *Clostridium perfringens*, *Pseudomonas aeruginosa*, *Shigella flexneri*, *Staphylococcus aureus*, and *Salmonella enterica*.

10.1.1.2 Virus

Generally, its resistance to disinfectants is stronger than that of bacteria. When chlorine is used for disinfection, the effective agent to inactive viruses is HOCl, which can damage the genome- and protein-mediated functions (Wigginton and Kohn, 2012). For virus inactivation, the contact time is usually 30–60 min for the disinfection of secondary effluent. When chlorine dose was higher than 10 mg/L with a contact time of 60 min, infectious rotaviruses can be effectively inactivated, and chlorination with chlorine doses of 16 and 8 mg/L and contact time of 30 min can reduce in indigenous enteroviruses of 1.2 and 0.35 logs, respectively. Increase of chlorine dose or extension of contact time can achieve a higher virus removal (Zhang et al., 2016a).

10.1.1.3 Parasitic protozoa

Although Cl_2 has a certain effect on parasitic protozoa, chlorine at the doses normally used for disinfection and sterilization in wastewater treatment plant cannot completely kill the parasitic protozoa in wastewater. Cl_2 concentration must be so high to completely inactivate parasitic protozoa, but high concentrations of Cl_2 may cause the secondary pollution of water and some health and environmental health concerns arise (Berglund et al., 2017). At present, there are fewer studies on the inactivation of parasitic protozoa with chlorination (Betancourt and Rose, 2004).

10.1.1.4 ARB and ARGs

Effective disinfection is a vital and regular process for disruption of ARB and ARGs. The mechanism of controlling ARB by chlorination is the same as that of pathogen, that HOCl can kill effectively the bacteria. The mechanisms of reducing ARGs are attributed to many reasons: killing the ARB carrying ARGs; suppressing frequency of ARGs transfers, or direct inactivation of ARGs. Some studies indicated that ARB resisting to cephalexin, ciprofloxacin, chloramphenicol, gentamicin, rifampicin, tetracycline, and vancomycin could be fully inactivated by 15 mg Cl_2 min/L, but sulfadiazine- and erythromycin-resistant bacteria were inactivated when the chlorine dose was more than 60 mg Cl_2 min/L (Yuan et al., 2015). After chlorination, 40% of erythromycin resistance gene (*ere(A)*, *ere(B)*, *erm(A)*, *erm(B)*) and 80% of tetracycline resistance genes (*tet(A)*, *tet(B)*, *tet(M)*, and *tet(O)*) in wastewater were removed. Another study showed that chlorine with doses up to 40 mg Cl min/L could promote the frequency of ARG conjugative transfers by 2–5-fold. However, with a chlorine dose higher than 80 mg Cl min/L, the frequency of ARG transfers was greatly suppressed (Guo et al., 2015).

10.1.1.5 Chlorine dioxide

Chlorine dioxide is another bactericidal agent whose disinfectant power is equal to or higher than chlorine. Chlorine dioxide is a yellow-green gas with a pungent smell, water-soluble, but very unstable. It is usually produced by sodium hypochlorite and hydrochloric acid according to the reaction:

$$5\ NaClO_2 + 4\ HCl \rightarrow 4\ ClO_2 + 5\ NaCl + 2\ H_2O$$

Chlorine dioxide is unstable and cannot be stored. It must be produced directly on site and added to water immediately after the production. Chlorine dioxide can be produced by using sodium chlorine combined with hydrochloric acid or chlorine gas (Kim et al., 2002).

Chlorine dioxide is characterized by high oxidizing power, which is the cause of its high germicidal potential. Due to the high oxidative power, possible bacterial elimination mechanisms may include inactivation of enzymatic systems or interruption of protein synthesis. Chlorine dioxide has been found to be more effective than chlorine in sewage effluent, and the rate of inactivation was found to be extremely rapid for chlorine dioxide (Collivignarelli et al., 2017).

10.1.1.6 Pathogenic bacteria

A previous study indicated that chlorine dioxide residual of less than 1 mg/L was effective against *Eberthella typhosa*, *Shigella dysenteriae*, and *Salmonella paratyphi B,* and 1–8 mg/L chlorine dioxide residual could inactivate 99.9% of *B. subtilis, B. mesentericus,* and *B. megatherium spores* (Aieta and Berg, 1986). Meanwhile, chlorine dioxide has been shown to be effective against *B. cereus, B. stearothermophilus,* and *Clostridium perjringens* spores. Another study examined the bactericidal effect of chlorine dioxide in untreated artificial and domestic wastewaters and secondary effluent of various organic loads, indicating that the inactivation of *Escherichia coli* in artificial wastewater is similar with that in real municipal wastewater (Amin et al., 2013).

10.1.1.7 Virus

Chlorine dioxide inactivates viruses primarily due to damage of viral capsid. Chlorine dioxide can adsorb and penetrate into the protein of capsomeres and react with internal RNA, resulting in the damage of the capacity of genetic group. Therefore, the surface of the virus adsorbs high concentration of chlorine dioxide. This strengthens the disinfection effect of chlorine dioxide considerably and results in eventual inactivation of viruses. With chlorine dioxide of 1.0 mg/L and contact time of 30 min, inactivation effect on Poliovirus-1, Coxsackie virus B_3, ECHO-11, Adenovirus-7, Herpes simplex virus-1, and Mumps virus can be achieved in water (Junli et al., 1997).

10.1.1.8 Parasitic protozoa

The control of pathogenic protozoa in wastewater is mainly to inactivate their oocysts. Although it was known that Giardia was much more resistant than bacteria to such disinfection, it is possible to kill the cysts given a high enough concentration of the disinfectant and contact time (Betancourt and Rose, 2004). Unfortunately, chlorination is generally not effective for inactivating *C. parvum* oocysts. Thus the primary target for effective disinfection for protozoa has been on *Cryptosporidium*. Chlorine dioxide has been found to be more effective in controlling *C. parvum* oocysts. A study report promising efficacy of ClO_2 against *C. parvum* with substantially lower contact time than those required for inactivation using free chlorine. For solutions containing 1.4 mg/L ClO_2, contact time of 857 min was required for a 3-log inactivation of *C. parvum* oocysts (Murphy et al., 2014).

10.1.1.9 ARB and ARGs

Disinfection with chlorine dioxide may increase the concentration of ARGs in wastewater. A study reported that chlorine dioxide preferentially increased the abundances of extracellular ARGs against macrolide (*ermB*), tetracycline (*tetA*, *tetB*, and *tetC*), sulfonamide (*sul1*, *sul2*, and *sul3*), β-lactam (*ampC*), aminoglycosides (*aph(2′)-Id*), rifampicin (*katG*), and vancomycin (*vanA*) by up to 3.8 folds. Similarly, the abundances of intracellular ARGs were also increased by up to 7.8 folds after chlorine dioxide (Liu et al., 2018).

10.1.2 Ozone

Ozone is an unstable gas produced by the dissociation of oxygen molecules in atomic oxygen. Ozone production can be done by electrolysis, photolytic reactions, and radiochemical reaction induced by electric shocks. Ozone is unstable and therefore must be generated *in situ*. Ozonation equipment includes air preparation equipment (ozone generator, contactor, and destruction unit), instrumentation, and controls. The capital costs of ozonation systems are relatively high, and operation and maintenance are relatively complex (Mezzanotte et al., 2007).

In water, ozone is unstable and undergoes decomposition to produce $^{\cdot}OH$, which is also a very strong oxidant that reacts quickly with a variety of molecules (Sharma et al., 2016). Ozone is primarily responsible for disinfection, while both O_3 and $^{\cdot}OH$ participate in oxidation reactions

$$O_3 \rightarrow O_2 + O.$$

Ozone is an extremely reactive oxidizing agent characterized by disinfection efficiencies higher than the disinfection with chlorine (Mezzanotte et al., 2007). The mechanisms of ozone sterilization process is a biochemical oxidation reaction due to many reasons: ozone can oxidize the enzymes required to break down the internal glucose of bacteria, causing the bacteria to inactivate and die; ozone directly acts with bacteria and viruses, destroying their organelles and DNA or RNA, destroying the metabolism of bacteria and causing bacterial death; and ozone penetrates the cell membrane tissue, invades the cells, acts on the outer membrane lipoprotein and the internal lipopolysaccharide, causing the bacteria to undergo permeability distortion and dissolve and die. Furthermore, ozonation is a promising technology for enhanced wastewater treatment to eliminate various organic micropollutants.

10.1.2.1 Pathogenic bacteria

Ozone can effectively kill Gram-positive (*Listeria monocytogenes*, *Staphylococcus aureus*, *Enterococcus faecalis*) and Gram-negative microorganisms (*Yersinia enterocolitica*, *Pseudomonas aeruqinosa*, *Salmonella typhimurium*), in both spores and vegetative cells. Disinfection of anaerobic sanitary wastewater effluent with ozone has been done at doses of 5.0, 8.0, and 10.0 mg O_3/L for contact time of 5, 10, and 15 min. The total coliform inactivation range is 2.00–4.06 logs, and the inactivation range for *Escherichia coli* is 2.41–4.65 logs (Amin et al., 2013). Another study shows that *Escherichia coli* can been deactivated in 1–3 orders of magnitude by ozone (Wysok et al., 2006).

10.1.2.2 Virus

Ozone treatment has been shown to reduce the concentrations of enteric viruses and bacteriophages. It was previously reported that a 6-log virus inactivation could be achieved at a residual 0.5 mg min/L and a 4-log virus inactivation at 1 mg min/L by ozonation, while conventional treatment reduced viral concentrations by 14 logs on norovirus, sapovirus, parechovirus, hepatitis E virus, astrovirus, picobirnavirus, parvovirus, and gokushovirus. Ozone treatment led to a further reduction by 1–2 log10 (Wang et al., 2018). Further ozonation reduced the amounts of several viruses to undetectable levels, indicating that this is a promising technique for reducing the transmission of many pathogenic human viruses (Wang et al., 2018).

10.1.2.3 Parasitic protozoa

Ozone may offer greater potential for parasite inactivation than chlorination. Ozone was proved to be efficient in the inactivation of *Cryptosporidium* for that 1 mg/L of ozone with contact time 5 min can inactivate more than 90% *Cryptosporidium* oocysts. However, a 99% inactivation of *C. parvum* oocysts was achieved with an ozone at 5 mg × min/L at 20°C. Ozone at 4 mg/L is able to inactivate *Schistosoma mansoni* eggs (Curtis, 2003).

10.1.2.4 ARB and ARGs

There is limited knowledge on the efficiency of ozonation to inactivate ARB and to remove ARGs. The literature findings available so far revealed that the process performance with respect to ARB and ARGs elimination is strongly depended on the susceptibility of the target bacterium/gene, ozone concentration, and contact time. ARB (*Enterococcus*, *P. aeruginosa*, *Staphylococcus*, and Enterobacteriaceae) and their associated ARGs (*vanA*, bla_{VIM}, *ermB*, and *ampC*) in wastewater effluents were damaged with ozone (0.9 g O_3 per 1 gDOC). *Enterococcus* was reduced by 98%, and *Staphylococcus* was reduced by 83%, but the abundance of *P. aeruginosa* was only altered slightly (Michael-Kordatou et al., 2018). Another study has shown that 3 mg/L ozone dose can decrease the levels of ARB (*E. coli* DH5α) and ARG (pB10) by more than 90%, and wastewater physicochemical properties can affect ozone effects on killing ARB and inactivating ARGs (Yuan et al., 2015).

10.1.3 UV

UV includes electromagnetic radiations between the X-rays and visible light in the range from 100 to 400 nm. The germicidal UV-ray portion falls in the range from 220 to 320 nm. The radiation penetrates the cell wall of the microorganisms and is absorbed by the nucleic acids, causing the inhibition of replication and the death of the cells (Hijnen et al., 2006). The generation of UV rays is carried out by means of lamps containing mercury vapors produced through an electric arc. The energy generated by the excitation of mercury vapors contained within the lamp results in the emission of UV radiation. A new UV source, ultraviolet light-emitting diode (UV-LED), has emerged in the past decade with a number of advantages compared to traditional UV mercury lamps. Studies on UV-LED water disinfection have increased during the past few years. UV-LEDs offer a variety of wavelengths, which is well aligned with the needs of efficient disinfection. Meanwhile, UV-LEDs possess several unique advantages such as environmental friendliness (no mercury), compactness and robustness (more durable), faster start-up time (no warm-up time), potentially less energy consumption, longer lifetime, and the ability to turn on and off with high frequency (Song et al., 2016).

In the secondary or tertiary treated effluents, UV radiation has an efficient disinfecting effect. However, the efficiency of UV disinfection can be influenced by suspended particles, particle sizes, or concentrations of dispersed microorganisms in wastewater. Besides, UV disinfection has one of the few problems, which cannot provide any residual disinfection once the treated effluent has left the UV photoreactor. This problem is compounded by the fact that microorganisms are able to repair damage and alterations induced to their DNA by UV treatment following the administration of a sublethal dose (Hijnen et al., 2006).

10.1.3.1 Pathogenic bacteria

UV disinfection is an effective measure to kill pathogenic bacteria in wastewater, It has been reported that 3-log inactivation of pathogenic bacteria like *E. coli*,

S. aureus, *S. sonnei*, and *S. typhi* can be achieved at 5 mJ/cm^2 dose of UV, and *S. faecalis* requires about a 1.4 times higher dose for 3 log units of inactivation (Chang et al., 1985). The germicidal efficiency of UV radiation is highly dependent on the wavelength. It has been found that 310-nm UV-LEDs result in 94.8 mJ/cm^2 per log inactivation for *E. coli* inactivation, which is much lower than the performance of 365-nm UVA-LEDs (229 mJ/cm^2). It has also been indicated that 275 nm UV is more effective than 255 nm UV for *E. coli* inactivation, which may be attributed to the higher output power of the 275-nm UV-LEDs, resulting in a higher fluence rate and shorter exposure time to reach the same UV dose (Song et al., 2016).

10.1.3.2 Virus

The use of commercial UV disinfection lamps is the most common form of disinfecting the effluent of wastewater treatment plants. In a range of the UV dose from 14 to 27 mJ/cm^2, 3-log inactivation can be achieved for enteroviruses including poliovirus, coxsackievirus, and echovirus (Zhang et al., 2016a). In the European standards for drinking water, a UV dose of 40 mJ/cm^2 is required to ensure 4-log virus inactivation, but this dose would not be sufficient to ensure 4-log inactivation for all viruses in the secondary effluent. Adenoviruses are known to be highly resistant to UV irradiation, and may need three times higher UV doses, namely 120 mJ/cm^2, to achieve 3-log inactivation, and 200 mJ/cm^2 or even higher to meet the 4-log virus removal requirement (Zhang et al., 2016a).

10.1.3.3 Parasitic protozoa

UV radiation is one of the most effective disinfection processes for the inactivation of the (oo) cyst of *Cryptosporidium* and *Giardia* in wastewater. At 100 mJ/cm^2 dose, UV radiation was rather effective in both feeds toward protozoan parasites such as *Giardia lamblia* cysts and *Cryptosporidium parvum* oocysts (approx. 60% and 65% removal respectively) in secondary effluents (Morita et al., 2002). UV radiation has also shown potential for inactivating protozoan cysts by 90%–99% depending on intensity and duration of exposure. Using low pressure collimated beam, a UV dose of 80 mJ/cm^2 was needed to achieve a 3-log inactivation of either rotavirus SA-11 or coliphage MS2 (Curtis, 2003).

10.1.3.4 ARB and ARGs

It has been reported the effective inactivation of a methicillin-resistant strain of *Staphylococcus aureus* and vancomycin-resistant *Enterococcus faecalis* can be achieved after UV disinfection at fluence of 77 mJ/cm^2. Furthermore, UV disinfection process can cause a total reduction of the proportion of *E. coli*, which is resistant toward amoxicillin, ciprofloxacin, and sulfamethoxazole after 60 min of irradiation (12.5 mJ/cm^2). In general, damage of ARGs requires much greater UV doses (200–400 mJ/cm^2 for 3- to 4-log reduction) than ARB inactivation (10–20 for 4–5-log reduction). With more than 10 mJ/cm^2 dose of UV irradiation, the frequency of ARG transfer was found to be largely suppressed (Manaia et al., 2016). However, UV disinfection can increase total relative abundance of ARGs in wastewater and the radiation at

500 mJ/cm^2 can obviously increase their total relative abundance during the same number of microbes. The genes *bacA* and RND-related ARGs mainly contribute to the ARG abundance enhancement (Hu et al., 2016).

10.1.4 PAA

PAA is the peroxide of acetic acid, a strong oxidant and disinfectant, with a wide spectrum of antimicrobial activity. In recent years, the use of peracetic acid as a disinfectant for wastewater effluents has been receiving more attention as a feasible alternative to wastewater chlorination. The PAA dosing system consists of a pump and a storage tank, which is simpler and safer than that of Cl_2 (Rossi et al., 2007).

PAA is produced by the reaction between H_2O_2 and AA as shown by the following equation:

$$CH_3CO_2H + H_2O_2 \rightarrow CH_3CO_3H + H_2O$$

Use of the PAA decomposes can result in the formation of acetic acid and oxygen. As an antimicrobial agent, PAA may be speculated that it functions much as other peroxides and oxidizing agents, and its disinfecting action is the release of active oxygen or the production of reactive hydroxyl radicals, which disrupts sulfhydryl (-SH) and sulfur (S–S) bonds within enzymes contained in the cell membrane. Thus, the bacterial cell wall is attacked to cause cell wall and membrane destruction, as well as certain enzymes and DNA damages.

10.1.4.1 Pathogenic bacteria

The use of PAA as a disinfectant for wastewater effluents has been investigated since the 1980s. A previous study suggested that PAA with dose of 5–10 mg/L and contact time of 15 min was optimal for secondary effluents, providing a reduction of more than 95% of total and fecal coliforms. For the secondary effluent, approximately 5 mg/L PAA residual reduced total coliform and fecal coliform about 4–5 logs after 20 min of contact time. A PAA dose of 25 mg/L provided about 5-log reductions for both *Escherichia coli* and *S. faecalis* within 5 min of contact time (Kitis, 2004).

10.1.4.2 Virus

The concentrations of PAA effective against bacteriophages MS2 in demineralized water is 15 mg/L, resulting in a log reduction of greater than 4 over 5 min of contact time. Poliovirus requires a much higher concentration of PAA, 750–1500 mg/L, to produce a 4-log reduction within 15 min (Baldry and French, 1989). Much higher doses or longer contact time are required for virus removals, especially highly resistant viruses (Kitis, 2004).

10.1.4.3 Parasitic protozoa

PAA shows low efficiency against some parasitic protozoas, such as *Giardia lamblia* cysts and *Cryptosporidium parvum* oocysts (Kitis, 2004).

10.1.4.4 ARB and ARGs

PAA can greatly damage many types of ARB and ARGs. After exposure to 20 mg/L PAA for 10 min, inactivation of ampicillin-resistant bacteria reaches 2.3 logs, but inactivation of tetracycline-resistant bacteria is significantly less efficient, reaching only 1.1 log, whereas their effect can result in a deactivation of bacterial cells (Huang et al., 2013). Some studies have shown that after exposure to 5 mg/L PAA for 20 min, *sul*1, *tet*(G), and BacHum can be reduced by 1.0–3.2 logs, while for 5 mg/L PAA and contact time 5–10 min, *sul*1, *tet*(G), and BacHum are only reduced by 0.6–1.6 logs. However, PAA disinfection is not effective in the removal of some ARGs (*amp*C, *mec*A, *erm*B, *sul*1, *sul*2, *tet*A, *tet*O, *tet*W, *van*A) in wastewater (Kampf, 2018).

Conventional disinfection technologies, especially chlorination, are still the most widely used processes for removal of bHRPs from wastewater. However, wastewater disinfection results in the harmful formation of disinfection by-products (DBPs), which are discharged from wastewater treatment plants and may impair aquatic ecosystems and downstream drinking water quality. Therefore, there is an urgent need to improve existing conventional disinfection or develop new technologies to control bHRPs while avoiding DBPs formation in wastewater.

10.2 Biological treatment progresses

Anaerobic and/or aerobic treatment reactors and membrane bioreactors are the most commonly used biological treatment processes in wastewater treatment plants, which are designed specifically to remove pollutants in wastewater, including chemical oxygen demand (COD), nitrogen, and phosphorus (Barancheshme and Munir, 2018). Meanwhile, these bioreactors can affect the fates of bHRPs in wastewater. In most wastewater treatment systems, any bHRPs removal that occurs is a fortuitous by-product of the principal design objective (usually COD removal).

10.2.1 Anaerobic and/or aerobic treatment reactors

Aerobic and anaerobic treatment processes are low energy and environmentally friendly strategies in wastewater treatment plants. The aerobic treatment processes occur in the presence of air and microorganisms that use oxygen to convert organic contaminants to carbon dioxide, water, and biomass. The anaerobic treatment processes, on the other hand, take place in the lack of air and microorganisms that do not require air to convert organic contaminants to methane and biomass.

10.2.1.1 Pathogenic bacteria

The underlying mechanisms for the removal of pathogenic bacteria in anaerobic and/or aerobic treatment reactors have not been rigorously investigated. However, they are likely to be related to the adsorption to solids, predation, or death. The opportunities for bacterial pathogen removal are limited. Between 90%, 99%, and 99%

removal of fecal coliforms and salmonellae, *Campylobacter* has been reported in activated sludge. The septic tank with a 3-day retention time can remove 50% −95% of the indicator bacteria, while the removal of *V. cholerae* O1 and *Salmonella* O1 may take longer time (12−15 days). UASB reactor can remove 67% of the fecal coliforms with an 8-h retention time (Curtis, 2003). Besides, some studies reported similar levels of removal of salmonellae, campylobacter, bifidobacteria, and fecal enterococci in anaerobic treatment reactors.

10.2.1.2 Virus

Viruses are too small to settle by themselves in wastewater treatment plants, but they may be removed via sedimentation if they associate with larger settleable solids. In previous reports, virus adsorption to solids and removal by sedimentation was thought to be an important virus removal mechanism in wastewater treatment plants. Approximately 90% of type 1 poliovirus and about 98% of Coxsackie A9 virus can be removed in activated sludge (Verbyla and Mihelcic, 2015). Poliovirus and HAV are predominantly adsorbed on the solid matters and viral concentrations in the liquid phase could decrease from 1.0 ± 0.4 logs to 2.2 ± 0.3 logs. The inactivation rate of three enteroviruses (polio 1, coxsackie B3, and echo 1) and a rotavirus (SA-11) under aerobic conditions was found to be significantly greater than the inactivation rate under anaerobic conditions (−0.77 log/day vs. −0.33 log/day) (Junli et al., 1997).

10.2.1.3 Parasitic protozoa

Different aerobic and anaerobic treatment processes have different removal efficiency for different parasitic protozoas (Christgen et al., 2015). During aerobic wastewater treatment, *Cryptosporidium* oocysts and *Giardia* cysts can be reduced by 2.96 logs and 1.40 logs, respectively. During anaerobic sludge digestion, no statistically significant reduction was observed for *Cryptosporidium* oocysts and *Giardia* cysts. These results demonstrate the relative persistence of the protozoa in sewage sludge during wastewater treatment. However, other study indicated that the total *C. parvum* oocyst removal in sewage treatment averaged 98.6%. Moreover, the anaerobic digestion process inactivated 90% of the *C. parvum* oocysts within 4 h of exposure. Nearly all the oocysts can be eliminated by anaerobic digestion for 24 h. This demonstrates that the activated sludge process and anaerobic digestion can be effective for the removal and inactivation of *C. parvum* oocysts.

10.2.1.4 ARB and ARGs

Samples of a municipal wastewater treatment plant have been studied to evaluate the variation of five ARGs (*tetG*, *tetW*, *tetX*, *sul1*, and *intI1*) in the influent and effluent of each treatment unit (Barancheshme and Munir, 2018). The concentration of ARGs in wastewater diminished in the anaerobic and anoxic effluent, while increment in the aerobic effluent was observed. It was concluded that anaerobic and anoxic treatments are much more successful in removing ARGs than aerobic treatment as microorganism has lower bioactivity under anaerobic condition, and the propagation of resistance genes is inhibited. Anaerobic−aerobic sequence bioreactors could remove more than 85% of ARGs in the influent, which means that it was more efficient compared with aerobic and anaerobic units (83% and 62%, respectively). Removal of antibiotics including sulfamethazine, sulfamethoxazole, trimethoprim, and lincomycin has been studied in five different wastewater treatment plants using aerobic/anaerobic treatment methods (Behera et al., 2011). It was found that the removal efficiency of antibiotics was low compared with other pharmaceutical compounds. The removal efficiencies of sulfamethazine, sulfamethoxazole, trimethoprim, and lincomycin were 13.1%, 51.9%, 69.0%, and 11.2%, respectively.

In another study, occurrence and release of tetracycline-resistant and sulfonamide-resistant bacteria, as well as three genes (*sul1*, *tetW*, and *tetO*) in the effluent of five wastewater treatment plants were studied, and the performance of different processes was compared. ARGs and ARB removal ranged from 2.57 logs to 7.06 logs in MBR, and 2.37 logs to 4.56 logs in activated sludge, oxidative ditch, and rotatory biological contactors (Munir et al., 2011).

10.2.2 MBR and membrane filtration

MBR is the combination of a membrane-based filtration process such as microfiltration or ultrafiltration systems with a suspended growth biological reactor. Essentially, the membrane system replaces the solids separation function of the secondary clarifiers in conventional activated sludge systems (Marti et al., 2011). The MBR technology combines the unit operations of aeration, secondary clarification, and filtration into a single process, producing a high quality effluent, suitable for any discharge and most reuse or recycle applications, while greatly reducing space requirements and under stringent norms. MBR is a modification of the activated sludge process in which separation of solids is achieved without the requirement of a secondary clarifier. Instead, this function is carried out by a membrane, which retains the particulate phase within the reactor and allows the treated, clarified effluent to pass to the next process or be discharged/reused. The small pore size of the membranes employed for separation of solids enables them to remove a wide range of microorganisms. There are two facets of treatment by MBR-activated sludge and membrane separation. The mechanisms of activated sludge in MBR on bHRPs are similar to that of aerobic treatment reactors. The role of membranes alone in removal of bHRPs is similar to that of membrane filtration (Hai et al., 2014).

Membrane filtration allows the separation through a physical barrier to the pollutants present in the water. With the passage through the membranes, there is almost complete removal of the bacteria, parasitic protozoa, and partial viruses. Membrane types are mainly divided into microfiltration (MF, 0.1−10 μm), ultrafiltration (UF, 0.01−0.1 μm), nanofiltration (NF, 0.001−0.01 μm), and reverse osmosis (RO, $<$0.001 μm). Whereas nanofiltration and reverse osmosis processes are effective in removing protozoan (oo) cysts, microfiltration and ultrafiltration are the most commonly applied/used technologies for microbial removal because of their cost-effectiveness. Microfiltration is able to ensure the removal of most bacteria and protozoa's cysts, but not providing adequate virus removal; ultrafiltration, on the other hand, allows complete removal of bacteria, viruses, and protozoa (Collivignarelli et al., 2017).

When well designed and operated, MBRs can consistently achieve efficient removals of suspended solids, protozoa, and coliform bacteria. Studies have reported efficient removal of indicator bacteria in MBR processes, 5 logs removal of coliforms and up to 7 logs removal of fecal enterococci. Although viruses are smaller than the pore sizes used in microfiltration processes, high removal rates of these have been reported but mainly after the build-up of a biofilm on the membrane. MBRs significantly reduced ARB abundances in the ranges of 2.80–3.54 logs, higher removals in conventional treatment plants (Bodzek et al., 2019).

10.3 AOPs

AOPs were first proposed for potable water treatment in the 1980s, which have been defined as the oxidation processes involving the generation of hydroxyl radicals ($OH\cdot$) in sufficient quantity to affect water purification. Later, the AOP concept has been extended to the oxidative processes with sulfate radicals ($SO_4^{\bullet-}$) (Deng and Zhao, 2015).

Hydroxyl radical is the most reactive oxidizing agent in water treatment, with an oxidation potential between 2.8 V (pH 0) and 1.95 V (pH 14). Hydroxyl radicals attack organic pollutants through four basic pathways: radical addition, hydrogen abstraction, electron transfer, and radical combination. Their reactions with organic compounds produce carbon-centered radicals ($R\cdot$ or $R\cdot{-}OH$). With O_2, these carbon-centered radicals may be transformed to organic peroxyl radicals ($ROO\cdot$). All of the radicals further react accompanied with the formation of more reactive species such as H_2O_2 and super oxide ($O_2^{\bullet-}$), leading to killing or inactivating the bHRPs in wastewater. Because hydroxyl radicals have a very short lifetime, they are only *in situ* produced during application through different methods, including a combination of oxidizing agents (such as H_2O_2 and O_3), irradiation (such as ultraviolet light or ultrasound), and catalysts (such as Fe^{2+}) (Deng and Zhao, 2015).

Persulfate ($S_2O_8^{2-}$) itself is a strong oxidant with a standard oxidation potential (E^o) of 2.01 V. Once activated by heat, ultraviolet (UV) irradiation, transitional metals, or elevated pH, $S_2O_8^{2-}$ can form more powerful sulfate radicals ($SO_4^{\bullet-}$, $E^o = 2.6$ V) to initiate sulfate radical-based advanced oxidation processes.

Despite the fact that AOPs are generally used for the oxidation of a wide range and variety of organic and inorganic compounds, they can be applied for the water disinfection, but be rarely used because these radicals have a very short half-life, so that the required detention times for disinfection are prohibitive due to extremely low radical concentrations. On other hand, all AOPs have a high operating cost.

There is a frequent debate on the role of $OH\cdot$ on pathogen disinfection. Some studies suggested that $OH\cdot$ plays an important role in inactivation of *E. coli* and *Bacillus subtilis spores.* $OH\cdot$ produced in a photocatalytic disinfection process have been found to provide residual effects that repress the post-UV reactivation of coliform bacteria. However, other studies have showed that OH· are not effective against *Giardia muris*. For the Fenton oxidation, under the optimal condition wherein Fe^{2+}/H_2O_2 can remove 2.58–3.79 logs of ARGs include *sul1*, *tetX*, and *tetG* within 2 h. For the UV/H_2O_2 process, all ARGs can achieve a reduction of 2.8–3.5 logs (Zhang et al., 2016b).

In recent years, a lot of researches have been conducted on the AOPs that can be driven by sunlight (Tsydenova et al., 2015). The use of renewable and free solar energy in such processes can substantially reduce treatment costs and is more favorable from an environmental perspective. The solar-enhanced methods seem to be particularly suitable for countries located in regions with abundant sunlight, which is the case of many developing countries with drinking water issues. Besides, the ability of AOPs to remove both pathogens and chemical pollutants could further help to improve the economic efficiency of water/wastewater treatment. TiO_2 photocatalysis and photo-Fenton are by far the most studied AOPs that have been shown to be capable of removing chemical pollutants and pathogens, including bacteria, viruses, fungi, and protozoa. In TiO_2 photocatalysis, free hydroxyl •OH radicals are generated upon irradiation of a catalytic semiconductor, such as TiO_2, with near-UV light of wavelengths <385 nm TiO_2 photocatalysis can remove 3 logs of *E. coli* and 1–2 logs *S. typhimurium* within 120 min. Photo-Fenton can remove 2.5–4 logs of *E. coli*, *S. typhimurium*, and *S. sonnei* within 40 min.

10.4 Natural disinfection

10.4.1 Constructed wetland

Constructed wetlands are small semiaquatic ecosystems, in which a great population of different microbial community multiplies and various physical–chemical reactions happen. Constructed wetlands containing floating, emergent and submergent aquatic plants, and other water-tolerant species have been found to economically reduce bHRPs in wastewater (Vacca et al., 2005; Karim et al., 2004). The main factors responsible to reduce bHRPs are physical processes such as mechanical filtration, sedimentation, and adsorption, chemical processes such as the release of oxygen from plants, and biological mechanisms such as natural death, predation and action of antibiotic substances released by macrophytes (Morató et al., 2014). Other studies have shown that many aquatic plants roots can secrete some acids such as tannic acid and gallic acid, which cause disinfection.

Some studies showed the reduction of 57% of total coliforms and 62% of fecal coliforms, 98% of *Giardia*, 87% of *Cryptosporidium*, and 38% of coliphage in a duckweed-based wetland system. In a multispecies wetland system, the total and fecal coliforms can be reduced by 98% and 93%, respectively, and *Giardia*, *Cryptosporidium*, and enteric viruses can be reduced by an average of 73%, 58%, and 98%, respectively. Reduction efficiencies of human enteric viruses by the surface flow wetlands ranged from 1 to 3 logs, including norovirus, adenovirus, Aichi virus 1, polyomaviruses, and enterovirus. Constructed wetlands can efficiently remove aqueous ARGs, but they can also act as reservoirs for specific ARGs. Vertical flow constructed wetlands can remove 33.2%–99.1% of tetracycline-resistance genes (*tet*) and integrase gene of Class 1 integrons (Huang et al., 2017).

10.4.2 Wastewater stabilization ponds

Wastewater stabilization ponds are often considered to be the most environmentally and economically sustainable technology for small, rural, or remote communities that require low-cost and low-maintenance wastewater treatment systems (Collivignarelli et al., 2017). This process consists of oxidation basins working in series with an aerobic final stage. Wastewater stabilization ponds have the capability to effectively attenuate organic and nutrient concentrations, as well as bHRPs in wastewater. A wide range of bHRPs such as bacterial, viral, protozoan, and helminthic pathogens can been removed in wastewater stabilization ponds systems. Removal of bHRPs is mostly due to natural die-off, which increases with time, pH, and temperature and other factors, including food shortages, predation, and algae adsorption (Curtis, 2003). When exposed to the direct rays of the sun, the temperature of wastewater stabilization ponds can reach up to 40°C. The pH at midday is commonly 9 or higher, which is attributed to the presence of algae, is an important factor for effective disinfection. Meanwhile, exposure to the ultraviolet rays of the sun may also play a role in removing the bHRPs in wastewater. Based on temperature and algal concentrations, removal efficiencies of bHRPs during wastewater treatment over the course of a year can be highly variable, where higher removal efficiencies would be expected in summer and fall seasons.

Some studies have indicated that wastewater stabilization ponds could remove up to 6 logs and 3 logs of bacteria and helminth eggs or protozoan (oo) cysts, respectively. For virus, 1-log reduction was achieved. Producing final effluents of wastewater stabilization ponds can meet World Health Organization guidelines for the use of treated wastewater in unrestricted agricultural irrigation. However, individual studies have shown that relative abundance of ARGs increased in one wastewater stabilization pond, due to that long-term stay in wastewater stabilization pond may promote transfer of ARGs among microorganisms (Liu et al., 2016).

10.5 Other technologies for bHRPs removal

10.5.1 Nanomaterials

Nanomaterials are considered new technologies for efficient disinfection and microbial control, which have demonstrated strong antimicrobial properties through diverse mechanisms including photocatalytic production of reactive oxygen species that damage cell components and viruses, compromising the bacterial cell envelope, interruption of energy transduction, and inhibition of enzyme activity and DNA synthesis. Some nanomaterials have also been shown to have strong antimicrobial properties, including chitosan, silver nanoparticles, nanosized ZnO, photocatalytic TiO_2, and carbon nanotubes (Li et al., 2008). Different from conventional chemical disinfectants, these antimicrobial nanomaterials are not strong oxidants and are relatively inert in water. Although some nanomaterials have been used as antimicrobial agents in consumer products including home purification systems as antimicrobial agents, the research of their potential for disinfection or microbial control in wastewater treatment is scarce.

Normally, antimicrobial nanomaterials are used in combination with existing disinfection technologies, like UV disinfection and membrane filtration. The photosensitive nanomaterials can be easily applied to UV reactors to improve UV disinfection, which can effectively kill cyst-forming protozoa such as *Giardia* and *Cryptosporidium*, and some pathogenic viruses such as adenoviruses. Composite membranes containing a 50-nm thick chitosan layer on a poly(acrylic acid)/poly(-ethylene glycol) diacrylate layer were found to exhibit potent antibacterial activity toward Gram-positive and Gram-negative bacteria, and the antibacterial activity of the membrane improved with increasing chitosan content. It is worth noting that the role of nanomaterials toward ARB and ARG in water and wastewater treatment is mostly elusive. A study conducted using nanoalumina showed significant promotion of the horizontal conjugative transfer of multidrug-resistance genes mediated by plasmids across genera (RP4, RK2, and pCF10) (Sharma et al., 2016). A study on photocatalytic activation used ultraviolet A (UV-A)/TiO_2 to oxidize methicillin-resistant *S. aureus*, multidrug-resistant *Acinetobacter baumannii*, and vancomycin-resistant *Enterococcus faecalis*. The photocatalytic process with TiO_2 reduced the numbers of bacteria by 3 logs.

10.5.2 Biochar

Biochar is a heterogeneous black carbon remnant of plant biomass pyrolysis, and it is a relatively new research field. Biochar is a porous material that has rich mineral elements and large specific surface area, and thus provides the most sites to be filled by sorption of contaminants. The main treatment mechanism of biochar is sorption. Moreover, electrostatic repulsion is another sorption mechanism that is attributed to complex properties of biochar (Gwenzi et al., 2017).

Biochar filters are more effective than other filters in removal of *Salmonella* spp. (3 logs reduction) but are less effective in removal of bacteriophages. The capacity of biochar to remove pathogens and indicator organisms depends on biochar type and operating conditions particularly hydraulic loading rates, and presence of potentially interfering concomitants in aqueous systems. The biochar filters can achieve a 1.4, 1.5, 2.0, and 1.3 logs reduction for *E. faecalis*, *E. coli*, *Saccharomyces cerevisiae yeast*, and Phix-174 phage, respectively, and the smaller the particle size of the biochar, the better the straining of bacterial and virus particles (Barancheshme and Munir, 2018).

10.5.3 Bacteriophage

Bacteriophages are viruses that infect and kill bacteria. Different from traditional broad-spectrum antibiotics, these viruses target specific bacteria without harming the body's normal microflora and play key roles in regulating the microbial balance in ecosystem. Phages can be bactericidal, and they can increase in number responding to the incidence of pathogens, tend to only minimally disrupt normal flora, are equally effective against ARB, are often easily discovered, seem to be capable of disrupting bacterial biofilms, and can have low inherent toxicities. The exploitation of phages as a realistic approach in the control of pathogens has attracted considerable interest in recent years because of the emergence of ARB. Bacteriophages are used to treat a bacterial infection, which is known as phage therapy. Phages have several characteristics including their effectiveness in killing their target bacteria, their specificity, adaptability, natural residence in the environment, and the fact that they are self-replicating and self-limiting (Jassim et al., 2016).

Some studies revealed the role of phages in water and wastewater treatment, especially disinfection. Successful phage treatment of wastewater bacterial pathogens would be dependent on the diversity of pathogen species within wastewater. There is potential for phage treatment to be used successfully in combination with biological sludge stabilization processes to reduce the abundance of specific pathogenic bacterial strains such as *E. coli* and *Salmonella*. Phage biocontrol is receiving greater concerns to mitigate the propagation of ARB. A study showed that polyvalent phage cocktails (PER01 and PER02) were significantly more effective than narrow host-range coliphage cocktails (MER01 and MER02) in suppressing a model ARB β-lactam-resistant *Escherichia coli* NDM-1, initially present at 6.2 logs. After 5 days, the NDM-1 concentration significantly decreased to 3.8 logs in the presence of the polyvalent phage cocktail, compared to 4.7 logs for the coliphage cocktail treatment (Yu et al., 2017).

10.6 Summary

This chapter presents an overview of the main control technologies for bHRPs in wastewater, and introduces the application, action mechanisms, and removal performance of the technologies. Conventional disinfection technologies (especially chlorination) are still commonly used in bHRPs' control progress, despite the fact that they may produce DBPs in the treated wastewater. The existing wastewater treatment technologies have been optimized to improve the removal efficiency of bHRPs without affecting that of chemical pollutants. The advanced technologies are very interesting but they are still in the research state. New effective technologies have to be developed or the multiple technologies can be combined to replace conventional disinfection technologies to reduce the production of DBPs and improve the removal of bHRPs.

References

Ahmed, M.B., Zhou, J.L., Ngo, H.H., Guo, W., Thomaidis, N.S., Xu, J., 2017. Progress in the biological and chemical treatment technologies for emerging contaminant removal from wastewater: a critical review. Journal of Hazardous Materials 323, 274–298.

Aieta, E.M., Berg, J.D., 1986. A review of chlorine dioxide in drinking water treatment. Journal American Water Works Association 78 (6), 62–72.

Amin, M.M., Hashemi, H., Bovini, A.M., Hung, Y.T., 2013. A review on wastewater disinfection. International Journal of Environmental Health Engineering 2 (1), 22.

Baldry, M., French, M., 1989. Activity of peracetic acid against sewage indicator organisms. Water Science and Technology 21 (12), 1747–1749.

Behera, S.K., Kim, H.W., Oh, J.-E., Park, H.-S., 2011. Occurrence and removal of antibiotics, hormones and several other pharmaceuticals in wastewater treatment plants of the largest industrial city of Korea. The Science of the Total Environment 409 (20), 4351–4360.

Barancheshme, F., Munir, M., 2018. Strategies to combat antibiotic resistance in the wastewater treatment plants. Frontiers in Microbiology 8, 2603.

Berglund, B., Dienus, O., Sokolova, E., Berglind, E., Matussek, A., Pettersson, T., Lindgren, P.-E., 2017. Occurrence and removal efficiency of parasitic protozoa in Swedish wastewater treatment plants. The Science of the Total Environment 598, 821–827.

Betancourt, W.Q., Rose, J.B., 2004. Drinking water treatment processes for removal of Cryptosporidium and Giardia. Veterinary Parasitology 126 (1–2), 219–234.

Bodzek, M., Konieczny, K., Rajca, M., 2019. Membranes in water and wastewater disinfection—review. Archives of Environmental Protection 45 (1), 3–18.

Chang, J.C., Ossoff, S.F., Lobe, D.C., Dorfman, M.H., Dumais, C.M., Qualls, R.G., Johnson, J.D., 1985. UV inactivation of pathogenic and indicator microorganisms. Applied and Environmental Microbiology 49 (6), 1361–1365.

Christgen, B., Yang, Y., Ahammad, S.Z., Li, B., Rodriquez, D.C., Zhang, T., Graham, D.W., 2015. Metagenomics shows that low-energy anaerobic-aerobic treatment reactors reduce antibiotic resistance gene levels from domestic wastewater. Environmental Science & Technology 49 (4), 2577–2584.

Collivignarelli, M., Abbà, A., Benigna, I., Sorlini, S., Torretta, V., 2017. Overview of the main disinfection processes for wastewater and drinking water treatment plants. Sustainability 10 (1), 86.

Curtis, T., 2003. Bacterial pathogen removal in wastewater treatment plants. The Handbook of Water and Wastewater Microbiology 477–490.

Deng, Y., Zhao, R., 2015. Advanced oxidation Processes (AOPs) in wastewater treatment. Current Pollution Reports 1 (3), 167–176.

Guo, M.-T., Yuan, Q.-B., Yang, J., 2015. Insights into the amplification of bacterial resistance to erythromycin in activated sludge. Chemosphere 136, 79–85.

Gwenzi, W., Chaukura, N., Noubactep, C., Mukome, F.N.D., 2017. Biochar-based water treatment systems as a potential low-cost and sustainable technology for clean water provision. Journal of Environmental Management 197, 732–749.

Hai, F., Riley, T., Shawkat, S., Magram, S., Yamamoto, K., 2014. Removal of pathogens by membrane bioreactors: a review of the mechanisms, influencing factors and reduction in chemical disinfectant dosing. Water 6 (12), 3603–3630.

Hijnen, W., Beerendonk, E., Medema, G.J., 2006. Inactivation credit of UV radiation for viruses, bacteria and protozoan (oo) cysts in water: a review. Water Research 40 (1), 3–22.

Hu, Q., Zhang, X.-X., Jia, S., Huang, K., Tang, J., Shi, P., Ye, L., Ren, H., 2016. Metagenomic insights into ultraviolet disinfection effects on antibiotic resistome in biologically treated wastewater. Water Research 101, 309–317.

Huang, J.J., Xi, J.Y., Hu, H.Y., Tang, F., Pang, Y.C., 2013. Inactivation and regrowth of antibiotic-resistant bacteria by PAA disinfection in the secondary effluent of a municipal wastewater treatment plant. Biomedical and Environmental Sciences 26 (10), 865–868.

Huang, X., Zheng, J., Liu, C., Liu, L., Liu, Y., Fan, H., 2017. Removal of antibiotics and resistance genes from swine wastewater using vertical flow constructed wetlands: effect of hydraulic flow direction and substrate type. Chemical Engineering Journal 308, 692–699.

Jassim, S.A., Limoges, R.G., El-Cheikh, H., 2016. Bacteriophage biocontrol in wastewater treatment. World Journal of Microbiology and Biotechnology 32 (4), 70.

Junli, H., Li, W., Nenqi, R., Li, L.X., Fun, S.R., Guanle, Y., 1997. Disinfection effect of chlorine dioxide on viruses, algae and animal planktons in water. Water Research 31 (3), 455–460.

Kampf, G., 2018. Biocidal agents used for disinfection can enhance antibiotic resistance in Gram-negative species. Antibiotics (Basel) 7 (4), 110.

Karim, M.R., Manshadi, F.D., Karpiscak, M.M., Gerba, C.P., 2004. The persistence and removal of enteric pathogens in constructed wetlands. Water Research 38 (7), 1831–1837.

Kim, B., Anderson, J., Mueller, S., Gaines, W., Kendall, A., 2002. Literature review—efficacy of various disinfectants against Legionella in water systems. Water Research 36 (18), 4433–4444.

Kitis, M., 2004. Disinfection of wastewater with peracetic acid: a review. Environment International 30 (1), 47–55.

Lazarova, V., Savoye, P., Janex, M., Blatchley III, E.R., Pommepuy, M., 1999. Advanced wastewater disinfection technologies: state of the art and perspectives. Water Science and Technology 40 (4–5), 203–213.

Li, Q., Mahendra, S., Lyon, D.Y., Brunet, L., Liga, M.V., Li, D., Alvarez, P.J., 2008. Antimicrobial nanomaterials for water disinfection and microbial control: potential applications and implications. Water Research 42 (18), 4591–4602.

Liu, L., Hall, G., Champagne, P., 2016. Effects of environmental factors on the disinfection performance of a wastewater stabilization pond operated in a temperate climate. Water 8 (1), 5.

Liu, S.S., Qu, H.M., Yang, D., Hu, H., Liu, W.L., Qiu, Z.G., Hou, A.M., Guo, J., Li, J.W., Shen, Z.Q., Jin, M., 2018. Chlorine disinfection increases both intracellular and extracellular antibiotic resistance genes in a full-scale wastewater treatment plant. Water Research 136, 131–136.

Manaia, C.M., Macedo, G., Fatta-Kassinos, D., Nunes, O.C., 2016. Antibiotic resistance in urban aquatic environments: can it be controlled? Applied Microbiology and Biotechnology 100 (4), 1543–1557.

Marti, E., Monclús, H., Jofre, J., Rodriguez-Roda, I., Comas, J., Balcázar, J.L., 2011. Removal of microbial indicators from municipal wastewater by a membrane bioreactor (MBR). Bioresource Technology 102 (8), 5004–5009.

Mezzanotte, V., Antonelli, M., Citterio, S., Nurizzo, C., 2007. Wastewater disinfection alternatives: chlorine, ozone, peracetic acid, and UV light. Water Environment Research 79 (12), 2373–2379.

Michael-Kordatou, I., Karaolia, P., Fatta-Kassinos, D., 2018. The role of operating parameters and oxidative damage mechanisms of advanced chemical oxidation processes in the combat against antibiotic-resistant bacteria and resistance genes present in urban wastewater. Water Research 129, 208–230.

Morató, J., Codony, F., Sánchez, O., Pérez, L.M., García, J., Mas, J., 2014. Key design factors affecting microbial community composition and pathogenic organism removal in horizontal subsurface flow constructed wetlands. The Science of the Total Environment 481, 81–89.

Morita, S., Namikoshi, A., Hirata, T., Oguma, K., Katayama, H., Ohgaki, S., Motoyama, N., Fujiwara, M., 2002. Efficacy of UV irradiation in inactivating *Cryptosporidium parvum* oocysts. Applied and Environmental Microbiology 68 (11), 5387–5393.

Munir, M., Wong, K., Xagoraraki, I., 2011. Release of antibiotic resistant bacteria and genes in the effluent and biosolids of five wastewater utilities in Michigan. Water Research 45 (2), 681–693.

Murphy, J.L., Haas, C.N., Arrowood, M.J., Hlavsa, M.C., Beach, M.J., Hill, V.R., 2014. Efficacy of chlorine dioxide tablets on inactivation of *Cryptosporidium* oocysts. Environmental Science & Technology 48 (10), 5849–5856.

Ottoson, J., Hansen, A., Björlenius, B., Norder, H., Stenström, T., 2006. Removal of viruses, parasitic protozoa and microbial indicators in conventional and membrane processes in a wastewater pilot plant. Water Research 40 (7), 1449–1457.

Rossi, S., Antonelli, M., Mezzanotte, V., Nurizzo, C., 2007. Peracetic acid disinfection: a feasible alternative to wastewater chlorination. Water Environment Research 79 (4), 341–350.

Sharma, V.K., Johnson, N., Cizmas, L., McDonald, T.J., Kim, H., 2016. A review of the influence of treatment strategies on antibiotic resistant bacteria and antibiotic resistance genes. Chemosphere 150, 702–714.

Song, K., Mohseni, M., Taghipour, F., 2016. Application of ultraviolet light-emitting diodes (UV-LEDs) for water disinfection: a review. Water Research 94, 341–349.

Tsydenova, O., Batoev, V., Batoeva, A., 2015. Solar-enhanced advanced oxidation processes for water treatment: simultaneous removal of pathogens and chemical pollutants. International Journal of Environmental Research and Public Health 12 (8), 9542–9561.

Vacca, G., Wand, H., Nikolausz, M., Kuschk, P., Kästner, M., 2005. Effect of plants and filter materials on bacteria removal in pilot-scale constructed wetlands. Water Research 39 (7), 1361–1373.

Verbyla, M.E., Mihelcic, J.R., 2015. A review of virus removal in wastewater treatment pond systems. Water Research 71, 107–124.

Wang, H., Sikora, P., Rutgersson, C., Lindh, M., Brodin, T., Björlenius, B., Larsson, D.J., Norder, H., 2018. Differential removal of human pathogenic viruses from sewage by conventional and ozone treatments. International Journal of Hygiene and Environmental Health 221 (3), 479–488.

Wigginton, K.R., Kohn, T., 2012. Virus disinfection mechanisms: the role of virus composition, structure, and function. Current Opinion in Virology 2 (1), 84–89.

Wysok, B., Uradziñski, J., Gomólka-Pawlicka, M., 2006. Ozone as an alternative disinfectant-a review. Polish Journal of Food and Nutrition Sciences 15 (1), 3.

Yu, P., Mathieu, J., Lu, G.W., Gabiatti, N., Alvarez, P.J., 2017. Control of antibiotic-resistant bacteria in activated sludge using polyvalent phages in conjunction with a production host. Environmental Science and Technology Letters 4 (4), 137–142.

Yuan, Q.-B., Guo, M.-T., Yang, J., 2015. Fate of antibiotic resistant bacteria and genes during wastewater chlorination: implication for antibiotic resistance control. PLoS One 10 (3), e0119403.

Zhang, C.-M., Xu, L.-M., Xu, P.-C., Wang, X.C., 2016a. Elimination of viruses from domestic wastewater: requirements and technologies. World Journal of Microbiology and Biotechnology 32 (4), 69.

Zhang, K., Farahbakhsh, K., 2007. Removal of native coliphages and coliform bacteria from municipal wastewater by various wastewater treatment processes: implications to water reuse. Water Research 41 (12), 2816–2824.

Zhang, Y., Zhuang, Y., Geng, J., Ren, H., Xu, K., Ding, L., 2016b. Reduction of antibiotic resistance genes in municipal wastewater effluent by advanced oxidation processes. The Science of the Total Environment 550, 184–191.

CHAPTER

Risk management policy for HRPs in wastewater

11

Kan Li, Hongqiang Ren, PhD

State Key Laboratory of Pollution Control and Resource Reuse, School of the Environment, Nanjing University, Nanjing, China

Chapter outline

Water is the essential resource of life and civilization, and the management of water usage, including wastewater treatment and discharge to prevent the contamination of drinking water, accompanies the whole history of civilization. As a more modern concept, the "water management" first appeared in Germany at the dawn of the 20th century for the purpose of agricultural irrigation projects (Supan, 1901; Ule, 1910), which then led to the management of wastewater produced by agriculture in late 1930s (Pruss, 1938). The management of wastewater produced by municipal or industrial usage was not documented until late 1960s (Stander, 1966; Symons, 1969).

Currently, the most important goal of successful water management is to eliminate or control the risk arising from water contamination to the human beings or ecosystem, focusing on the decrease of high-risk pollutants (HRPs) to reach acceptable levels. The knowledge of pollutants and their risks to both the ecosystem and human beings has been growing throughout the long history of civilization, especially for the health hazards possibly caused by waterborne pathogens or certain chemicals. In the historical book of "Lv Shi Chunqiu" first published more than 2600 years ago, the occurrence of several endemic diseases, such as ulcerate, humpback, and swollen feet, were documented, and their correlations with water quality were proposed. The use of filtration for wastewater treatment has been recorded simultaneously in "Zuo Shi Chunqiu," and the remaining of such filtration facility can be found in Yang City Relics built during the same period. However, the correlation of HRPs in wastewater with certain diseases was not reported until 1940s

High-Risk Pollutants in Wastewater. https://doi.org/10.1016/B978-0-12-816448-8.00011-3

(Subrahmanyan and Bhaskaran, 1950; Ward and Turner, 1942). The correlation of biological HRPs bacteria and other microbes with jaundice, diarrhea, or other gastrointestinal symptoms were observed for the first time. In mid-1950s, the outbreak of Itai-itai disease (1955) and Minamata disease (1956) in Japan with a large number of victims and heavy casualty shocked the world (Kobayashi, 1970; McAlpine and Araki, 1958). After etiology investigation indicated that these two diseases resulted from mercury and cadmium pollution, a wide range of regulations and laws concerning the standard of heavy metals in wastewater and surface water was drafted and validated across the world (Funabashi, 2006). Afterward, the knowledge of risk management of HRPs in wastewater has grown rapidly, and the publications focusing on the theme of "high risk," "pollutant," "management," and "wastewater" have a hit of 1772 (Figs. 11.1−11.3) in the combined database of Web of Knowledge (https://apps.webofknowledge.com/).

Among the different research areas, toxicity (81.2% of all publications), epidemiology (97.0%), and ecological (99.4%) investigation are the top three fields concerning HRP management in wastewater (Fig. 11.2). This is mainly attributed to that all the management of HRPs should depend on the risk evaluation results. Due to the growing demands of human and ecoenvironmental health, a continuous invest has been put into this research field, and the publications increased greatly in the past 2 decades (Fig. 11.2).

Further analysis of correlation of the publications with the citations indicates that the research focus of the risk management of HRPs in wastewater has changed along with the extension of knowledge regarding pollutants risk control (Fig. 11.3), from the traditional strategy of "temporal dynamics" to elimination utilizing technologies, such as "Fenton process" and "BP-UV filter," from a specific location of "hospital

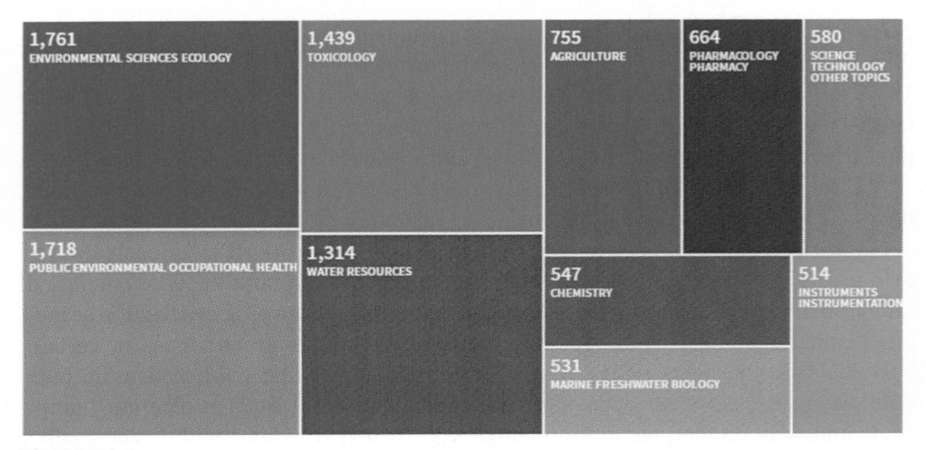

FIGURE 11.1

Summary of the research fields regarding the publications in "high risk," "pollutant," and "wastewater management," generated from Web of knowledge (https://apps.webofknowledge.com/).

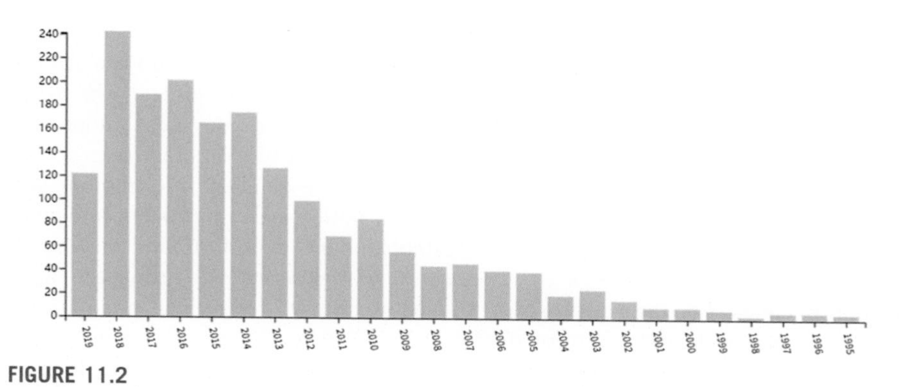

FIGURE 11.2

The annual number of publications in research fields of "high risk," "pollutant," and "wastewater management," generated from Web of knowledge (https://apps.webofknowledge.com/).

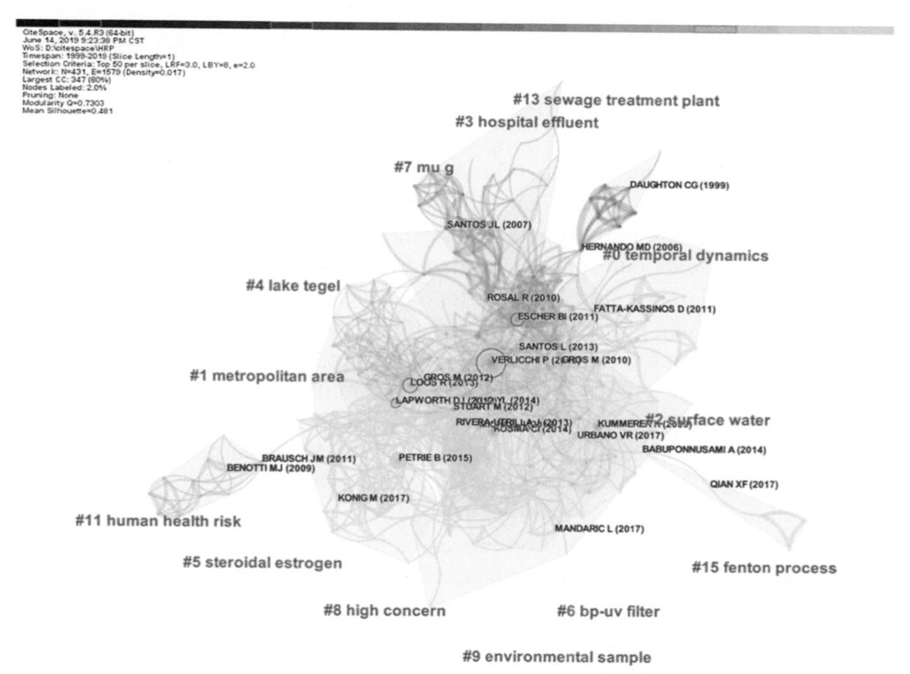

FIGURE 11.3

The evolution of research area during the past 3 decades of publications of "high risk," "pollutant," and "wastewater management." Note: each knot of the citation net is one article or patent; the color of the knot indicates the publication year of this article or patent; and the red phrase with a number is the hot topic concerning these key words.

This picture is generated from citation analysis using Citespace Chen, C.M., 2004. Searching for intellectual turning points: Progressive knowledge domain visualization. Proceedings of the National Academy of Sciences of the United States of America 101, 5303–5310.

effluent" or "sewage treatment plant" to the whole "metropolitan area" combined with all "surface water," with an emphasis on the "high concern" chemicals or biological contaminants such as pollutants with "steroidal estrogen" effects. The trend of the studies concerning HRPs in wastewater tends to cover more areas, more pollutants, as well as to provide permanent solutions instead of temporary ones.

This tremendous growth of research brought more detailed information into views of scientists and other stakeholders, such as policy makers, officers, and general public; thus, an acceleration of the establishment, evolution, and mature of management system of HRPs in wastewater was experienced during the past several decades. For the early age of mid-20th century, the general populations to a large part were not aware of the hazards of HRPs and were later informed through the epidemiology publications of scientists or researchers. The people usually looked for help from government and officers, for example, the Minamata disease and Itai-itai disease mentioned earlier, as well as perfluoroalkyl substances pollution in Mid-Ohio Valley (Sun et al., 2017). However, this procedure caused great health damage of residents on local and regional scales, even global scale, and the solution of the penalty or compensation is lagged due to the lack of laws, standards, or regulations. For example, the symptoms of Itai-itai disease were actually first recorded as early as the 19th century and reported to the local government no late than 1912, but did not draw enough attention until 1955, when no law or regulations concerning wastewater safety were settled yet. The nitrogen fertilizer production caused the spread of Minamata disease through the continuous discharge of wastewater into Minamata River for 12 years after the outbreak of disease, and the "Niigata Minamata Disease" broke out in Niigata Prefecture nearly 10 years later in 1965, which contributed three of the "Four Major Pollution-Related Diseases" in Japan. After identification of the diseases as a direct consequence of certain pollutants discharged by industry manufacture, conflict of these companies with local residents heated up, and drew attention of the whole nation, which finally prompted the pass of Pollution-Related Health Damage Compensation Law of Japan nearly 2 decades later in 1973. Although more than 1000 victims were compensated annually after the law was put into effect, all the law suits were not finished until 2013, and lots of the victims passed away without compensation (Hino et al., 2018; Iguchi and Koga, 2015; Voulvoulis et al., 2013). These risk remediation processes then evolved into a more sophisticated risk management procedure in recent decades, with the combination of conventions, laws, standards, and registration regulations, and a batch of developed nations has established a system for the management of HRPs in wastewater. Due to the public awareness of correlation between environment protection and HRPs management, especially after the economic globalization, the industry distribution has been largely transferred.

11.1 Risk thresholds and criteria for different disposal or reuse purposes

Currently, the global government structure of HRPs management incudes four levels: global level, regional level, national level, and local level.

11.1.1 The United Nations environment program

The severe ecoenvironment accidents that caused disease, huge casualty and environment damage during 1950–1970s finally pushed the cooperation of three suborganization of the United Nations, the International Labor Organization, the Food and Agriculture Organization, and the World Health Organization, to promote the 1972 UNz. Conference on Human Environment to control the pollution caused by the industrialization met the urgent requirements of international laws system to efficiently reduce the potential hazards, which then led to the milestone settle up of United Nations Environment Programme (UNEP, https://www.unenvironment.org/) in January 1973.

As its founding, the UNEP has identified "management of chemicals and wastes" as one of its Safeguard Standards, and put its effort in minimizing globalized ecoenvironment hazards, through the establishment of global environment legal system. For example, the Stockholm Convention contains obligations to eliminate or severely restrict the production and use of a number of the persistent organic pollutants, which are persistent, toxic, and bioaccumulative in particular food chain, is prone to long-range environmental transport (Lallas, 2001). After signature of over 90 nations in 2001, the global manufacture of chemicals, such as dichloro-diphenyl-trichloroethane (DDT), lindane, polychlorinated dibenzo-p-dioxins/dibenzofurans, and perfluorooctane sulfonate (PFOS), contained in Stockholm Convention is largely restrained or even forbidden (Fiedler, 2007; Hardy and Maguire, 2010; Vijgen et al., 2011; Wang et al., 2009). These organic HRPs were phased out through a global schedule, and well controlled right now (Fiedler et al., 2019). In 2013, another convention concerning the production and usage of mercury, the Minamata Convention, was signed by more than 130 nations under United Nations (Kessler, 2013), and has already showed its impact on reducing mercury emissions and global deposition (Giang et al., 2015).

11.1.2 Risk thresholds and criteria for different disposal or reuse purposes on regional, national, or local scale

The management of HRPs has to fulfill the obligation on global scale, thus the supporting policies must be issued to effectively execute the global conventions. These policies normally followed or detailed the requirements of international conventions, for example, the Water Framework Directive (WFD), the Urban Wastewater Treatment Directive (UWWTD) and the Integrated Pollution Prevention and Control (IPPC) Directive of European Union (EU), in combination tends to reach the good chemical status of waters, and with the limits on concentrations in surface waters of 41 dangerous chemical substances to human or ecosystem. The convention obligation of EU in both Minamata Convention and Stockholm Convention was put into effect, which successfully pushed the reuse of wastewater in arid and semiarid areas for irrigation of agricultural crops (Fuerhacker, 2009).

However, the imbalance of nature conditions or global economic development exists in individual regions or nations, so there are always needs for specific control of HRPs. As on the global level, the banned or phase out strategy were applied in several cases on regional or national scale. For example, a homolog of PFOS been banned in Stockholm Convention, the perfluorooctanoate (PFOA) was discovered to cause carcinogen in rodents (Li et al., 2017), which can lead to a phase out of its production leading by the U.S. Environment Protection Agency, the "2010/2015 PFOA Stewardship Program," to reduce the production of PFOA by 95% in 2010, and totally eliminated from emissions and products in 2015 (https://www.epa.gov/assessing-and-managing-chemicals-under-tsca/fact-sheet-20102015-pfoa-stewardship-program). This program has largely reduced the PFOA wastewater volume (Hu et al., 2016), as well as the serum concentration of PFOA in the US citizens (Olsen et al., 2017).

However, this is not the only option for HRPs management, especially for many HRPs, such as heavy metals or biological pathogens. It is extremely hard to totally eliminate these HRPs; thus, the standards of reasonable limits are more practical, and popularly accepted. As mentioned earlier, due to the occurrence of HRPs-induced diseases, the industrialized nations all put an effort to control and minimize the potential hazards of HRPs to reach an acceptable level during the early age of 1960–1970s. However, the globalization and industries transfer largely changed the situation across the world, and the developing countries now afford more manufacture tasks in global economic system (Szirmai and Verspagen, 2015; Timmer et al., 2015), which leads to more burden of ecoenvironment system. Due to the population density and lower developmental levels on hygiene conditions, there is an even serious situation faced by these nations. Therefore, began in late 20th century, the HRP controlling system started to emerge in developing countries as well. Table 11.1 shows the HRPs included in the standards of several developing countries. The most of the criteria are concerning industrial wastewater; however, several HRPs also have thresholds in discharged municipal wastewater, especially for heavy metals and several organic pollutants, such as DDT and phenols. Of course, the inclusion of all HRPs in one standard may largely overlook the specific situation of distinguished industries. The early industrialized nations such as the United States tend to apply the discharge or reuse standard targeting certain industry, which could be more specific and effective. However, the general standards of HRPs issued by the developing countries still notes an effort for facing the problems of HRPs. The heavy metals are the most frequently regulated HRP species, and organic pollutants and disinfection residues are also emphasized by these standards; however, most of these standards show lack of control of biological HRPs, and only China and India include limits of *E. coli*, thus the wastewater effluent from spots like hospitals or

Table 11.1 The HRPs already in control of six developing nations.

China	Thailand	Cambodia	Malaysia	Bangladesh	India
Integrated wastewater effluent	Industrial wastewater effluent	Wastewater effluent to public water area and sewer	Wastewater effluent discharge	Industrial wastewater discharge	Effluent discharged to inland surface water
Zinc (Zn)	Zinc (Zn)	Zinc (Zn)	Zinc	Zinc (Zn)	Zinc (Zn)
Chromium (Hexavalent)	Chromium (Hexavalent)	Chromium (Hexavalent)	Chromium (Hexavalent)	Chromium, (Hexavalent)	Chromium (Hexavalent)
	Chromium (trivalent)	Chromium (trivalent)	Chromium (trivalent)		
Arsenic (As)	Arsenic (As)	Arsenic (As)	Arsenic	Arsenic (As)	Arsenic (As)
Copper (Cu)	Copper (Cu)	Copper (Cu)	Copper	Copper (Cu)	Copper (Cu)
Mercury (Hg)	Mercury (Hg)	Mercury (Hg)	Mercury	Mercury (Hg)	Mercury (Hg)
Cadmium (Cd)	Cadmium (Cd)	Cadmium (Cd)	Cadmium	Cadmium (Cd)	Cadmium (Cd)
	Barium (Ba)	Barium (Ba)	Barium		
Selenium (Se)	Selenium (Se)	Selenium (Se)	Selenium (Se)	Selenium (Se)	Selenium (Se)
Lead (Pb)	Lead (Pb)		Lead	Lead (Pb)	Lead (Pb)
Nickel (Ni)	Nickel (Ni)	Nickel (Ni)	Nickel(Ni)	Nickel (Ni)	Nickel (Ni)
Manganese (Mn)	Manganese (Mn)	Manganese (Mn)	Manganese (Mn)	Manganese (Mn)	Manganese (Mn)
		Iron (Fe)	Iron	Iron (Fe)	Iron (Fe)
Silver (Ag)		Silver (Ag)	Silver (Ag)		
		Molybdenum (Mo)			
			Vanadium (V)		Vanadium (V)
		Tin (Sn)	Tin (Sn)		
Cyanide	Cyanide (as HCN)	Cyanide (CN)	Cyanide	Cyanide (CN)	Cyanide (CN)
	Formaldehyde		Formaldehyde		
	Phenols	Phenols	Phenols	Phenols	
	Free chlorine	Free chlorine	Chlorine (free)		
		Ion chlorine			
Total residual chlorine			Chloride	Chloride	Total residual chlorine

Continued

Table 11.1 The HRPs already in control of six developing nations.—*cont'd*

China	Thailand	Cambodia	Malaysia	Bangladesh	India
Alpha emitters			Alpha emitters	Alpha emitters	
Beta emitters			Beta emitters	Beta emitters	
DDT		DDT			DDT
Polychlorinated biphenyl		Polychlorinated biphenyl			Polychlorinated biphenyl
Carbon tetrachloride		Carbon tetrachloride			Carbon tetrachloride
Hexachlora benzene		Hexachlora benzene			Hexachlora benzene
Endrin		Endrin			Endrin
Dieldrin		Dieldrin			Dieldrin
Aldrin		Aldrin			Aldrin
Isodrin		Isodrin			Isodrin
Perchloro ethylene		Perchloro ethylene			Perchloro ethylene
Hexachloro butadiene		Hexachloro butadiene			Hexachloro butadiene
Chloroform		Chloroform			Chloroform
1,2-Dichloro ethylene		1,2-Dichloro ethylene			1,2-Dichloro ethylene
Trichloro ethylene		Trichloro ethylene			Trichloro ethylene
Trichloro benzene		Trichloro benzene			Trichloro benzene
O-dichlorobenzene					
P-dichlorobenzene					
P-nitrochlorobenzene					
2,4-Dinitrochlorobenzene					
Phenol					
M-cresol					
2,4-Dichlorobenzene					
2,4,6-Trichlorophenol					
Dibutyl phthalate					
Dioctyl phthalate					
Fecal coliform					Fecal coliform

pharmaceutical manufacturer could cause big trouble if it is mixed with the drinking water resources downstream. Another challenge of the HRPs in wastewater in all countries is that although the thresholds and criteria of several HRPs were clarified, the management of wastewater treatment plants (WWTPs) inner or outside the waste effluent producer can largely affect the effects of the practice of whole standard, underlying that discharges and excessive emissions both can largely reduce the efforts to minimize HRPs effluents from wastewater. Besides, although the health and ecological risk of HRPs can finally affect the interest of all stakeholders, several parties may benefit themselves at a relatively early stage. For example, farms located alongside one WWTP may use wastewater for irrigation, due to the high trophic contents of the wastewater. This is extremely common for some of emerging economies with limited farmlands and density population, such as African nations and India (Qadir et al., 2010; Shuval et al., 1997). The HRPs such as heavy metals and organic pollutants can thus enter into the food chain of citizens who consume these crops or vegetables (Qadir et al., 2010; Shuval et al., 1997).

Therefore, to make the standards and regulations of wastewater management system to work more efficiently, especially in developing countries, not only the entrepreneur producing or selling HRPs or products containing HRPs should be considered, but also all the stakeholders parties should be participants in the whole management process.

11.2 Risk management policies and regulations

Although the conventions, laws, regulations, or standards have gain control of most of the traditional HRPs, there are still a potential of hazards to a large extent, because with the development of material and biology science, especially as the acceleration of science and technology, new kinds or forms of chemicals or biological products were continuously synthesized or developed. These new developed chemicals could possibly cause great hazards to the human beings or the environment, just as what happened during the 1950–1970s. To fit in with this new situation, a remarkable policy, the "Registration, Evaluation, Authorization and Restriction of Chemicals (REACH, Fig. 11.4, Table 11.2)," was issued by the European Union, following the rule "No data no market" which was never applied before. The REACH successfully pushed the manufacturer of the new chemicals to compulsively provide the data for registration to evaluate the potential hazards, and other stakeholders, such as importers and downstream users to be aware of the responsibility, so as to avoid the situation of untraceable hazards. After registration, REACH worked through a procedure of screening and regulatory management option analysis, requiring that the manufactured or imported chemicals with annual amount of 10 tons or more, or showing biocidal activity need to undergo persistent, bioaccumulative and toxic (PBT) assessment, or very persistent and very bioaccumulative (vPvB) assessment, and subsequent chemical safety assessment. The REACH Regulation pays specific attention to the PBT/vPvB concern. One aim of the REACH Regulation is the

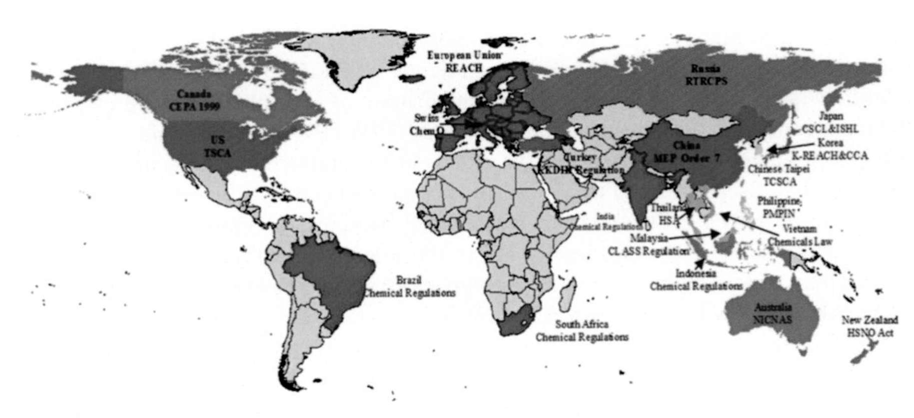

FIGURE 11.4

The laws or regulations concerning chemicals risk management in different nations across the world.

minimization and final substitution of PBT and vPvB substances when suitable technically and economically viable alternatives are available. Another kind of high-risk substances specifically managed by REACH was endocrine disruptor substances. The success of REACH later inspired a wide range of nations to modify or newly draft their own REACH like policies or regulations (Table 11.2), such as the US Toxic Chemical Substance Control Act (TSCA) and Japan Chemical Substance Control Law (CSCL). Due to the successful application of these chemical control policies, the potential hazards of HRPs has been largely reduced, and they can be traced and recalled from the market if necessary immediately.

11.3 Problems and future development of risk management of HRPs

With the regulations or laws like REACH, the potential hazards of new HRPs have been largely limited, and may be identified and eliminated as soon as consequential diseases occurred, but the existing policies and management tools are still imperfect. The most obvious deficiency of the standards or regulations is that all these policies are mostly focusing on chemicals, but neglected the potential hazards of biological pathogens. However, as the WHO notes in "Guidelines for the safe use of wastewater, excreta and greywater" (Organization, 2006), the potential hazards of biological pathogens cannot be neglected, as the pathogens in the treated or reused wastewater can affect human health via different contact pathways. Waterborne pathogens can be classified into categories as bacteria, protozoa, and viruses, such as *salmonella*, *cryptosporidium*, and Rotavirus, and it is currently estimated that there are 1407 species of pathogens capable of infecting human beings (Bitton, 2014).

Table 11.2 A brief introduction of the laws or regulations concerning HRPs across the world.

Nations/ Regions	Laws or regulations	Brief introduction	Website
Europe Union	EU REACH Regulation	"Registration, Evaluation, Authorization and Restriction of Chemicals" (REACH) is a regulation of the European Union, adopted to improve the protection of human health and the environment from the risks that can be posed by chemicals, … It also promotes alternative methods for the hazard assessment of substances to reduce the number of tests on animals.	http://ec.europa.eu/environment/chemicals/reach/reach_en.htm
United States of America	US Toxic Chemical Substance Control Act (TSCA)	The Toxic Substance Control Act (TSCA) is the most important chemical control law in the United States. It authorized environmental protection agency (EPA) to take certain regulatory actions against both new and existing chemical substances. In 2016, TSCA was amended by "The Frank R. Lautenberg Chemical Safety for the 21st Century Act" The toxicology data are not mandatory in TSCA, but must be submitted if available.	https://www.epa.gov/tsca-inventory
Swiss	Swiss Chemicals Ordinance (ChemO)	Although the ChemO is the substitute of REACH in Swiss, it the ChemO governs not only new substances, but also the whole determination and assessment of dangers and risks that substances and preparations may pose to human life and health and to the environment. This includes the handling of substances and preparations that may endanger people or the environment, e.g. biocidal products and the active substances contained therein, and to plant protection products and the active substances and coformulants contained therein.	https://www.admin.ch/opc/en/classified-compilation/20141117/index.html
Canada	Canadian Environmental Protection Act (Taylor and Chénier, 2003)	The Canadian Environmental Protection Act 1999 (Taylor and Chénier, 2003) is the most important chemical control law in Canada. It is a broad act covering a number of subjects, that is, chemicals, living organisms, marine environment, and hazardous wastes. The CEPA controls the HRPs through its Toxic Substance List.	http://www.ec.gc.ca/lcpe-cepa

Continued

Table 11.2 A brief introduction of the laws or regulations concerning HRPs across the world.—*cont'd*

Nations/ Regions	Laws or regulations	Brief introduction	Website
China	China MEP Order 7 (China REACH)	The Measures for Environmental Administration of New Chemical Substances (China MEP Order 7) was issued in January 2010 by the Chinese Ministry of Environmental Protection (recently changed into the Ministry of Ecology and Environment) and came into force on October 15, 2010. It regulates HRPs basically the same way as EU REACH regulation, and is also known as "China REACH."	http://www.gov.cn/flfg/2010-02/04/content_1528001.htm
Japan	Japan Chemical Substance Control Law (CSCL)	The Act on the Evaluation of Chemical Substances and Regulation of Their Manufacture, etc. (hereinafter the "Chemical Substances Control Law") was firstly enacted in 1973 to prevent environmental pollution by chemical substances that pose a risk to human health or the environment, however, not until the latest amendment made in 2009 did it includes the substances of newly synthesized. It controls the manufacture and import amounts of both persistent HRP chemicals, as well as chemicals could causing long-term toxicity for humans or for predator animals at higher trophic level.	http://www.meti.go.jp/english/policy/safety_security/chemical_management/index.html
Korea	Korea Act on Registration and Evaluation, etc. of Chemical Substances (K-REACH)	The amended Act on Registration and Evaluation, etc. of Chemical Substances in South Korea (also known as K-REACH) was promulgated in March 2018 and will come into force on January 1, 2019. It is called K-REACH because of its similarity of REACH; however, as the new rules regulates, the quantity limitation of registration is only 100 kg.	http://www.chemsafetypro.com/Topics/Korea/K-REACH_Amendment.html
	Korea Chemicals Control Act (CCA)	The CCA (previously toxic chemicals control act or TCCA) passed national assembly in May in 2013 and it comes into force in January 2015. It is a new law focusing on chemical reporting and chemical accident prevention.	http://elaw.klri.re.kr/eng_service/lawView.do?hseq=34829&lang=ENG

Chinese taipei	Taiwan Toxic Chemical Substance Control Act (TCSCA)	Toxic Chemical Substances Control Act (TCSCA) is the most important chemical control law in taiwan. TCSCA requires enterprises who manufacture, import, export, sale, transport, use, storage, or discarding chemical substances certain controlled toxic chemical substances to apply for permits, registration, or approval and comply with relevant management measures.	http://law.moj.gov.tw/eng/news/news_detail.aspx?SearchRange=G&id=7444&k1=toxic
Thailand	Thailand Hazardous Substance Act	The current Hazardous Substance Act B.E. 2535 was issued in 1992. It is the most important chemical control law in Thailand. The purpose of the Act is to regulate the importation, production, marketing, and possession of all hazardous chemicals (including industrial chemicals, pesticides, and biocides) used in Thailand.	https://www.chemsafetypro.com/Topics/Thailand/Thailand_Hazardous_Substance_Act_BE_2535.pdf
Turkey	Turkish REACH - KKDIK Regulation	The KKDIK regulation came into force on December 23, 2017. "KKDIK" are the first letters of REACH written in Turkish. Similar to EU REACH regulation, the KKDIK regulation requires companies to register all substances manufactured in Turkey or imported into Turkey with volume above 1 t/y.	http://www.resmigazete.gov.tr/eskiler/2017/06/20170623M1-18.htm
Russia	Russian technical regulation for chemical product safety	October 7, 2016, the Russian Federation approved its technical regulation for chemical product safety via government decree no. 1019. The new technical regulation will come into force on 1 July 2021. The new regulation makes GHS labels and safety data sheets compulsory for chemical products in Russia and sets out requirements on new substance notification and chemical product registrations.	http://www.mintest-russia.com/news/technical-regulations-on-the-safety-of-chemical-products-/
Australia	Australia NICNAS	NICNAS assesses industrial chemicals that are new to Australia for their health and environmental effects before they are introduced to Australia. All new industrial chemicals (i.e., those not listed on AICS) must be notified to NICNAS and assessed before their import or manufacture in Australia unless they are exempt.	http://www.nicnas.gov.au/nicnas-handbook/handbook-main-content
New Zealand	New Zealand HSNO Act	New Zealand's Hazardous Substances and New Organisms Act, known as HSNO Act, came into force for new organisms on July 29, 1998 and for hazardous substances on July 2, 2001. The HSNO manages the risks that hazardous	http://www.legislation.govt.nz/act/public/1996/0030/latest/DLM381222.html

Continued

Table 11.2 A brief introduction of the laws or regulations concerning HRPs across the world.—*cont'd*

Nations/ Regions	Laws or regulations	Brief introduction	Website
		substances and new organisms may pose to human health and the environment in New Zealand.	
Malaysia	Malaysia CLASS Regulations	A manufacturer or importer is required to prepare an inventory of hazardous chemicals, imported or supplied in a quantity of 1 ton per year or above for each calendar year mandatorily notify of hazardous chemicals.	http://www.dosh.gov.my/ICOP
Philippine	Philippine PMPIN	Pre-Manufacture and Pre-Importation Notification (PMPIN) for new substances was authorized by the Republic Act 6969Toxic Substances and hazardous and Nuclear Waste Control Act of 1990. Manufacturers and importers of a new substance that is not listed on PICCS are required to notify DENR-EMB of their intent to manufacture or import the new substance.	http://www.emb.gov.ph/portal/chemical/Permitings/Pre-ManufacturePre-ImportationNotification.aspx
Vietnam	Vietnam Chemicals Law	Government Regulation Number 74 year 2001 regarding hazardous and Toxic Material Management. The Ministry of Industry and trade is leading chemical management in Vietnam, the Chemical Law provides regulations on chemical handling, safety in chemical handling, right and obligations of organizations and individuals engaged in chemical handling, and state management of chemical handling.	https://www.chemsafetypro.com/Topics/Vietnam/Vietnam_Reference_Resource.html
India	Chemical Regulations in India	The regulation was firstly enacted in 1989 by the Ministry of Environment & Forests (MoEF) and later amended in 1994 and 2000. It regulates the manufacture, storage and import of hazardous chemicals in India. The transport of hazardous chemicals must meet the provisions of the Motor Vehicles Act, 1988.	http://nagarikmancha.org/images/MANUFACTURE,%20STORAGE%20AND%20IMPORT%20OF%20HAZARDOUS%20CHEMICAL%20RULES,%201989.pdf
Indonesia		The Government Regulation Number 74 year 2001 regarding hazardous and Toxic Material Management is the most a	

	Chemical Regulations in Indonesia	substance may be regarded as B3 if it is explosive, oxidizing, extremely flammable, highly flammable, flammable, extremely toxic, highly toxic, moderately toxic, harmful, corrosive, irritant, dangerous to the environment, carcinogenic, teratogenic, or mutagenic.	http://portal.fiskal.depkeu.go.id/dbpkppim/index.php?r=dokumen/preview&id=194
Brazil	Chemical Regulations in Brazil	In Oct 2018, the National Chemicals Safety Commission (Comitê Nacional sobre segurança Química, or CONASQ), along with the Ministry of Environment (Ministério do Meio Ambiente, or MME), published a Preliminary Bill for the Inventory, Evaluation, and Control of Chemical Substances.	https://chemicalwatch.com/71197/brazil-releases-draft-law-for-chemical-management
South Africa	Chemical Regulations in South Africa	In South Africa, industrial chemicals are mainly controlled by National Environmental Management Act (1998) and hazardous Substances Act (1973). The National Environmental Management Act authorizes DEA to prohibit or control certain substances or chemicals that pose threat to the environment and human health, and hazardous Substances Act is the most important chemical regulation in South Africa. It controls the production, import, use, handling, and disposal of hazardous substances.	https://www.environment.gov.za/sites/default/files/legislations/nema_amendment_act107.pdf; https://www.acts.co.za/hazardous-substances-act-1973/index.html

The U.S. EPA has shown its leading role on these aspects: the "Drinking Water Contaminant Candidate List (CCL)" addressed a special attention on the pathogens in aquatic systems, and identified more than 500 potential aquatic pathogens in drinking water (http://www.epa.gov/safewater/ccl/pdfs/report_ccl4_microbes_universe.pdf). These efforts also brought changes to the management policies. For example, concerns have also been focused on biopathogens in many policies nowadays, such as Swiss ChemO, Canadian CEPA, and New Zealand's HSNO Act (Table 11.2). The lack of control of the biological HRPs can also be attributed to the current shortage of the reliable detection methods, as the culture methods always take time, and the accuracy is affected by the complex composition of the microbes in wastewater (Ince et al., 2012; Johnson et al., 2006); however, with the development of biological hazard evaluation model, such as epidemiological assessment (EA) and quantitative microbial risk assessment (QMRA), there should be significant improvement of biological HRPs detection, evaluation, and management in the near future (Bouwknegt et al., 2013).

The second deficiency of the management system is that it is still vulnerable when dealing with combined effects of pollutants, which in many cases, are the realistic composition of wastewater. For example, the relative environmental legislations, in most countries except New Zealand and Australia, have set the maximum acceptable heavy metal concentrations in the aquatic environment for each heavy metal (Gikas, 2008). It is well known that the combined toxicity of multiple pollutants can be additive, synergism, or antagonism, but the calculation models currently apply a simple addition of the concentrations of multiple HRPs, or effect summation for the multiple HRPs at normalized doses. This is inconsistent with the real situation of HRPs in real wastewater, as the interaction among the HRPs and other chemicals can be quite complicated. For example, the pharmaceutical and personal care products and heavy metals in hospital, animal breeding, or pharmaceutical wastewater may interact with biological pathogens and push them to evolve into antibiotic-resistant bacteria, which could potentially cause an even bigger problem afterward (Zhang et al., 2011; Zhang and Zhang, 2011).

11.4 Summary

The first generation of management system of HRPs in wastewater was established in 1970s to prevent the occurrence of pollution hazards to human health or ecosystem. This system combined the standards, regulations, laws, and conventions as management tools, and later grew into a legal structure fully covering the management from a city or a town scale to the whole world. It has successfully dealt with traditional HRPs in the past several decades in the 20th century. However, to avoid "dealing after happening," the European Union established a more efficient management system of all chemicals, the REACH, to force all the stakeholders to participate in tracing and identifying new HRPs, which was then adopted by many nations across the world. This system, combined with the "Drinking Water

Contaminant Candidate List" proposed by the U.S. EPA for management of biological pathogens in wastewater based on QMRA assessment, has been successfully utilized to control the hazards of HPRs in wastewater in the past decade. In the future, with more efforts addressed on the combination effects of HRPs in real wastewater, more effective management can be achieved, and the long-term goal of safe utilization of chemicals and biological materials can be fulfilled.

References

Bitton, G., 2014. Biological Treatment and Biostability of Drinking Water. John Wiley & Sons, Inc.

Bouwknegt, M., Knol, A.B., Jp, V.D.S., Evers, E.G., 2013. Uncertainty of population risk estimates for pathogens based on QMRA or epidemiology: a case study of Campylobacter in the Netherlands. Risk Analysis 34 (5), 847.

Chen, C.M., 2004. Searching for intellectual turning points: progressive knowledge domain visualization. Proceedings of the National Academy of Sciences of the United States of America 101, 5303–5310.

Fiedler, H., 2007. National PCDD/PCDF release inventories under the Stockholm convention on persistent organic pollutants. Chemosphere 67 (9), S96–S108.

Fiedler, H., Kallenborn, R., de Boer, J., Sydnes, L.K., 2019. The Stockholm convention: a tool for the global regulation of persistent organic pollutants. Chemistry International 41 (2), 4–11.

Fuerhacker, M., 2009. EU Water Framework Directive and Stockholm Convention Can we reach the targets for priority substances and persistent organic pollutants? Environmental Science and Pollution Research 16, 92–97.

Funabashi, H., 2006. Minamata disease and environmental governance. International Journal of Japanese Sociology 15 (1), 7–25.

Giang, A., Stokes, L.C., Streets, D.G., Corbitt, E.S., Selin, N.E., 2015. Impacts of the minamata convention on mercury emissions and global deposition from coal-fired power generation in Asia. Environmental Science and Technology 49 (9), 5326–5335.

Gikas, P., 2008. Single and combined effects of nickel (Ni(II)) and cobalt (Co(II)) ions on activated sludge and on other aerobic microorganisms: a review. Journal of Hazardous Materials 159 (2–3), 187–203.

Hardy, C., Maguire, S., 2010. Discourse, field-configuring events, and change in organizations and institutional fields: narratives of DDT and the Stockholm Convention. Academy of Management Journal 53 (6), 1365–1392.

Hino, O., Yan, Y., Ogawa, H., 2018. Environmental pollution and related diseases reported in Japan: from an Era of "risk evaluation" to an Era of "risk management". Juntendo Medical Journal 64 (2), 122–127.

Hu, X.C., Andrews, D.Q., Lindstrom, A.B., Bruton, T.A., Schaider, L.A., Grandjean, P., Lohmann, R., Carignan, C.C., Blum, A., Balan, S.A., 2016. Detection of poly-and perfluoroalkyl substances (PFASs) in US drinking water linked to industrial sites, military fire training areas, and wastewater treatment plants. Environmental Science and Technology Letters 3 (10), 344–350.

Iguchi, M., Koga, M., 2015. 12 the state of environment and energy governance in Japan. Environmental Challenges and Governance: Diverse Perspectives from Asia 37, 219.

Ince, O., Basak, B., Ince, B.K., Cetecioglu, Z., Celikkol, S., Kolukirik, M., 2012. Effect of nitrogen deficiency during SBR operation on PHA storage and microbial diversity. Environmental Technology 33 (16), 1827–1837.
Johnson, G., Wilks, M., Warwick, S., Millar, M.R., Fan, S.L.-S., 2006. Comparative study of diagnosis of PD peritonitis by quantitative polymerase chain reaction for bacterial DNA vs culture methods. Journal of Nephrology 19 (1), 45.
Kessler, R., 2013. The Minamata Convention on Mercury: A First Step toward Protecting Future Generations. National Institute of Environmental Health Sciences.
Kobayashi, J., 1970. Relation Between the'itai-itai' Disease and the Pollution of River Water by Cadmium from a Mine.
Lallas, P.L., 2001. The Stockholm convention on persistent organic pollutants. American Journal of International Law 95 (3), 692–708.
Li, K., Gao, P., Xiang, P., Zhang, X., Cui, X., Ma, L.Q., 2017. Molecular mechanisms of PFOA-induced toxicity in animals and humans: implications for health risks. Environment International 99, 43–54.
McAlpine, D., Araki, S., 1958. Minamata disease - an unusual neurological disorder caused by contaminated fish. Lancet 2 (SEP20), 629–631.
Olsen, G.W., Mair, D.C., Lange, C.C., Harrington, L.M., Church, T.R., Goldberg, C.L., Herron, R.M., Hanna, H., Nobiletti, J.B., Rios, J.A., 2017. Per-and polyfluoroalkyl substances (PFAS) in American Red Cross adult blood donors, 2000–2015. Environmental Research 157, 87–95.
Organization, W.H, 2006. Guidelines for the Safe Use of Wastewater, Excreta and Greywater. World Health Organization.
Pruss, M., 1938. The agricultural wastewater management - it's importance for the further production. Zeitschrift des Vereines Deutscher Ingenieure 82, 42-42.
Qadir, M., Wichelns, D., Raschid-Sally, L., McCornick, P.G., Drechsel, P., Bahri, A., Minhas, P., 2010. The challenges of wastewater irrigation in developing countries. Agricultural Water Management 97 (4), 561–568.
Shuval, H., Lampert, Y., Fattal, B., 1997. Development of a risk assessment approach for evaluating wastewater reuse standards for agriculture. Water Science and Technology 35 (11–12), 15–20.
Stander, G.J., 1966. Water pollution research - a key to wastewater management. Journal Water Pollution Control Federation 38 (5), 774–&.
Subrahmanyan, K., Bhaskaran, T.R., 1950. The risk of pollution of ground water from borehole latrines. Indian Medical Gazette 85 (9), 418–420.
Sun, Z., Zhang, C., Yan, H., Han, C., Chen, L., Meng, X., Zhou, Q., 2017. Spatiotemporal distribution and potential sources of perfluoroalkyl acids in Huangpu River, Shanghai, China. Chemosphere 174, 127–135.
Supan, 1901. The water management template. Petermanns Mitteilungen 47 (12). A93-A93.
Symons, G.E., 1969. Industrial wastewater management – a look ahead. Water and Wastes Engineering 6 (1), A11–&.
Szirmai, A., Verspagen, B., 2015. Manufacturing and economic growth in developing countries, 1950–2005. Structural Change and Economic Dynamics 34, 46–59.
Taylor, K.W., Chénier, R., 2003. Introduction to Ecological Risk Assessments of Priority Substances Under the Canadian Environmental Protection Act, 1999. Human and Ecological Risk Assessment: An International Journal 9 (2), 447–451.
Timmer, M., de Vries, G.J., De Vries, K., 2015. Routledge Handbook of Industry and Development. Routledge, pp. 79–97.

Ule, 1910. Earth is drying up. Germany's mistakes in Water Management. Petermanns Mitteilungen 56 (5), 282-282.

Vijgen, J., Abhilash, P., Li, Y.F., Lal, R., Forter, M., Torres, J., Singh, N., Yunus, M., Tian, C., Schäffer, A., 2011. Hexachlorocyclohexane (HCH) as new Stockholm Convention POPs—a global perspective on the management of Lindane and its waste isomers. Environmental Science and Pollution Research 18 (2), 152–162.

Voulvoulis, N., Skolout, J.W., Oates, C.J., Plant, J.A., 2013. From chemical risk assessment to environmental resources management: the challenge for mining. Environmental Science and Pollution Research 20 (11), 7815–7826.

Wang, T., Wang, Y., Liao, C., Cai, Y., Jiang, G., 2009. Perspectives on the Inclusion of Perfluorooctane Sulfonate into the Stockholm Convention on Persistent Organic Pollutants. ACS Publications.

Ward, T.G., Turner, T.B., 1942. Study of certain epidemiological features of leptospiral jaundice in Baltimore. American Journal of Hygiene 35 (1), 122–133.

Zhang, T., Zhang, X.-X., Ye, L., 2011. Plasmid metagenome reveals high levels of antibiotic resistance genes and mobile genetic elements in activated sludge. PLoS One 6 (10), e26041.

Zhang, X.-X., Zhang, T., 2011. Occurrence, abundance, and diversity of tetracycline resistance genes in 15 sewage treatment plants across China and other global locations. Environmental Science & Technology 45 (7), 2598–2604.

Abbreviations

2, 4-D	2,4-dichlorophenoxyacetic acid
A/A/O	anoxic-aerobic-aerobic
A/O	anaerobic/aerobic
A^2/O	anaerobic/anoxic/aerobic
AAS	atomic absorption spectrometry
AB	adsorption biodegradation
AC	activated charcoal
ACE	acetaminophen
AchE	acetylcholinesterase
ADI	acceptable daily intake
AF	assessment factor
AFS	atomic fluorescence spectrometry
Ag	silver
AHs	aromatic hydrocarbons
AOPs	advanced oxidation processes
ARB	antibiotic resistant bacteria
ARGs	antibiotic resistance genes
As	arsenic
ASV	anodic stripping voltammetry
AZI	azithromycin
BAC	biological activated carbon
BAF	biological aerated filter
BALF	bronchioalveolar lavage fluid
BEO	bioelectrochemical oxidation
bHRPs	biological high-risk pollutants
BioMAC	biofilm-assisted carbon filtration
BOD	biochemical oxygen demand
BPA	bisphenol A
CAS	chemical abstract service
CAS	cyclic activated sludge
CB	conduction band
CBZ	carbamazepine
Cd	cadmium
CFN	caffeine
CFO NPs	Cu—Fe—O nanoparticles
ChemO	chemicals ordinance
COD	chemical oxygen demand
COPD	chronic obstructive pulmonary disease
CoV	coronavirus
CR	carcinogenic risk
Cr	chromium
CSCL	chemical substance control law

CTA	cellulose triacetate
Cu	copper
CW	constructed wetland
DBPs	disinfection by-products
DCF	diclofenac
DDT	dichloro-diphenyl-trichloroethane
DIU	diuron
DMP	dimethyl phthalate
DOM	dissolved organic matter
DPASV	differential pulse voltammetry
E1	estrone
E2	estradiol
EA	epidemiological assessment
EC_{50}	concentration for 50% of maximal effect
ED	electrodialysis
ED-VMD	electrodialysis-vacuum membrane distillation
EDC	1-(3-dimethylaminopropyl)-3-ethylcarbodiimide hydrochloride
EDCs	endocrine disrupting chemicals
EE2	ethinyl estradiol
EGSB	expanded granular sludge bed
EPA	environmental protection agency
ERA	ecological risk assessment
ERY	erythromycin
EU	European Union
FLX	fluoxetine
FO	forward osmosis
GAC	granular activated carbon
GC	gas chromatography
GC−MS	gas chromatography−mass spectrometry
GFP	green fluorescent protein
HAV	hepatitis A virus
HBV	hepatitis B virus
HCS	high-content screening
HCV	hepatitis C virus
HDV	hepatitis D virus
HEV	hepatitis E virus
Hg	mercury
HI	hazard index
HIV	human immunodeficiency virus
HPLC	high-performance liquid chromatography
HQ	hazard quotient
HRA	health risk assessment
HRI	health risk index
HRPs	high-risk pollutants
HRT	hydraulic retention time
HS-SPME	headspace solid-phase microextraction
HSNO Act	Hazardous Substances and New Organisms Act

IBP	ibuprofen
ICP-MS	inductively coupled plasma-mass spectrometry
ICP-OES	inductively coupled plasma-optical emission spectrometry
ICTV	international committee on taxonomy of viruses
IL-6	interleukin-6
IPPC	integrated pollution prevention and control
ISO	International Organization for Standardization
LacZ	β-galactosidase
LC	liquid chromatography
LC_{50}	concentration for 50% of lethal effect
LCA	life cycle assessment
LDH	lactate dehydrogenase
LIBS	laser induced breakdown spectroscopy
Luc	firefly luciferase
Lux	bacteria luciferase
MBR	membrane bioreactor
MCE	mineralization current efficiency
MEC	measured environmental concentration
MERS-CoV	Middle East respiratory syndrome coronavirus
MF	microfiltration
MFC	microbial fuel cell
Mg	manganese
MoA	mode of action
MRL	method reporting limits
MRL	minimal risk level
MWCNTs	multiwalled carbon nanotubes
MWTPs	municipal wastewater treatment plants
ND	not detected
NDM-1	New Delhi metallo-beta-lactamase-1
NDMA	N-nitrosodimethylamine
NF	nanofiltration
Ni	nickel
NOEC	no-observed effect concentration
NOM	natural organic matter
NOR	norfloxacin
NPDES	national pollutant discharge elimination system
NPX	naproxen
OCPs	organic pesticides
OECD	organization for economic cooperation and development
OF	ofloxacin
OPPs	organophosphorus pesticides
PAA	peracetic acid
PAC	powdered activated carbon
PAHs	polycyclic aromatic hydrocarbons
Pb	lead
PBT	persistent, bioaccumulative, and toxic
PCBs	polychlorinated biphenyls
PCDD	polychlorinated dibenzo-*p*-dioxins

PCDF	polychlorinated dibenzo-*p*-dibenzofurans
PCR	polymerase chain reaction
PDCoV	porcine deltacoronavirus
PEC	predicted environmental concentration
PEF	photo-electro fenton
PFASs	perfluoroalkyl substances
PFOA	perfluorooctanoic acids
PFOS	perfluorooctane sulfonate
PMN	polymorphonuclear
PNEC	predicted no-effect concentration
POPs	persistent organic pollutants
PPCPs	pharmaceuticals and personal care products
PTFE	polytetrafluorethylene
QMRA	quantitative microbial risk assessment
qPCR	quantitative PCR
REACH	registration, evaluation, authorization, and restriction of chemicals
RI	risk index
RMOA	regulatory management option analysis
RO	reverse osmosis
ROX	roxithromycin
RT-PCR	reverse transcription-PCR
RVFCW	recirculating vertical flow constructed wetlands
SARS-CoV	severe acute respiratory syndrome coronavirus
SBR	sequencing batch reactor
SCCPs	short-chain chlorinated paraffins
SI-ASV	sequential injection anodic stripping voltammetry
SMX	sulfamethoxazole
SPF	sono-photo-Fenton
SPIs	Salmonella pathogenicity islands
SRT	sludge retention time
SSF	subsurface-flow
STPs	sewage treatment plants
TBA	tert-butanol
TFC	thin-film composite
TMP	trimethoprim
TNF-α	tumor necrosis factor-α
TSCA	toxic chemical substance control act
TSS	total suspended solids
UASB	up-flow anaerobic sludge blanket
UF	ultrafiltration
UNEP	United Nations environment program
UV	ultraviolet
UV-LED	ultraviolet light-emitting diode
UWWTD	urban wastewater treatment directive
VB	valence band
VOCs	volatile organic compounds
vPvB	very persistent and very bioaccumulative
WAS	waste activated sludge

WFD water framework directive
WHO World Health Organization
WWTPs wastewater treatment plants
Zn zinc

Index

'*Note*: Page numbers followed by "f" indicate figures and "t" indicate tables.'

B

C

D

E

F

G

H

M

N

Q

R